Copper Proteins and Copper Enzymes

Volume I

Editor

René Lontie, D.Sc.

Professor
Faculty of Sciences
Laboratorium voor Biochemie
Katholieke Universiteit
Louvain, Belgium

CRC Press, Inc.
Boca Raton, Florida

Library of Congress Cataloging in Publication Data
Main entry under title:
Copper proteins and copper enzymes.

  Includes bibliographies and indexes.
  1. Copper proteins. 2. Copper enzymes.
I. Lontie, René, 1920-
QP552.C64C663 1984   574.19′24   82-24366
ISBN 0-8493-6470-1 (v. 1)
ISBN 0-8493-6471-X (v. 2)
ISBN 0-8493-6472-8 (v. 3)

To my parents

PREFACE

These volumes of *Copper Proteins and Copper Enzymes* are intended to describe the contemporary spectroscopy and other biophysical chemistry now being applied to copper proteins in order to determine the structures of their active sites. Several chapters of the treatise describe the functional understanding which is emerging from the new work. The authors are all major contributors to research progress on copper proteins and the volumes will be found to be definitive and authoritative.

The subject, copper proteins and copper enzymes, is a very lively one and is best considered in the broadest biological and chemical contexts as it continues to develop. Copper itself, born in the dust of the cosmos and comprising perhaps 0.007% of the earth's crust, occurs in the biosphere as about two dozen families of copper proteins which serve to transport O_2, to activate it toward reaction with organic molecules, and to transfer electrons between donors and acceptors. The families of copper proteins include the azurins, plastocyanins, metallothioneins, superoxide dismutases, ceruloplasmins, laccases, ascorbate oxidases, cytochrome *c* oxidases, monoamine oxidases, diamine oxidases, galactose oxidase, hexose oxidase, urate oxidases, polyphenol oxidases, phenol *o*-hydroxylases, *p*-coumarate 3-monooxygenase, dopamine β-monooxygenase, lysine protocollagen oxygenase, quercetin 2,3-dioxygenase, and the arthropodan and molluscan hemocyanins. The number of types of copper-binding domains in these proteins appears to be very limited, perhaps three or four (blue type-1 mononuclear copper sites; type-2 mononuclear copper sites; diamagnetic binuclear type-3 copper sites), and it is probable that there are very few evolutionary prototypes from which the existing families sprung.

The structures of the copper-binding domains are being rapidly worked out, as these volumes will demonstrate. However, the overall three-dimensional protein structures are in general not known, although progress is being made. The three-dimensional structures will represent major opportunities to understand the chemical biology of these proteins because their functional properties almost certainly depend upon the presence of structural domains other than the copper-binding ones; e.g., the affinity of the copper sites for oxygen may be strongly affected by the kind and disposition of amino-acid residues. In the case of hemoglobin, almost 90 abnormal hemoglobins are known to arise from amino-acid substitutions, which show altered O_2 affinities. Since the K_m (O_2) of copper enzymes is an adaptive property, it is likely that the structures of the active sites will vary in space even though the ligands remain the same for each type.

Another interesting problem against which the information in these volumes should be weighed lies in the fact that each of the functions served by copper proteins is also served by families of iron-, heme-, and flavin-containing proteins. Why then was copper selected when other prosthetic groups were available?

Howard S. Mason
Department of Biochemistry
School of Medicine
The Oregon Health Sciences University
Portland, Oregon

THE EDITOR

René Lontie, D. Sc., is Head of the Laboratorium voor Biochemie and Professor in the Faculty of Sciences, Katholieke Universiteit te Leuven, Louvain, Belgium.

He was born in Louvain in 1920 and educated there. He received his doctorate in physical chemistry (with Professor J. C. Jungers) from the Katholieke Universiteit in 1942. He was trained in protein chemistry as a Research Assistant and Senior Research Assistant of the National Fund for Scientific Research (Belgium) at the Laboratorium voor Biochemie in Louvain under Professor P. Putzeys. As a Graduate Fellow of the Belgian American Educational Foundation he was a Research Fellow in Physical Chemistry at the Department of Physical Chemistry, Harvard Medical School, Boston, Mass. (Professor E. J. Cohn, Professor J. T. Edsall, Professor J. L. Oncley).

He was President of the Vlaamse Chemische Vereniging and of the Belgische Vereniging voor Biochemie — Société Belge de Biochimie. He is a fellow of the American Association for the Advancement of Science, a member of the Royal Society of Sciences of Uppsala, Sweden, the New York Academy of Science, the American Chemical Society, the Biochemical Society (London), the Société de Chimie biologique (Paris), the Society of the Sigma Xi. He is member of the Advisory Board of the *European Journal of Biochemistry* and of *Inorganica Chemica Acta,* Bioinorganic Chemistry, Articles and Letters.

His major research interests, which included milk and barley proteins, are focused now on copper proteins, mainly on the structure, function, and biosynthesis of hemocyanins.

CONTRIBUTORS

Luciana Avigliano
Associate Professor of Molecular Biology
Institute of Biological Chemistry
University of Rome
Rome, Italy

John F. Boas
Australian Radiation Laboratory
Yallambie
Victoria, Australia

Gerhard Buse
Professor of Molecular Biology
RWTH Aachen
Abteilung Physiologische Chemie
Aachen, West Germany

Anthony E. G. Cass
Lecturer in Applied Enzymology
Centre for Biotechnology
Imperial College
London, England

Man Sung Co
Graduate Student
Department of Chemistry
Stanford University
Stanford, California

Murray J. Ettinger
Department of Biochemistry
State University of New York at Buffalo
Buffalo, New York

Ole Farver
Associate Professor
Department of Chemistry AD
The Royal Danish School of Pharmacy
Copenhagen, Denmark

E. Martin Fielden
Professor
Head, Division of Molecular Processes
Medical Research Council
Radiobiology Unit
Harwell, Didcot
England

Constant Gielens
First Assistant
Faculty of Sciences
Laboratorium voor Biochemie
Katholieke Universiteit te Leuven
Louvain, Belgium

Barry Halliwell
Lecturer in Biochemistry
Department of Biochemistry
University of London King's College
London, England

Hans-Jürgen Hartmann
Senior Research Associate
Anorganische Biochemie
Physiologisch-Chemisches Institut
 der Universität Tübingen
Tübingen, West Germany

H. Allen O. Hill
University Lecturer
Inorganic Chemistry Laboratory
University of Oxford
Fellow and Praelector
The Queen's College
Oxford, England

Keith O. Hodgson
Associate Professor
Department of Chemistry
Stanford University
Stanford, California

Peter F. Knowles
Reader in Biophysical Chemistry
Astbury Department of Biophysics
University of Leeds
Leeds, England

Daniel J. Kosman
Professor of Biochemistry
Department of Biochemistry
School of Medicine
State University of New York at Buffalo
Buffalo, New York

Torbjørn Ljones
Professor of Chemistry
Department of Chemistry
University of Trondheim
Dragvoll, Norway

Thomas M. Loehr
Professor of Chemistry
Department of Chemical and Biochemical
  Sciences
Oregon Graduate Center
Beaverton, Oregon

René Lontie
Professor
Faculty of Sciences
Laboratorium voor Biochemie
Katholieke Universiteit te Leuven
Louvain, Belgium

Bruno Mondovì
Professor of Biochemistry
Applied Biochemistry
University of Rome
Rome, Italy

Israel Pecht
Jacques Mimran Professor of Chemical
  Immunology
Department of Chemical Immunology
The Weizmann Institute of Science
Rehovot, Israel

Gisèle Préaux
Professor
Faculty of Sciences
Laboratorium voor Biochemie
Katholieke Universiteit te Leuven
Louvain, Belgium

Bengt Reinhammar
Lecturer in Biochemistry
Department of Biochemistry and
  Biophysics
Chalmers Institute of Technology
University of Göteborg
Göteborg, Sweden

Donald A. Robb
Lecturer in Biochemistry
Department of Bioscience and
  Biotechnology
University of Strathclyde
Glasgow, Scotland

Giuseppe Rotilio
Professor of Biological Chemistry
Faculty of Sciences
University of Rome
Rome, Italy

Lars Rydén
Lecturer in Biochemistry
Department of Biochemistry
Biomedical Center
Uppsala University
Uppsala, Sweden

Joann Sanders-Loehr
Professor of Chemistry
Chemistry Department
Portland State University
Portland, Oregon

Tore Skotland
Nyegaard & Co. A/S
Oslo, Norway

Ulrich Weser
Professor of Inorganic Biochemistry
Physiologisch-Chemisches Institut
  der Universität Tübingen
Tubingen, West Germany

Kapil D. S. Yadav
Lecturer
Department of Chemistry
University of Gorakhpur
Gorakhpur, India

COPPER PROTEINS AND COPPER ENZYMES

René Lontie

Volume I

Volume II

Volume III

TABLE OF CONTENTS

Volume I

Chapter 1

INTRODUCTION

René Lontie

Copper is an essential trace element, although nearly all organisms have access to only very minute amounts. Notable exceptions are bacteria involved in the leaching of copper from low-grade ore[1] and the "copper flowers" in Central Africa.[2] These species raise an interesting problem, as they have to cope with otherwise toxic amounts of the metal.

Several books and conferences have been devoted to trace elements[3-7] and to copper and copper proteins.[8-18] Many chapters on copper and copper proteins are also to be found, e.g., in more general treatises and proceedings of symposia,[19-23] like in serial publications.[24-28]

Four chapters describe spectroscopic techniques which have contributed so much to the understanding of the active sites: electron paramagnetic resonance, nuclear magnetic resonance, X-ray absorption, and resonance Raman. Absorption and circular dichroic spectra are presented in the chapters on the individual proteins.

The blue oxidases contain three types of copper: "blue" (meaning strongly blue), "nonblue" (weakly colored), and diamagnetic pairs,[25] respectively named, type 1, type 2, and type 3. Most authors also use these designations for proteins with only one type. A different nomenclature is advocated in Chapter 3 of Volume I.

The polypeptide chains show a great diversity in size from the smaller ones of the small blue proteins and the copper-thioneins to those of the gastropodan hemocyanins, which are among the largest in nature. Special precautions are needed with the longer polypeptide chains in order to avoid proteolytic cleavage during their isolation, as shown for molluscan hemocyanins and ceruloplasmin (Chapter 6 of Volume II and Chapter 2 of Volume III).

Several chapters illustrate the great effort which went into the determination of amino-acid sequences, like those in Volume I on the small blue proteins, in Volume II on Cu/Zn-superoxide dismutases, hemocyanins, and tyrosinase, and in Volume III on ceruloplasmin, cytochrome *c* oxidase, and metallothioneins.

Conformations were only determined at a high resolution for two small blue proteins and for bovine Cu/Zn-superoxide dismutase (respectively, Chapters 6 and 7 of Volume I, and Chapter 2 of Volume II).

The first steps in the study of the biosynthesis of copper proteins are also presented: the isolation of mRNA's for hemocyanins and for ceruloplasmin (Chapter 6 of Volume II and Chapter 2 of Volume III) and the consideration of copper-thioneins as possible copper donors (Chapter 5 of Volume III).

While the type-1 proteins appear as a rather homogeneous group, the type-2 enzymes (Chapters 1, 2, 4, and 5 of Volume II) show a great diversity in reactivity. Type-3 copper in the blue oxidases (Chapters 1 to 3 of Volume III), by the presence of type-1 and type-2 copper, differs also quite much from the isolated type-3 sites in hemocyanins and tyrosinase (Chapters 6 and 7 of Volume II).

It has not been possible to cover all the copper proteins and copper enzymes. Typical representatives are missing, like urate oxidase (uricase) (EC 1.7.3.3)[29,30] and nitrite reductase (EC 1.7.99.3).[31,32]

Copper does not seem essential for some enzymes, which have been claimed to contain this metal. A crystalline preparation of ribulose 1,5-bisphosphate carboxylase (EC 4.1.1.39) from tobacco did not seem to contain appreciable amounts of copper,[33] in contrast with an earlier report on a preparation from spinach.[34] With parsley the ribulose 1,5-bisphosphate oxygenase seemed a separate enzyme, which contained copper.[35] Indoleamine 2,3-dioxygenase did not contain significant amounts of copper, as a copper-rich protein was eliminated in the last step of the purification.[36] With L-tryptophan 2,3-dioxygenase (EC 1.13.11.11) of *Pseudomonas acidovorans*, which contains protoheme IX, it was shown that copper was not essential for the catalytic activity, although some preparations contained firmly bound copper.[37]

While this heme enzyme did not need copper, soluble guanylate cyclase from bovine lung contained 1 mol of copper and 1 mol of heme (ferroprotoporphyrin IX) per mol of enzyme, which could be important for the regulation of the activity of the enzyme.[38] A protein with unknown function with 2 copper and 2 heme *b* per M_r of 400,000 was isolated from bovine erythrocytes.[39]

The binding of copper to several proteins has also been the object of many investigations and copper(II), not unlike cobalt(II), has been used as a probe for the binding site.[40]

Chapter 3 of Volume II and the last Chapter of Volume III deal with physiological aspects of copper and copper proteins and enzymes: the so debated role of superoxide dismutases and the metabolism of copper, respectively.

There remain the inevitable problems of nomenclature. The terms type-1, type-2, and type-3 copper seem to have lost their initial meaning. Several of the small blue proteins have been named prematurely by combining the name of a plant, a relative or a town, e.g., with ''cyanin''. When more is known of their structure and function a more general name with the indication of the species would be preferable (cf. ''phytocyanin'' in Chapter 6 of Volume I). A similar situation prevailed with erythrocuprein, hemocuprein, and hepatocuprein, which via cytocuprein were finally named Cu/Zn-superoxide dismutase. For semantic reasons terms like Semi-Met seem preferable to Half-Met for the hemocyanins (Chapter 6 of Volume II).

The astute reader will discover shades of opinion and outright contradictions, which illustrate the complexity of the subject. Many more amino-acid sequences and especially protein conformations are needed to understand the evolution of these fascinating proteins and the structure of their active sites.

REFERENCES

1. **Brierley, C. L.,** Microbiological mining, *Sci. Am.*, 247 (2), 42, 1982.
2. **De Plaen, G., Malaisse, F., and Brooks, R. R.,** The ''copper flowers'' of Central Africa and their significance for prospecting and archaeology, *Endeavour, New. Ser.*, 6, 72, 1982.
3. **O'Dell, B. L. and Campbell, B. J.,** Trace elements: metabolism and metabolic function, in *Metabolism of Vitamins and Trace Elements, Comprehensive Biochemistry*, Vol. 21, Florkin, M. and Stotz, E. H., Eds., Elsevier, Amsterdam, 1970, chap. 2.
4. **Underwood, E. J.,** *Trace Elements in Human and Animal Nutrition*, 4th ed., Academic Press, New York, 1977.
5. **Flodin, N. W.,** *Vitamin/Trace Mineral/Protein Interactions, Annual Research Reviews*, Vol. 1, Eden Press, Westmount, Quebec, 1979.
6. **Prasad, A. S.,** *Trace Elements and Iron in Human Metabolism*, John Wiley & Sons, Chichester, 1979.

7. **Levander, O. A. and Cheng, L., Eds.,** Micronutrient interactions: vitamins, minerals, and hazardous elements, *Ann. N.Y. Acad. Sci.,* 355, 1980.

8. **McElroy, W. D. and Glass, B., Eds.,** *Copper Metabolism, A Symposium on Animal, Plant and Soil Relationships,* Johns Hopkins Press, Baltimore, 1950.

9. **Peisach, J., Aisen, Ph., and Blumberg, W. E., Eds.,** *The Biochemistry of Copper,* Academic Press, New York, 1966.

10. *Biological Roles of Copper,* Ciba Foundation Symp. No. 79, Excerpta Medica, Amsterdam, 1980.

11. **Nriagu, J. O.,** *Copper in the Environment Part 1: Ecological Cycling,* Wiley-Interscience, New York, 1980.

12. **Nriagu, J. O.,** *Copper in the Environment Part 2: Health Effects,* Wiley-Interscience, New York, 1980.

13. **Loneragan, J. F., Robson, A. D., and Graham, R. D., Eds.,** *Copper in Soils and Plants,* Academic Press, New York, 1981.

14. **Sigel, H., Ed.,** *Properties of Copper, Metal Ions in Biological Systems,* Vol. 12, Marcel Dekker, New York, 1981.

15. **Sigel, H., Ed.,** *Copper Proteins, Metal Ions in Biological Systems,* Vol. 13, Marcel Dekker, New York, 1981.

16. **Spiro, Th. G.,** *Copper Proteins, Metal Ions in Biology Series,* Vol. 3, Wiley-Interscience, New York, 1981.

17. **Owen, Ch. A., Jr.,** *Copper Deficiency and Toxicity: Acquired and Inherited, in Plants, Animals, and Man,* Noyes, Park Ridge, N.J., 1981.

18. **Owen, Ch. A., Jr.,** *Biochemical Aspects of Copper,* Noyes, Park Ridge, N.J., 1982.

19. **Eichhorn, G. L., Ed.,** *Inorganic Biochemistry,* Vols. 1 and 2, Elsevier, Amsterdam, 1973.

20. **Yasunobu, K. T., Mower, H. F., and Hayaishi, O., Eds.,** *Iron and Copper Proteins,* Advances in Experimental Medicine and Biology, Vol. 74, Plenum Press, New York, 1976.

21. **Brill, A. S.,** *Transition Metals in Biochemistry,* Springer-Verlag, Berlin, 1977, chap. 3.

22. **Weser, U., Ed.,** *Metalloproteins: Structure, Molecular Function and Clinical Aspects,* Thieme, Stuttgart, 1979.

23. **Karcioğlu, Z. A. and Sarper, R. M., Eds.,** *Zinc and Copper in Medicine,* Charles C Thomas, Springfield, Ill., 1980.

24. **Freeman, H. C.,** Crystal structures of metal-peptide complexes, *Adv. Protein Chem.,* 22, 257, 1967.

25. **Malkin, R. and Malmström, B. G.,** The state and function of copper in biological systems, *Adv. Enzymol.,* 33, 177, 1970.

26. **Österberg, R.,** Models for copper-protein interaction based on solution and crystal structure studies, *Coord. Chem. Rev.,* 12, 309, 1974.

27. **Karlin, K. D. and Zubieta, J.,** Coordination chemistry of copper-sulphur complexes of physical relevance, *Inorg. Perspect. Biol. Med.,* 2, 127, 1979.

28. **Österberg, R.,** Physiology and pharmacology of copper, *Pharmacol. Ther.,* 9, 121, 1980.

29. **Mahler, H. R.,** Uricase, in *The Enzymes,* Vol. 8, Boyer, P. D., Lardy, H., and Myrbäck, K., Eds., Academic Press, New York, 1963, 285.

30. **Farina, B., Faraone Mennella, M. R., and Leone, E.,** Main properties of ox kidney uricase, *Ital. J. Biochem.,* 28, 270, 1979.

31. **Iwasaki, H. and Matsubara, T.,** A nitrite reductase from *Achromobacter cycloclastes, J. Biochem. (Tokyo),* 71, 645, 1972.

32. **Kakutani, T., Watanabe, H., Arima, K., and Beppu, T.,** Purification and properties of a copper-containing nitrite reductase from a denitrifying bacterium, *Alcaligenes faecalis* strain S-6, *J. Biochem. (Tokyo),* 89, 453, 1981.

33. **Chollet, R., Anderson, L. L., and Hovsepian, L. C.,** The absence of tightly bound copper, iron, and flavin nucleotide in crystalline ribulose 1,5-bisphosphate carboxylase-oxygenase from tobacco, *Biochem. Biophys. Res. Commun.,* 64, 97, 1975.

34. **Wishnick, M., Lane, M. D., Scrutton, M. C., and Mildvan, A. S.,** The presence of tightly bound copper in ribulose diphosphate carboxylase from spinach, *J. Biol. Chem.,* 244, 5761, 1969.

35. **Brändén, R.,** Ribulose-1,5-diphosphate carboxylase and oxygenase from green plants are two different enzymes, *Biochem. Biophys. Res. Commun.,* 81, 539, 1978.

36. **Hirata, F., Shimizu, T., Yoshida, R., Ohnishi, T., Fujiwara, M., and Hayaishi, O.,** Copper content of indoleamine 2,3-dioxygenase, in *Iron and Copper Proteins,* Advances in Experimental Medicine and Biology, Vol. 74, Yasunobu, K. T., Mower, H. F., and Hayaishi, O., Eds., Plenum Press, New York, 1976, 354.

37. **Ishimura, Y., Makino, R., Ueno, R., Sakaguchi, K., Brady, F. O., Feigelson, Ph., Aisen, Ph., and Hayaishi, O.,** Copper is not essential for the catalytic activity of L-tryptophan 2,3-dioxygenase, *J. Biol. Chem.,* 255, 3835, 1980.

38. **Gerzer, R., Böhme, E., Hofmann, F., and Schultz, G.,** Soluble guanylate cyclase purified from bovine lung contains heme and copper, *FEBS Lett.,* 132, 71, 1981.
39. **Sellinger, K.-H. and Weser, U.,** Erythrocyte Cu$_2$(haem$_b$)$_2$ protein, *FEBS Lett.,* 133, 51, 1981.
40. **Bertini, I. and Scozzafava, A.,** Copper(II) as probe in substituted metalloproteins, in *Properties of Copper, Metal Ions in Biological Systems,* Vol. 12, Sigel, H., Ed., Marcel Dekker, New York, 1981, chap. 2.

Chapter 2

ELECTRON PARAMAGNETIC RESONANCE OF COPPER PROTEINS

John F. Boas

TABLE OF CONTENTS

I. INTRODUCTION

Copper proteins play an important role in both plant and animal physiology, e.g., hemocyanin (Hc) is the oxygen-carrying protein in the hemolymph of many molluscs and arthropods; cytochrome *c* oxidase (which also contains iron) is the terminal oxidase in the respiratory chain; ascorbate oxidase catalyses the aerobic oxidation of L-ascorbate; and tyrosinase catalyses the ortho-hydroxylation of monophenols to diphenols and the subsequent dehydrogenation of *o*-diphenols to *o*-quinones. In some cases, such as ceruloplasmin, stellacyanin (St), and the azurins (Az), the physiological role of the copper protein may not be as clear-cut, as will be seen from the discussion in other chapters in these volumes and from earlier reviews.[1-5]

A characteristic of many copper proteins is their intense blue color, which is due to an optical absorption at around 600 nm with a molar absorption coefficient of between 1,000 and 10,000 M^{-1} cm^{-1}. This absorption coefficient is some 100 times larger than that found for simple copper amino-acid or peptide complexes.[1-3] A second characteristic of many copper proteins, and one usually associated with the intense blue color, is an unusual electron paramagnetic resonance (EPR) spectrum which has a hyperfine splitting when the magnetic field is parallel to the z or symmetry axis of around 0.0080 cm^{-1} or approximately one half of that of the simple copper complexes.[1-6] As shown by Malmström and co-workers in the early 1960s, both the blue color and the unusual EPR spectrum were associated with the activity of the multicopper enzymes ceruloplasmin and laccase.[6,7] Similar results were found for "blue" copper proteins containing only one copper atom.[1-4,8]

In general, copper proteins in their native state may contain either or both copper(I) or copper(II). Since copper(I) is not a paramagnetic ion, this form will only concern us in passing in this review. However, the copper(II) ion is paramagnetic and should give an EPR signal. The copper(II) ions in proteins have been broadly classified as type 1, type 2, and type 3 on the basis of their EPR signals, and this classification has been reviewed recently by Fee[3] and Boas et al.[4] Type 1 and type 2 are distinguished by their EPR spectral behavior, while type 3 does not give an EPR signal in the native state for reasons which we discuss below. Appendix 1 lists some of the naturally occurring copper proteins and gives representative values of the **g** and hyperfine parameters associated with their EPR spectra.

Type-1 copper(II) is associated with the intense blue color, and is found in the single copper proteins, such as the Azs and plastocyanin (Pc), and in the multicopper enzymes laccase, ceruloplasmin, and ascorbate oxidase.[3] As shown in Appendix 1, the EPR spectrum of type-1 copper is characterized by **g**-values similar to those of simple copper(II) chelates and by a value of the main hyperfine splitting approximately one half of that found for the simple complexes. The origin of this small hyperfine splitting is still one of the contentious issues in EPR studies of copper proteins. As pointed out by Malmström and Vänngåard in 1960, it is difficult to explain this with the same model as that used for the simple complexes.[6] As we shall see, various theoretical models have been proposed for the blue copper site but, until recently, there has been little experimental evidence as to the nature and symmetry of the site and the surrounding ligands.

Type-2 copper gives EPR signals similar to those obtained for the simple copper(II) complexes. It is observed in association with the type-1 copper in ceruloplasmin, laccase, and ascorbate oxidase, and on its own in superoxide dismutase, the monoamine oxidases, galactose oxidase, and some other systems (e.g., see Boas et al.[4]). It has been argued that the label "type 2" should be reserved for those proteins where type-2 copper appears in association with type 1 (e.g., see Fee[3]). However, we will continue to refer to type-2 copper only on the basis of the EPR signals. The type-2 copper site appears to have an approximately square planar configuration, with nitrogen and oxygen being the coordinating ligands.[4,9]

Type-3 copper ions are characterized by an absorption band in the 300-nm region and the absence of an EPR spectrum.[2,4] They are found in ceruloplasmin, laccase, and ascorbate oxidase in association with type-1 and type-2 copper, and as the only form of copper in OxyHc and tyrosinase. The absence of an EPR spectrum led many workers to suggest that the copper was either copper(I) and hence diamagnetic, or was in a mixed valence complex with the copper(I) and copper(II) ions in close association (e.g., see the papers in Reference 1). An opposing school of thought held that the copper ions were copper(II) but were so close together that strong exchange coupling occurred, giving neither an observable paramagnetic susceptibility nor an EPR signal.[10] In the case of Hc, the recent evidence from resonance Raman spectroscopy,[11-13] extended X-ray absorption fine structure (EXAFS),[14-16] and from the magnetic measurements described in this chapter has tended to support the latter view. As in the case of the type-1 copper, the nature of the sites is controversial. There is still little direct evidence as to the type and symmetry of the surrounding ligands, although some indirect evidence is discussed in this chapter and in other chapters in these volumes. In most cases, an X-ray structure determination is still required. Much of the speculation surrounding the type-1 and type-3 copper sites has arisen because it has proved very difficult to reproduce their magnetic and spectral properties in the laboratory.

In the clarification of the nature and function of the active sites of the copper proteins, EPR has had some successes, but has often not lived up to the expectations held for it. In principle, as pointed out by Brill and Venable,[17] one may obtain the following information from EPR studies:

1.   The number of distinct magnetic centers
2.   The orientation with respect to the crystal axes of the symmetry axes of the magnetic centers, when single crystals are studied
3.   The symmetries of the protein environments of the magnetic centers, and the ground states of the transition metal ions
4.   The delocalization of the electrons in the metal ion-protein bonds
5.   Identification of the protein atoms bound to the metal ions if the former have nuclear magnetic moments
6.   An estimate of the distance between paramagnetic metal ions

To these we may add:

7.   An estimate of the proportion of copper ions in each type of magnetic center and of the number of metal ions not in an EPR-detectable center
8.   Details of the chemical changes which take place either when the protein is modified or which occur during a chemical reaction

In practice such richness of detail is often not obtained due to difficulties such as those of obtaining large enough single crystals, of obtaining samples sufficiently free of extraneous copper or other impurities, of obtaining sufficient resolution in the EPR spectra, and of interpreting the data.

While conventional EPR alone has perhaps been of limited value in providing definitive answers on the nature of the copper sites in proteins, it has proved to be a more powerful technique when used in conjunction with magnetic susceptibility measurements and electron resonance techniques such as electron nuclear double resonance (ENDOR) and electron spin-echo spectrometry. When the results from these measurements are combined with those obtained from other techniques such as X-ray diffraction, resonance Raman spectroscopy, nuclear magnetic resonance (NMR), X-ray absorption spectroscopy, and photoelectron spectroscopy, a detailed picture of the copper sites is expected to emerge. Some of these techniques and the results obtained for various copper-protein systems are discussed in other chapters in these volumes.

This chapter will examine some of the general aspects of the application of EPR and its related techniques such as ENDOR and spin-echo spectroscopy to studies of copper proteins, and discuss their advantages and limitations. Since much of the available EPR data has been discussed in earlier reviews,[2-5] we will concentrate on what, to the author, are the more recent highlights, rather than attempt to give a totally complete and comprehensive review.

II. BASIC PRINCIPLES OF EPR

The copper(II) ion has the $3d^9$ electronic configuration with a single unpaired electron. The effective spin is therefore equal to the actual spin of the free ion, namely $S = 1/2$, so that a Kramers doublet is the lowest energy level. In a static magnetic field of flux density B, the doublet is split into two levels which are labeled by the z component of the spin quantum number, M, as $M = -1/2$ and $M = +1/2$. The separation of the levels is given by:

$$E = g\beta B_0 \tag{1}$$

where β is the Bohr magneton and g is the g-factor, a quantity which takes into account the orbital contribution to the spin magnetic moment.[18] For a completely free electron, $g = 2.0023$. B_0 is the magnitude of $\boldsymbol{B}$.

If an alternating magnetic field of frequency v is applied perpendicular to the static magnetic field, magnetic dipole transitions are allowed between the lower level, labeled as $M = -1/2$ and the upper level, labeled as $M = +1/2$, provided that:

$$h v = g\beta B_0 \tag{2}$$

where h is Planck's constant. The EPR transition for a two-level system with $S = 1/2$ is shown in Figure 1. Because the Einstein transition probabilities for stimulated absorption and emission are equal and the probability for spontaneous emission is quite small, there will only be a net absorption of energy from the alternating field if the population of the lower level is greater than that of upper level. The population difference in an applied static field in thermal equilibrium and in the absence of an alternating field is given by the ratio:

$$\frac{N_+}{N_-} = e^{-g\beta B_0/kT} \tag{3}$$

where the N_+ and N_- refer to the population of the $M = +1/2$ and $M = -1/2$ levels, respectively, T is the temperature, and k the Boltzmann constant. Provided that the magnitude of the alternating field is not too large, this population difference is maintained because electrons in the $M = +1/2$ state lose energy to their surroundings and return to the $M = -1/2$ state. There will therefore be a net absorption of energy from the alternating field.

In practice, the frequency of the alternating field is fixed, and the static or d.c. field is swept slowly through the resonance condition given by Equation (2). For reasons we will discuss later in this chapter, the energy levels are not sharply defined, so that the sample absorbs energy over a range of field values, giving an absorption line shape. The absorption of energy can be detected experimentally and its intensity depends on a number of factors, including the sample temperature, the number of electron spins in the sample, and the intensity of the applied alternating field. The source of the alternating radiation fields is usually a klystron, operating in the microwave frequency region between 1 and 100 GHz. The klystron is coupled by a waveguide to the resonant cavity in which the sample is placed and is locked electronically to the resonant frequency. A circulator is often incorporated between the klystron and the cavity to ensure optimum transfer of the microwave power from the klystron to the cavity and from the cavity to the detector. The cavity is coupled to the waveguide with a small iris, whose effective diameter can be controlled by the movement of a device such as a metal pin or a dielectric plunger. When the iris diameter is such that all the incident microwaves are absorbed by the cavity, the cavity is said to be critically coupled. The spectrometer is usually operated close to this condition. Inside the cavity the electromagnetic energy density is much higher than in the attached waveguide. The microwave energy absorbed by the sample at resonance depends on this energy density, expressed as a quality factor, Q, and on the size of the sample in relation to the cavity dimensions given by a filling factor, η.

The absorption of microwave power by the sample at resonance changes the effective coupling of the cavity to the waveguide and hence the amount of microwave power reflected

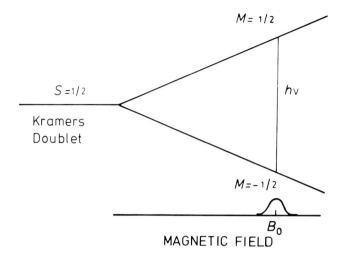

FIGURE 1.   EPR in a Kramers doublet, where $h\nu = g\beta B_0$.

from the cavity. There will therefore be a change in the microwave-power level at the detector. In the earliest EPR spectrometers, this change was observed directly by measuring the current through the crystal detector with a meter or an oscilloscope. However, direct detection is not very sensitive and nowadays the most common method of detection is to modulate the d.c. or slowly varying external magnetic field at 100 kHz and employ phase sensitive detection. The output of the detector-amplifier system is then the first derivative of the absorption and is usually observed on a chart recorder as a function of the external magnetic field. A basic EPR spectrometer is shown in Figure 2.

Typical microwave frequencies are around 9 GHz (3 cm wavelength), so that for copper(II) ions, which have typical *g*-values of around 2.2, magnetic field strengths of approximately 0.3 T (3000 gauss) are required. These are usually provided by an electromagnet, although some EPR spectrometers operating at very high frequencies ($\approx$70 GHz) use a superconducting magnet. The magnetic field is swept linearly through the resonance at a rate which depends on the characteristics of both the sample and the spectrometer. A typical field sweep rate may be around 20 mT (200 gauss)/min. Some present day spectrometers have provision for repetitive sweeping of the magnetic field through the resonance, allowing the employment of a computer of average transients (CAT) and hence greater sensitivity, since the signals add proportionately to the number of sweeps while the noise adds as the square root. The experimental techniques and the factors affecting EPR spectrometer sensitivity are discussed by numerous authors, and the interested reader may refer to the books by such authors as Poole,[19] Alger,[20] and Ingram[21] for more details. A little later in this chapter we will discuss some of the factors affecting EPR spectral sensitivity and resolution which are of specific relevance to studies of copper proteins.

The observation of an EPR spectrum alone, while giving information about the valence state of the copper site, does not give directly any other information of the type outlined by Brill and Venable.[17] To obtain such information it is necessary to represent the spectrum mathematically and find values of parameters which can be related to structural details such as site symmetry and bonding. The spin Hamiltonian is a convenient means of representing an EPR spectrum and is simply the sum of a number of terms involving these parameters and the electron and nuclear spin operators.

If a spin Hamiltonian can be fitted to the experimental spectrum, we are in a position to use the parameters so obtained to make deductions about the properties of the copper-ion sites. It should be noted here that the spin Hamiltonian is a quantum mechanical operator

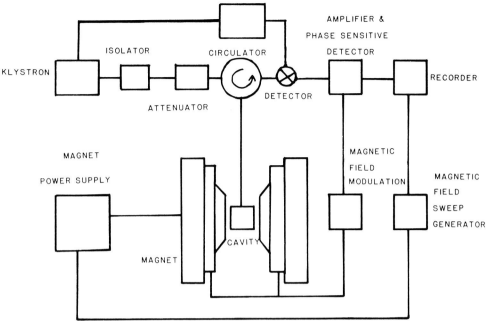

FIGURE 2.   A basic EPR spectrometer.

which, when applied to the wave functions of the appropriate levels, gives their energies in terms of the spin Hamiltonian parameters, the applied magnetic field, and the appropriate quantum numbers.

In the next two sections we discuss the development of the appropriate spin Hamiltonians and their application in the interpretation of the EPR spectra of copper proteins. A full account of the spin Hamiltonian and its use is given by Abragam and Bleaney in their most comprehensive treatise, and we will draw heavily on their treatment of this subject.[18]

III. THE ANALYSIS AND INTERPRETATION OF THE EPR SPECTRA OF TYPE-1 AND TYPE-2 COPPER(II) IONS

A. The Spin Hamiltonian for Mononuclear Copper(II) Ions

Classically, we write the energy of a dipole with a magnetic moment $\boldsymbol{\mu}$ in a field $\boldsymbol{B}$ as:

$$E = \boldsymbol{\mu}\cdot\boldsymbol{B} \tag{4}$$

which may be related to the Hamiltonian operator in the quantum mechanical formulation as:

$$\hat{\mathcal{H}} = \beta(\boldsymbol{B}\cdot\mathbf{g}\cdot\hat{\boldsymbol{S}}) \tag{5}$$

where Hamiltonian $\mathcal{H}$ is the operator which, when applied to the wave functions of the system, gives the energy eigenvalues, β is the Bohr magneton, equal to $-e/2mc$ where e and m are the electronic charge and mass respectively and c is the velocity of light. The spin operator vector, $\hat{\boldsymbol{S}}$, refers to the effective spin of magnitude 1/2 for the copper(II) ion. $\mathbf{g}$ is the $\mathbf{g}$ matrix, which describes the interaction of the orbital and spin angular momenta via the spin-orbit coupling. $\mathbf{g}$ is not a tensor even though it has some tensor properties and

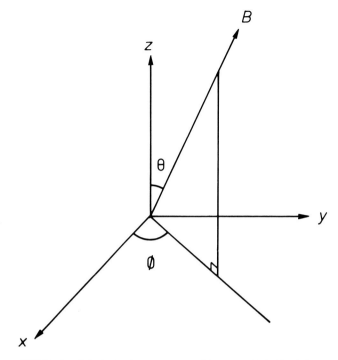

FIGURE 3.   Spherical polar coordinates of the magnetic field relative to
the principal axes of the spin Hamiltonian.

is often referred to as one.[22] For a free electron, **g** has no orbital angular momentum
component and is isotropic, with a value of $g = 2.0023$. An electron bound to an ion will
of course have both orbital and spin angular momentum, although for 3d transition series
ions such as copper(II), the orbital angular momentum is very largely "quenched" by the
crystal field. Hence the electron magnetic moment is regarded as spin only, with the remaining orbital contribution being represented by the **g** matrix. In general, this has nine
components, due to the dependence of the amount of orbital angular momentum admixture
on the surroundings, and we must write the Hamiltonian (Equation 5) as:[23]

$$\hat{\mathcal{H}} = \beta \left\{ g_{xx} B_x \hat{S}_x + g_{yy} B_y \hat{S}_y + g_{zz} B_z \hat{S}_z + g_{xy} B_x \hat{S}_y + g_{yx} B_y \hat{S}_x \right.$$
$$\left. + g_{yz} B_y \hat{S}_z + g_{zy} B_z \hat{S}_y + g_{zx} B_z \hat{S}_x + g_{xz} B_x \hat{S}_z \right\} \tag{6}$$

In all but a few cases where the symmetry of the surrounding ions is very low, $g_{xy} = g_{yx}$, $g_{yz} = g_{zy}$, and $g_{zx} = g_{xz}$, and the Hamiltonian may be diagonalized by an appropriate
choice of the x, y, and z directions to give:

$$\hat{\mathcal{H}} = \beta \left(g_x B_x \hat{S}_x + g_y B_y \hat{S}_y + g_z B_z \hat{S}_z \right) \tag{7}$$

The quantities g_x, g_y, and g_z are the principal g-values and the x, y, and z axes are the
principal axes of the **g** matrix. The g value at which resonance occurs (see Equation 1) for
a particular orientation of the magnetic field with respect to the principal axes is given by:[24]

$$g^2 = l^2 g_x^2 + m^2 g_y^2 + n^2 g_z^2 \tag{8}$$

where l, m, and n are the direction cosines of the magnetic field with respect to the principal
axes. This expression contains the two polar angles θ and ϕ as shown in Figure 3.

The familiar relationship:

$$g^2 = g_{\parallel}^2 \cos^2\theta + g_{\perp}^2 \sin^2\theta \qquad (9)$$

arises when the **g** matrix has axial symmetry ($g_x = g_y$) and θ is the angle between the magnetic field and the z axis which is taken to be the symmetry axis.

In practice, the microwave frequency is fixed, and from Equations (2) and (8) we see that resonance will occur at a magnetic field which varies with the angle between the field direction and the directions of the principal axes of the **g** matrix.

As shown in Appendix 1, the g-values of copper(II) ions in proteins generally show axial or near axial symmetry, with $g_{\parallel}$ (the g-value along the z or symmetry axis) ≈ 2.3 and $g_{\perp}$ (the g-value in the x-y plane) ≈ 2.05. The spectra of powders or frozen solutions generally show prominent peaks only at the magnetic fields corresponding to the principal g-values.

A number of additional interactions must be considered in the discussion of the spin Hamiltonian for copper(II). The most important of these is the nuclear hyperfine interaction, which arises from the interaction between the nuclear magnetic moment and the magnetic field at the nucleus generated by the magnetic electrons.[25] Since the copper nucleus has a spin $I = 3/2$, each electron spin level is split into four components, with values of the z component of the nuclear spin quantum number m of $-3/2$, $-1/2$, $+1/2$, and $+3/2$. Thus the electron resonance line is split into four approximately equally spaced components, as allowed transitions take place between pairs of states having the same values of m. As shown in Figure 4, the splitting of the hyperfine components is $A/g\beta$ as a first approximation, where A is the magnitude of the nuclear hyperfine interaction. Because A has a directional dependence, the hyperfine interaction is more generally written as:[26]

$$\hat{\mathcal{H}}_{hf} = \hat{\boldsymbol{S}} \cdot \mathbf{A} \cdot \hat{\boldsymbol{I}} \qquad (10)$$

where **A** is another interaction matrix which can be diagonalized by an appropriate choice of axes. $\hat{\boldsymbol{I}}$ is the nuclear spin operator vector. The general form of $\hat{\boldsymbol{S}} \cdot \mathbf{A} \cdot \hat{\boldsymbol{I}}$ is

$$A_{xx}\hat{S}_x\hat{I}_x + A_{yy}\hat{S}_y\hat{I}_y + A_{zz}\hat{S}_z\hat{I}_z + A_{xy}\hat{S}_x\hat{I}_y + A_{yx}\hat{S}_y\hat{I}_x$$
$$+ A_{yz}\hat{S}_y\hat{I}_z + A_{zy}\hat{S}_z\hat{I}_y + A_{zx}\hat{S}_z\hat{I}_x + A_{xz}\hat{S}_x\hat{I}_z \qquad (11)$$

where x, y, and z are the principal axes of the **g** matrix. Provided that the principal axes of the hyperfine matrix are the same as those of **g**, we may write Equation (11) in the form:

$$\hat{\mathcal{H}}_{hf} = A_x\hat{S}_x\hat{I}_x + A_y\hat{S}_y\hat{I}_y + A_z\hat{S}_z\hat{I}_z \qquad (12)$$

or, for axial symmetry:

$$\hat{\mathcal{H}}_{hf} = A_{\parallel}\hat{S}_z\hat{I}_z + A_{\perp}(\hat{S}_x\hat{I}_x + \hat{S}_y\hat{I}_y) \qquad (13)$$

Cases where the principal axes of **g** and **A** are not coincident in copper(II) complexes are now relatively well documented, e.g., in the review by Pilbrow and Lowrey.[22] To date noncoincidence of **g** and **A** has not been observed in copper proteins, probably because the effects are obscured by the rather broad lines and the random orientations of the copper complexes. Single crystal studies may well reveal noncoincidence. In contrast a study of the cobalt(II) ion in vitamin B-12r complexes, in which frozen solutions were used, did show that **g** and **A** were noncoincident.[27] However, the linewidths were somewhat less than those found in the copper(II) proteins.

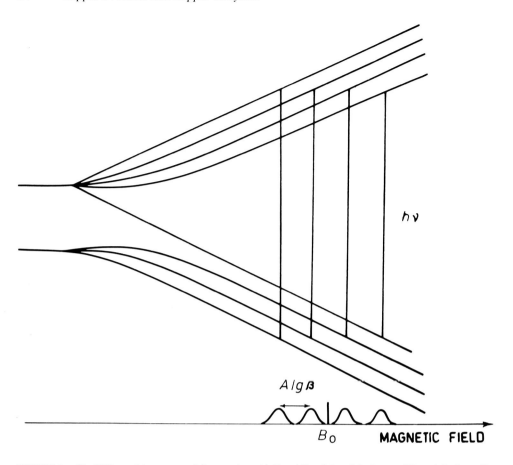

FIGURE 4. The EPR transitions expected from an ion with $S = 1/2$ and $I = 3/2$. $h\nu = g\beta B_0$, and the hyperfine splitting is $A/g\beta$.

To first order in perturbation theory, assuming coincidence, the hyperfine splitting is given by:[28]

$$g^2 A^2 = l^2 g_x^2 A_x^2 + m^2 g_y^2 A_y^2 + n^2 g_z^2 A_z^2 \tag{14}$$

where l, m, and n are the direction cosines of Equation (8). The hyperfine splitting is seen to be angular dependent, and in a powder or frozen solution the random orientations of the magnetic complexes will tend to smear out the hyperfine structure, except for those orientations where the external magnetic field is close to a magnetic axis. Furthermore, in many copper proteins, the linewidths are so large that hyperfine structure is only resolved for those orientations where the magnetic field is parallel to the axis along which the hyperfine splitting is the largest. This is often the axis for which the g-value is largest, and is chosen as the z axis. Hyperfine structure is often not resolved in the directions perpendicular to the z axis, i.e., along the x and y directions. These effects were first noted for copper proteins by Malmström and Vänngård.[6] As recorded in Appendix 1, typical values of A_z (or $A_{\parallel}$) for the "nonblue" or type-2 copper ions are in the range 0.015 to 0.020 cm^{-1}, while for the "blue" or type-1 copper ions they are in the range 0.003 to 0.01 cm^{-1}. Typical values of A_x and A_y (or $A_{\perp}$) are 0.003 cm^{-1} or less, and we note that these are often not as accurately determined because of the lack of resolution, as described above.

An additional complication which should be noted here is that copper has two isotopes, ^{63}Cu and ^{65}Cu, each with $I = 3/2$, but with slightly different magnetic moments. The isotope

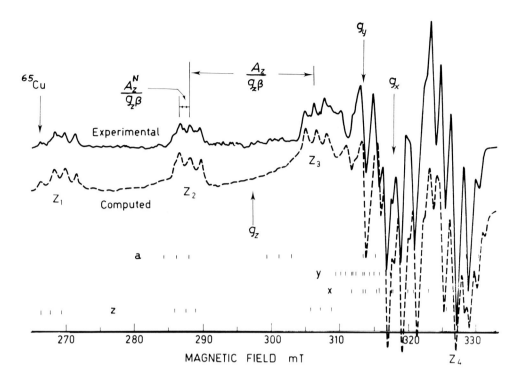

FIGURE 5. EPR spectrum of copper(II) thiosemicarbazone in dimethylformamide at 123 K and a microwave frequency of 9,139 MHz. The stick spectra labeled *x*, *y*, and *z* refer to the field positions for transitions along the *x*, *y*, and *z* axes, respectively, while *a* refers to the high field angular anomaly. Z_1 to Z_4 refer to spectral components for the magnetic field along the molecular *z* axis. Other features to note are the partial resolution of the ^{63}Cu and ^{65}Cu resonances, the hyperfine splitting due to the copper nucleus ($= A_z/g_z\beta$), and the superhyperfine splitting due to a single ^{14}N nucleus ($= A_z^N/g_z\beta$). (From Boas, J. F., Pilbrow, J. R., and Smith, T. D., *Magnetic Resonance in Biological Systems*, Berliner, L. J. and Reuben, J., Eds., Plenum Press, New York, 1978, chap. 7. With permission.)

effect can be seen in cases where the resonance lines are exceptionally narrow (e.g., see Figure 5), but usually only results in a line broadening. Some authors have reduced this broadening to some extent by isotopic substitution of the less abundant ^{65}Cu with ^{63}Cu.[29,30]

Three further interactions need to be considered in our discussion of the spin Hamiltonian for copper(II) ions. The nuclear quadrupole interaction arises from the interaction of the nuclear quadrupole moment (since, for the copper nucleus, $I = 3/2$) with the electric field gradient. The interaction shifts the relative positions of the nuclear hyperfine levels and thus alters the observed hyperfine splittings. It also "mixes" the hyperfine levels, so that "forbidden" transitions can take place in which the quantum number *m* can change by 1 or 2. The quadrupole interaction is a true tensor, and its contribution to the spin Hamiltonian is written as:[31]

$$\hat{\mathcal{H}}_q = \hat{I} \cdot \mathbf{P} \cdot \hat{I} \qquad (15)$$

This can be treated in a similar manner to the **A** and **g** matrices, as **P** will generally have the same principal axes as **A**. In this case we may write the interaction in the form:

$$\hat{\mathcal{H}}_q = P_x \hat{I}_x^2 + P_y \hat{I}_y^2 + P_z \hat{I}_z^2$$

or as
$$\hat{\mathcal{H}}_q = P_{\parallel} \left\{ \left[\hat{I}_z^2 - \frac{1}{3} I(I+1) \right] + \frac{1}{3}\eta \left(\hat{I}_x^2 - \hat{I}_y^2 \right) \right\}$$

where
$$P_{\parallel} = \frac{3}{2} P_z \text{ and } \eta = (P_x - P_y)/P_z \qquad (16)$$

When **P** is axial, Equation (16) becomes

$$\hat{\mathcal{H}}_q = P_\parallel \left[\hat{I}_z^2 - \frac{1}{3} I(I+1) \right]$$ (17)

Both forbidden transitions and shifts in line positions due to quadrupole interactions have been observed in single crystals of a number of copper complexes (e.g., see Bleaney et al.,[32] Maki and McGarvey,[33] White and Belford[34]). Quadrupole effects are less readily observed in powder and frozen solution spectra, and to date, none have been observed by EPR in the spectra of copper proteins. However, the quadrupole coupling tensor has been determined for copper(II) in stellacyanin (St) by means of ENDOR.[35] Typical values of **P** for simple copper(II) systems range from 0.0008 cm^{-1} (Reference 33) when oxygen is one of the coordinating ligands to 0.00007 cm^{-1} in some dithiocarbamate complexes.[34] A case where a substantial quadrupole effect was thought to have been observed in a frozen solution was that of cobalt(II) 4,4',4'',4'''-tetrasulfophthalocyanine.[36] However, it was later shown that the original interpretation was not correct,[37] due to an error in the calculation, and in fact the results were best fitted by a value of **P** = 0.

The interaction of the magnetic moment of the copper nucleus with the external magnetic field also needs to be considered. However, its effect is expected to be quite small and has not been observed in any of the EPR or ENDOR experiments conducted to date on the copper proteins. For the record, the term in the Hamiltonian is written as:

$$\hat{\mathcal{H}}_N = \beta_n (\boldsymbol{B} \cdot \boldsymbol{g}^I \cdot \hat{\boldsymbol{I}})$$ (18)

where $\boldsymbol{g}^I$ is a matrix similar to **g**.[38] The effects of this term are less than 1/1,000 that of the term in **g**. β_n is the nuclear magneton.

The final interaction to be considered arises from the interaction of the magnetic moments of ligand nuclei with the electron spin. Once again, the interaction is a matrix, but it may have different principal axes and different components for each ligand nucleus. This may lead to very complex spectra, particularly if the ligands are not identical, and we will not discuss the detailed interpretation of such spectra here. Nevertheless, the interpretation of the ligand hyperfine structure (generally referred to as superhyperfine structure) is a key element in descriptions of the binding sites and reaction mechanisms of copper proteins.

The treatment of the superhyperfine part of the Hamiltonian follows that used for the copper hyperfine interaction (Equations 10 to 14), although first-order perturbation theory is generally sufficient when the energies are calculated. The inclusion of these interactions into the spin Hamiltonian requires a term of the form:[39]

$$\hat{\mathcal{H}}_L = \sum_i (A_{ix} \hat{S}_x \hat{I}_{ix} + A_{iy} \hat{S}_y \hat{I}_{iy} + A_{iz} \hat{S}_z \hat{I}_{iz})$$ (19)

where the sum over i is over all ligand nuclei with a nuclear spin I_i and hyperfine interaction constants A_i. The expression (Equation 19) assumes that the axes of the superhyperfine interaction matrix are coincident with those of **g** and **A**. However, this may not always be the case. An example of noncoincident hyperfine and superhyperfine matrix axes is that of copper(II) phthalocyanine. As pointed out by Guzy et al.[40] (and recently reiterated by Brown and Hoffman[41]), the principal axes of the superhyperfine matrices are along the copper-nitrogen bond direction, and not along the molecular axis as assumed by Harrison and Assour.[42]

The most common nucleus giving rise to superhyperfine structure in the EPR spectra of copper proteins is ^{14}N. This has a nuclear spin $I = 1$, and thus each hyperfine line is split into three approximately equally spaced components of equal intensity. There are a number of references in the literature to the observation of nitrogen superhyperfine structure effects, both for copper proteins in their natural state (e.g., see Bereman and Kosman[30]) and for copper proteins after chemical treatment with imidazole[30] and such ions as azide,[43] cyanide,[44] or nitrite.[45] Interaction with fluoride ions has also been observed in some copper proteins.[30,46,47] Interactions with other halide ions, thiocyanate, and acetate have been observed to result in changes in the EPR spectra of copper proteins, even though the superhyperfine splitting is not observed directly.[47] Substitution of isotopes with a nuclear magnetic moment sometimes results in a broadening of the EPR signals, e.g., the substitution of ^{17}O for ^{16}O (see Aasa et al.[48]) and of ^{13}C for ^{12}C (e.g., see Marwedel et al.[44]).

To summarize the above discussion, the spin Hamiltonian for mononuclear copper(II) ions may be written as:

$$\hat{\mathcal{H}} = \beta(g_x B_x \hat{S}_x + g_y B_y \hat{S}_y + g_z B_z \hat{S}_z) \quad : \quad \text{the electron Zeeman interaction}$$

$$+ A_x \hat{S}_x \hat{I}_x + A_y \hat{S}_y \hat{I}_y + A_z \hat{S}_z \hat{I}_z \quad : \quad \text{the hyperfine interaction}$$

$$+ P_{\parallel}\left\{\left[\hat{I}_z^2 - \frac{1}{3} I(I+1)\right] + \frac{1}{3}\eta(\hat{I}_x^2 - \hat{I}_y^2)\right\} \quad : \quad \text{the quadrupole interaction}$$

$$- \beta(g_x^I B_x \hat{I}_x + g_y^I B_y \hat{I}_y + g_z^I B_z \hat{I}_z) \quad : \quad \text{the nuclear Zeeman interaction}$$

$$+ \sum_i (A_{ix}\hat{S}_x \hat{I}_{ix} + A_{iy}\hat{S}_y \hat{I}_{iy} + A_{iz}\hat{S}_z \hat{I}_{iz}) \quad : \quad \text{the ligand superhyperfine interaction} \quad (20)$$

This expression assumes that all tensor and matrix axes are coincident. It should be emphasized that the spin Hamiltonian is a quantum mechanical operator, and includes terms containing the electron spin operators $\hat{S}_x$, $\hat{S}_y$, and $\hat{S}_z$, the nuclear spin operators $\hat{I}_x$, $\hat{I}_y$, and $\hat{I}_z$ and the nuclear spin operators of the ligands, namely $\hat{I}_{ix}$, $\hat{I}_{iy}$, and $\hat{I}_{iz}$. It also involves the components of the applied magnetic field, namely B_x, B_y, and B_z. The magnetic properties of the ion are determined by the parameters of the spin Hamiltonian, namely the components of the **g, A, g^I, and A$_i$** matrices and the **P** tensor. However, before discussing the relationship between these parameters and the bonding and symmetry properties of the ion, it is necessary to discuss how they are determined from the experimental spectrum.

B. Determination of the Spin Hamiltonian Parameters

To determine the spin Hamiltonian parameters, the experimental spectrum is compared to a simulated spectrum. The latter is calculated using trial parameters and the expressions for the magnetic fields at which transitions occur. A new set of trial parameters is determined from the comparison of experimental and simulated spectra, and the process repeated until satisfactory agreement is obtained between the two.

The theoretical expressions for the resonance fields are obtained from the expressions for the energies of the states between which the transitions occur. In turn, these expressions for the energy are found by applying the spin Hamiltonian operator to the wave functions of the appropriate states, which may be written in terms of the magnetic quantum numbers of the electron, central nucleus, and ligand nuclei, namely, M, m, and m', respectively. EPR transitions are allowed between pairs of states labeled as $|M,m,m'>$ and $|M+1,m,m'>$, and the transition energy is calculated by subtracting the energies of the states between which the transition occurs. The allowed transitions are given by $\Delta M = 1$, $\Delta m = 0$, and $\Delta m' = 0$, but forbidden transitions, generally of much lesser intensity, may also be observed when $\Delta M = 1$, $\Delta m \neq 0$, and $\Delta m' \neq 0$.

The most convenient method of calculating the energy of a state is by using perturbation theory, a full account of which is given by Abragam and Bleaney[49] and will not be reproduced here. Second-order perturbation theory is usually sufficient, although ENDOR experiments may require the use of exact matrix diagonalization procedures.

When single crystals are studied, the experimental spectrum can be related relatively simply to the expressions for the magnetic fields at which transitions occur. The directions of the magnetic axes and the principal values of the **g** and hyperfine matrices can be found by rotating the crystal with respect to the magnetic field direction. The relationship between the crystal and magnetic axes is also easily determined, at least in principle.[50] An example of this type of study is that of copper(II) in insulin single crystals.[51]

Unfortunately most copper proteins cannot be obtained as single crystals large enough for EPR studies, so that powders or frozen solutions must be used. Spectra of chemically modified proteins are also obtained from frozen solutions. In such systems, the copper(II) sites are randomly oriented with respect to the magnetic field, and the spectrum is the sum of the contributions from all orientations. Consideration of Equations (8) and (14) shows that the peaks in the experimental spectrum will occur at values of the magnetic field which correspond to the field being along the principal directions of the **g** and hyperfine matrices, at least to first order. This is because the change in g and A in these equations per unit angle is least rapid in the vicinity of these directions.[52] In practice, determination of the principal g and hyperfine values by direct inspection from the experimental results is often neither possible nor accurate, so that a computer simulation needs to be performed using trial values of the spin Hamiltonian parameters. The simulated spectrum can then be compared with the experimental spectrum, new trial parameters chosen, and the spectrum simulated again. This process can be repeated until satisfactory agreement is found between experimental and simulated spectra. Because of the random distribution of the orientations of the complexes with respect to the magnetic field, an integration over all angles must be performed. Since this cannot be done analytically, the usual method is to approximate the integral by using the trapezoidal rule and summing over finite angular intervals. Detailed discussions of the procedure are given by such authors as Venable,[53] Boas et al.,[54] and Boyd et al.,[55] and only an outline is given below.

C. Computer Simulation of Spectra from Mononuclear Copper(II) Ions

The EPR signal intensity of a transition labeled as n at a magnetic field B_0 which is directed at polar angles θ and ϕ to the axes of the magnetic complex is given by:

$$I(B_0) = P(n,\theta,\phi) \, G \, (B_0,n,\theta,\phi) \tag{21}$$

where $P(n,\theta,\phi)$ is the transition probability for the nth transition at the angle (θ,ϕ) and G (B_0,n,θ,ϕ) is the lineshape function. The total EPR signal intensity at the field B_0 is given by the sum of the contributions from all transitions and all angles, as given by Equation (22), namely:

$$F(B) = \sum_{n=1}^{N} \sum_{\theta=0}^{\theta'} \sum_{\phi=0}^{\phi'} P(n,\theta,\phi) \, G \, (B_0,n,\theta,\phi) \, (\Delta \cos\theta) \, \Delta\phi \tag{22}$$

where N is the number of transitions, θ' and ϕ' are the upper limits of the polar angles θ and ϕ, and $(\Delta \cos\theta)\,\Delta\phi$ is the solid angle segment for the copper(II) sites whose symmetry axes lie at an angle to the magnetic field in the range $(\theta, \theta + \Delta\theta; \phi, \phi + \Delta\phi)$.

The full simulated spectrum is obtained by the calculation of $F(B_0)$ at intervals of the magnetic field over the range of interest, e.g., a spectrum such as that shown in Figure 5 was calculated at intervals of 0.2 mT (2 gauss) over a range of magnetic field from 260 mT (2,600 gauss) to 340 mT (3,400 gauss).

For a copper(II) ion, ignoring ligand superhyperfine interactions, there are four allowed transitions, labeled as $< -1/2, m| \rightarrow <1/2, m|$, where m takes the values of $\pm 3/2$ or $\pm 1/2$. Each transition has the same probability at a given value of θ and ϕ, so that $P(n,\theta,\phi)$ only incorporates the effect of the g anisotropy. As shown by Aasa and Vänngård,[56] the g anisotropy factor for axial symmetry in the field-swept domain is

$$\frac{g_\perp^2}{g} \left[\frac{g_\parallel^2}{g^2} + 1 \right] \tag{23}$$

with a more complicated expression for lower than axial symmetry. These authors point out that previous expressions for the g anisotropy factor were derived for the frequency-swept domain, and have to be corrected by a factor of $1/g$ if used in the field-swept situation applicable to conventional EPR spectrometers (e.g., see Bleaney,[57] Kneubühl and Natterer,[58] and Pilbrow[59]). Bonamo and Pilbrow[60] have shown that it is important to use the expressions of Aasa and Vänngård,[56] such as Equation (23), if the intensities of mononuclear copper(II) systems are to be simulated correctly.

The line shape function $G(B_o,n,\theta,\phi)$ is usually either Gaussian or Lorentzian, and is expressed as the first derivative to facilitate comparison with experiment. The Gaussian first derivative function is

$$G'(B_O,n,\theta,\phi) = \left\{ [B_O - B_O(n,\theta,\phi)] / \sigma^3 \right\} \exp \left\{ -[B_O - B_O(n,\theta,\phi)]^2 / 2\sigma^2 \right\} \tag{24}$$

where the linewidth parameter σ is the standard deviation of a Gaussian distribution and corresponds to the peak-to-peak half width of the first derivative. A similar expression exists for Lorentzian lines, but in the author's experience, Gaussian lineshapes have generally proved adequate in computer simulation of experimental spectra of copper(II) systems.

The linewidth parameter may also be anisotropic, and can be written as (cf. Equation 14):

$$\sigma^2 = (\sigma_x^2 g_x^2 \sin^2\theta \cos^2\phi + \sigma_y^2 g_y^2 \sin^2\theta \sin^2\phi + \sigma_z^2 g_z^2 \cos^2\theta)/g^2 \tag{25}$$

or, for axial symmetry:

$$\sigma^2 = (\sigma_\parallel^2 g_\parallel^2 \cos^2\theta + \sigma_\perp^2 g_\perp^2 \sin^2\theta)/g^2 \tag{26}$$

Examples of cases where an anisotropic linewidth was required to simulate the spectrum are given by Schoot Uiterkamp et al.[61]

The computed line shape is obtained at equal increments of the magnetic field over the required range by summing the contribution from each transition as it is calculated. Typically, one may use 40 intervals for θ and 20 for ϕ, depending on the overall symmetry, the need to economize on computer time, and the accuracy and degree of resolution required. If the overall symmetry is axial, θ is taken from 0 to $\pi/2$ and of course $\phi = 0$. Orthorhombic symmetry requires $0 \leqslant \theta \leqslant \pi/2$ and $0 \leqslant \phi \leqslant \pi/2$; monoclinic symmetry requires $0 \leqslant \theta \leqslant \pi$ and $0 \leqslant \phi \leqslant \pi/2$; and triclinic symmetry requires $0 \leqslant \phi \leqslant \pi$ and $0 \leqslant \phi \leqslant \pi$. To save computer time, the contribution of a given transition when the line shape is Gaussian can often be truncated at $\pm 3\sigma$ without noticeably affecting the simulations. Fortran programs for mononuclear spectra can be executed in a few seconds on medium-sized computers, and an example of the type of information that can be obtained is shown in Figure 5. It should be pointed out that such well-resolved spectra are generally not obtained from copper(II) proteins, and that it is correspondingly more difficult to extract the parameters required.

D. The Interpretation of EPR *g*-Values and Hyperfine Constants
1. The Total Hamiltonian and the Reduced Hamiltonian

The fitting of the parameters of the spin Hamiltonian to an experimental spectrum, while often a task of considerable complexity, does not give any direct insight into the structural and bonding parameters of the copper(II) ion. To obtain this insight it is necessary to relate the *g* and *A* values to the qualities of ultimate interest, namely the wave functions of the copper ion and its near neighbors. In order to do so, one must consider the derivation of the spin Hamiltonian from the total Hamiltonian as outlined by Pryce[62] and Abragam and Pryce.[63,64] An extensive discussion is given by Abragam and Bleaney.[65] The aim is to derive analytic expressions for the *g*-values and hyperfine constants in terms of the splittings between the energy levels and the admixtures of the appropriate wave functions, a process achieved by the application of a reduced form of the total Hamiltonian of the system to the wave functions as outlined below.

The total Hamiltonian comprizes terms arising from the following interactions, listed in decreasing order of magnitude:

1. The interaction of the electrons with the Coulombic field of the nucleus, modified by the repulsive field of the other electrons; these electrostatic interactions produce splittings of the energy levels of around 10^5 cm^{-1}
2. Spin-orbit coupling, i.e., the coupling of spin and orbital angular momenta; this produces splittings of between 10^2 and 10^3 cm^{-1}
3. The crystal- or ligand-field interaction, i.e., the electrostatic interaction of the electrons of the central ion with its surroundings, which produces splittings of between 10^2 and 10^3 cm^{-1}
4. Magnetic spin-spin interactions between the electrons of the ion
5. Interactions with the external magnetic field
6. Nuclear hyperfine interactions due to the nucleus of the central ion and surrounding ions

For iron-group ions such as copper(II), the crystal-field interaction is generally much larger than the spin-orbit coupling, and can be decomposed into two parts, namely a dominant part which has octahedral or tetrahedral symmetry and a weaker part which has lower symmetry and describes the distortions from octahedral or tetrahedral symmetry. The effect of the dominant part of the crystal-field interaction is to split the energy levels of the free ion ground state which have been determined using Hund's rules. In the case of copper(II) the free ion ground state is $^2D_{5/2}$, and this is split into a triplet and a doublet. Distortions from octahedral or tetrahedral symmetry, which must occur for the copper(II) ion since it is an odd electron system, split the triplet or doublet to leave an orbital singlet as the ground state.

As discussed by Abragam and Pryce,[63] the remaining interactions can be regarded as forming a reduced Hamiltonian, containing terms which express changes from the lowest energy value of the ground state and second-order terms such as those due to spin-orbit coupling. The reduced Hamiltonian is then applied to the wave functions of the ground and excited states, and since the orbital magnetic moment is very largely quenched by the crystal field, a Hamiltonian can be written for the ground state in terms of the spin only. The constants of this "spin only" Hamiltonian, namely **g, A, P**, etc., which can be determined experimentally by EPR, are expressed in terms of the wave functions of the ground and excited states and the splitting between these. The interpretation of the spin Hamiltonian parameters thus requires a decision as to which wave functions should be used. The problem has been approached with various degrees of sophistication but as we shall see, the more sophisticated calculations do not always lead to a better understanding.

2. *Ligand-Field and Molecular Orbital Calculations*

In principle, it would be desirable to include in the calculations all the metal orbitals and all the ligand orbitals, using known structural parameters and wave functions, and thus determine theoretical linear combination of atomic orbitals (LCAO) coefficients and molecular orbital (MO) energies. The g and hyperfine constants can then be evaluated by applying the appropriate Hamiltonian and compared with the experimentally determined values. An example of this type of calculation for copper(II) in dialkyldiselenocarbamates is given by Keijzers and de Boer,[66] who used the known crystal structure, Slater-type orbitals, and the LCAO-MO extended Hückel method. They found good agreement between experimental and calculated values of g and A. However, they concluded that neither the principal values of **g** nor the hyperfine couplings could be used to calculate the LCAO coefficients directly from experimental data because of the large number of unknown coefficients and energies.

Unfortunately, in the copper proteins the required experimental parameters are not always determined and more important, the nature of the coordinating ligands and their geometry are generally not known. The construction of the appropriate molecular orbital scheme is therefore a matter of speculation, and one may remark that for the copper site in proteins, a calculation of the type performed by Keijzers and de Boer[66] is of limited value and considerable difficulty. A qualitative molecular orbital treatment is given by Nickerson and Phelan,[67] but they do not attempt any calculation of g or A values. Furthermore, their basic assumptions of a square planar local symmetry and of binding of the copper to four deprotonated peptide backbone nitrogens, while possibly valid for ribonuclease, do not appear to be valid for other copper(II) proteins, particularly not for the ''blue'' copper proteins.

Less sophisticated attempts to introduce molecular orbital coefficients were first made by Maki and McGarvey in 1958.[33] The approach has generally been to write the wave functions in terms of orbital admixtures, apply the Hamiltonian to derive expressions for the g and A values, and then calculate the bonding parameters using the experimental data. These calculations all presuppose a knowledge of the ground state and the adjoining ligands, and while they have been successfully applied to a number of relatively well-known and simple complexes, their application to copper proteins, particularly the blue copper proteins, is not likely to give much useful information in the absence of structural details. Details of this type of analysis are given, e.g., by Buluggiu et al.,[68] Swalen et al.,[69] Harrison and Assour,[42] Weeks and Fackler,[70] and Kivelson and Neiman.[71]

3. *Crystal-Field Calculations*

a. *Introduction*

In calculations using the crystal-field approximation, the only wave functions considered are those of the copper(II) ion, with the sole effect of the ligands being on the separation and ordering of the orbitals. The simplest approach is to consider the 3d orbitals only, while a more sophisticated approach is to consider the admixtures of 4s and 4p character into the 3d wave functions.

The crystal field acting on a 3d transition series ion can be resolved into an octahedral or a tetrahedral component and a smaller component of lower symmetry describing the distortion from these symmetries.[63] The application of either an octahedral or a tetrahedral crystal field to the 3d orbital set of the copper(II) ion results in the splitting of the orbital levels into a triplet and a doublet.[63,72,73] Since the $3d^9$ electronic configuration of the copper(II) ion can be regarded as a hole in an otherwise full shell, the unpaired electron will reside in one of the d-type orbitals which can be labeled as $|xy>$, $|yz>$, $|xz>$, $|x^2-y^2>$, $|3z^2-r^2>$. It is easily shown that if the major component of the crystal field has octahedral symmetry the doublet is lowest, so that the unpaired electron resides in either the $|x^2-y^2>$ or the $|3z^2-r^2>$ orbital. Similarly, if the major component of the crystal field has tetrahedral symmetry, the triplet labeled as $|xy>$, $|yz>$, and $|xz>$ is lowest, and the unpaired electron of the copper(II) ion resides in one of these orbitals (e.g., see Carrington and McLachlan[74]). Distortions from pure octahedral or tetrahedral symmetry, which must occur as a result of the Jahn-Teller effect since $3d^9$ is an odd electron system, leave an orbital singlet state lowest.[75]

Which one of the five d-orbital states is the ground state is determined not only by whether the predominant term in the crystal field has octahedral or tetrahedral symmetry, but also by the distortions from these symmetries. When the symmetry of the crystal field is low, i.e., when the distortion from the simple octahedral or tetrahedral symmetry is very large, the splitting of the levels may give a different orbital set for the ground states. An example is the case of C_{2h} symmetry,[76] where the lowest orbital set is $|x^2 - y^2>$, $|xy>$, and $|3z^2 - r^2>$ and the upper orbital set is $|xz>$ and $|yz>$. It should be noted that C_{2h} symmetry is a distortion of square planar symmetry D_{4h} in which the orbitals separate into four groups,[77] namely $|3z^2 - r^2>$; $|x^2 - y^2>$; $|xy>$ and $|xz>$, $|yz>$.

It should be noted that although the simplest approach is to assume that the ground orbital singlet is pure, i.e., has no admixture of other d orbitals, this is not strictly correct. As we shall see, admixtures of the other d orbitals into the ground state are required to obtain reasonable agreement between theoretical predictions and the EPR experimental results. The relatively new technique of polarized neutron diffraction[78] can yield the unpaired electron populations of orbitals of both the metal and ligand atoms in transition metal complexes. Although systems containing copper(II) are difficult to study by this technique, an analogous study has been performed on phthalocyaninatocobalt(II). Here the spin-orbital populations were nonintegral and different from the usually accepted figure of unity for the ground state, namely $|3z^2 - r^2>$ or $3d_{z^2}^{1.0}$. Williams et al.[79] found that the ground state was in fact $3d_{xy}^{0.4}$ $3d_{xz,yz}^{0.17}$, $3d_{z^2}^{0.79}$, $3d_{x^2-y^2}^{0.21}$, $4s^{-0.14}$, showing that although the ground state was predominantly $|3z^2 - r^2>$, there was considerable admixture of other states. For simplicity, and because useful information can be obtained, we will outline the main results from the simplest model first, before discussing the more complicated cases.

b. Predominantly Octahedral Crystal Fields

The most common case considered for copper(II) ions is an octahedral field with an elongated tetragonal, rhombic, or trigonal bipyramidal distortion. Here the ground state is predominantly $|x^2 - y^2>$, and in the case of axial symmetry, the application of the reduced Hamiltonian to the ground and excited 3d orbital wave functions gives the following expressions for the spin Hamiltonian parameters:[80]

$$g_\parallel = 2 - \frac{8\lambda}{\Delta_0}$$

$$g_\perp = 2 - \frac{2\lambda}{\Delta_1}$$

$$A_\parallel = \rho\left(-\kappa - \frac{4}{7} - \frac{6\lambda}{7\Delta_1} - \frac{8\lambda}{\Delta_0}\right)$$

$$A_\perp = \rho\left(-\kappa + \frac{2}{7} - \frac{11\lambda}{7\Delta_1}\right)$$

$$P_\parallel = -\frac{3e^2 q}{7I(2I-1)} <r_q^{-3}> \tag{27}$$

where the wave functions of the lowest doublets are

$$| + > = |x^2 - y^2, + > - \frac{\lambda}{\Delta_0} |xy, + > - \frac{\lambda}{\sqrt{2\Delta_1}} |zx, - >$$

$$| - > = |x^2 - y^2, - > + \frac{\lambda}{\Delta_0} |xy, - > - \frac{\lambda}{\sqrt{2\Delta_1}} |yz, + > \quad (28)$$

In these equations $+$ and $-$ refer to the electron spins of $+1/2$ and $-1/2$, respectively, Δ_0 and Δ_1 are the energy level splittings, λ is the spin-orbit coupling constant and has a value of about -830 cm^{-1} for the free ion copper(II), κ is the Fermi contact term which describes the s-electron density at the copper nucleus, $\rho = 2g_n \beta \beta_n <r^{-3}>$ where $<r^{-3}>$ is the mean inverse cube radius of the distances of the electrons from the nucleus,[81] q is the quadrupole moment of the nucleus, and $<r_q^{-3}>$ is related to the electron density at the nucleus in a similar way to $<r^{-3}>$, but is modified to take account of the distortion of the charge cloud of the inner electrons by the outer electrons and the ligands.[82]

The following deductions may be made from Equation (27). First, since λ is negative and Δ_0 is less than Δ_1, $g_\parallel$ is greater than $g_\perp$, and both are always greater than 2.0. Second, $A_\parallel$ will be greater in magnitude than $A_\perp$.

If values of Δ_0, Δ_1, λ, ρ, and κ, typical of simple copper complexes, are inserted into the equations, the values of the parameters obtained are $g_\parallel \approx 2.3$, $g_\perp \approx 2.05$, $A_\parallel = 0.018$ cm^{-1}, and $A_\perp = 0.001$ cm^{-1}. These are representative of the values found for most simple copper(II) complexes, and are also typical of those for the type-2 copper sites in proteins, as shown in Appendix 1. The results of more exact calculations (e.g., see Pilbrow and Spaeth[81]) are similar to those given above. Likewise, the introduction of rhombic symmetry, while adding extra terms to the spin Hamiltonian, does not alter the basic picture.

One may therefore assert that the gross EPR spectral features of most simple copper complexes and of the type-2 copper ions in proteins can be explained using a crystal-field model and a ground state which is essentially $|x^2 - y^2>$. However, an inspection of Appendix 1 shows that the EPR spectra of type 1 are not satisfactorily explained in this way, as the values of A_z or $A_\parallel$ are approximately one half those of type 2. As was first pointed out by Malmström and Vänngård,[6] unrealistically low values of ρ are required if one attempts to fit the experimental spectra using the relations derived for the $|x^2 - y^2>$ ground state. An alternative ground state must therefore be considered, and it is this that has led to much of the debate regarding the nature of the blue copper proteins. One may remark that although the prime area of controversy has been over the blue copper sites in laccase, ceruloplasmin, ascorbate oxidase, and the single copper proteins (the Az's, St, etc.) the copper of cytochrome *c* oxidase offers an even more extreme situation.

In an essentially octahedral representation, an alternative ground state is $|3z^2 - r^2>$. This may arise if the distortion from cubic is compressed rhombic, tetragonal, or trigonal bipyramidal, rather than elongated. The appropriate formulae for the g and hyperfine constants have been calculated, and typical values for axial symmetry and a $|3z^2 - r^2>$ ground state are $g_\parallel \approx 2.0$, $g_\perp \approx 2.2$ (i.e., $g_\parallel < g_\perp$) and both $A_\parallel$ and $A_\perp$ are around 0.007 cm^{-1}. Examples of EPR spectra attributed to copper(II) ions in this ground state are given in the papers by Schatz and McMillan,[84] De Bolfo et al.,[85] and Pilbrow and Spaeth.[81] However, it is quite clear that this case does not fit any of the copper protein systems.

c. Predominantly Tetrahedral Crystal Fields

If the major component of the crystal field is tetrahedral, the triplet levels $|xy>$, $|yz>$, and $|xz>$ are lower than the doublet states. Departures from pure tetrahedral symmetry determine which of the triplet levels is the ground state. A simple calculation, considering only the states within the manifold of the triplet, spin-orbit coupling and an axial distortion from original tetrahedral symmetry gives $g_\parallel$ and $g_\perp = 2$ for the $|xy>$ state, $g_\parallel = 0$ and $g_\perp = 0$ for the $|xz>$ state, and $g_\parallel = -4$ and $g_\perp = 0$, respectively, for the $|yz>$ state. The sign of the g-values depends on the phase of the wave functions and is not important for our purposes. These results indicate that the $|xy>$ ground state is a possibility and may explain the g and hyperfine values of the copper proteins if the symmetry of the copper site is basically tetrahedral.

Several authors have attempted more complete calculations of the g- and A-values for the $|xy>$ ground state of the copper(II) ion in a tetragonally distorted tetrahedral crystal field. Bates et al.[86] and Bates[87] included admixtures of 4s and 4p in the calculations. Bates et al.[86] were able to explain the observed g-values and the absence of an observable hyperfine interaction in copper(II) ($\alpha\alpha'$-bromo)dipyrromethane (g_x = 2.084, g_y = 2.069, g_z = 2.283; no resolved hyperfine structure) by assuming an admixture of the $3d^8 4p$ configuration into the $3d^9$ ground state and a tetragonal distortion from tetrahedral symmetry. In the case of Cu^{2+} in ZnO ($g_\parallel$ = -0.74, $g_\perp$ = 1.531; $|A_\parallel|$ = 0.0195 cm^{-1}, $|A_\perp|$ = 0.0231 cm^{-1}, but see Hausmann and Schreiber[88]), Bates[87] assumed a trigonal distortion from tetrahedral symmetry. However, as pointed out by Parker,[89] Bates[87] calculated the expressions for a predominantly $|yz>$ state, rather than an $|xy>$ state.

Sharnoff[90] also included s- and p-orbital admixtures in an attempt to calculate the g- and A-values for the copper(II) ion in tetrachlorocuprates. However, as pointed out by Parker,[89] Sharnoff[90] obtained parameters for the $|yz>$ state. In addition, he used the hyperfine parameters obtained by Bates et al.[86] for the $|xy>$ ground state, so that the g and hyperfine values refer to different states.

The most complete calculation for an $|xy>$ ground state is that of Parker,[89] who included admixtures not only of s and p states, but also of the other d states in a tetragonally distorted tetrahedral field. His calculations were for the case of copper(II) substituted into NH$_4$F, where the experimental results gave g_x = 2.0071, g_y = 2.1040, g_z = 2.4702, and $A < 0.0004$ cm^{-1}. The sizes of the hyperfine interactions were found to be determined by the algebraic sums of the contributions from the s, p, and d electrons. In the calculation of $A_\parallel$, the d and s electron contributions add, while contributions from the p electrons and the orbital magnetic moments both have the opposite sign. In the calculation of $A_\perp$, the d-electron contribution has the opposite sign to the sum of the p and s contributions. Parker[89] was able to show that the very small hyperfine parameters observed for copper(II) in NH$_4$F could be obtained theoretically, using appropriate values for the other parameters in the equations. Although Parker[89] did not consider how his results might be applied to the type-1 site in the blue copper proteins, they indicate that the possibility of an $|xy>$ ground state, arising from distortions from tetrahedral symmetry, must be considered.

4. Crystal-Field Calculations for the "Blue Copper" Site

As discussed above, a predominantly $|x^2 - y^2>$ ground state does not explain the low values of A_z or $A_\parallel$ associated with the type-1 copper site. The alternative ground state, suggested by the calculations described above, is $|xy>$ and this possibility has been discussed by several authors.

In a paper which appears to have been overlooked by many authors, Brill and Bryce[91] developed a theoretical model with a ground state primarily $|xy>$ in character but including admixtures of both p orbitals and other 3d orbitals. They assume a tetrahedral distortion from a square planar configuration, which mixes the $3d_{xy}$ and 4p orbitals in the ground state. The higher orbital states are admixtures of the appropriate 3d, 4s, and 4p wave functions.

Applying the Hamiltonian in the same fashion as Bates,[87] Brill and Bryce[91] calculated expressions for the EPR g and hyperfine constants and for the oscillator and rotational strengths of the optical transitions. From either of the two sets of six pieces of experimental information (i.e., six from the EPR and six from the optical data) they were able to calculate the isotropic contact term and the hybridization coefficients of the orbitals. The values of the other set of six parameters could then be deduced. Excellent agreement was found between the experimental and calculated results for the two Az's studied, e.g., $A_\parallel$ calculated $= -0.0059$ cm^{-1}, and $A_\parallel$ experimental $= -0.0058$ or -0.0060 cm^{-1}. The ground state orbital was found to be $0.90 \, |xy> + 0.44 |4p>$, which led to an angle of distortion from square planar towards tetrahedral, above and below the $x-y$ plane, of 12.5°. Brill and Bryce[91] contended that their theory would have general application to the other blue copper proteins.

A similar argument is used by Greenaway et al.[92] in the case of cytochrome c oxidase. They conclude that the small value of A_z can best be explained by admixture of metal 4s and 4p character into the 3d orbitals and a $3d_{xy}$ (i.e., $|xy>$) ground state. Greenaway et al.[92] also comment that the inclusions of metal-ligand overlap and charge-transfer contributions in the calculations are insufficient to account for the low value of g_x, which can only be accounted for by admixture of metal 3d and 4p character in the wave functions.

Two recent papers by Solomon et al.[93] and Penfield et al.[94] have described modified crystal field-type calculations of the EPR parameters for the blue copper sites in St, Pc, and Az. Both papers draw heavily on the X-ray structural data for Pc[95] and Az,[96] which have shown that the copper ion is coordinated to two nitrogen and two sulfur ligands in a distorted tetrahedral geometry. The most detailed structure to date is that for Pc, and Penfield et al.[94] conclude that the appropriate description is that of an elongated tetrahedron with significant rhombic distortions, i.e., C_{3v} with rhombic distortions. The relative energies and the linear combinations of the d orbitals for the ground and excited states were calculated using the individual ligand strengths, distances, and angles. The ground state is found to be almost purely $|x^2-y^2>$, with the first excited state being $|xy>$. The calculated g-values for Pc are in good agreement with those found experimentally. However, Penfield et al.[94] did not calculate the hyperfine parameters.

Similar calculations have been outlined by Dooley et al. for laccase[97] and Dawson et al. for ceruloplasmin.[98]

The calculations of Solomon et al.[93] also resulted in a $|x^2-y^2>$ ground state for Pc, Az, and St. These authors found good agreement between the calculated and observed g-values for all three proteins, but very poor agreement for the values of $A_\parallel$ (or A_z), which were calculated using the theoretical expressions of Bates et al.[86]

The assignment of an $|x^2-y^2>$ orbital as the ground state is in conflict with most of the previous EPR studies on distorted tetrahedral-like copper(II) systems, which have used a ground state orbital of $|xy>$.[86-92,99,100] While both sets of calculations result in agreement between the calculated and experimental g-values, the values of $A_\parallel$ calculated by Solomon et al.[93] are a factor of 2 too large. This contrasts with the excellent agreement between theory and experiment obtained by Brill and Bryce,[91] who assumed that the ground state was $|xy>$ but with an admixture of the $|4p_z>$ orbital.

The reason for the discrepancy appears to lie in the neglect of p-orbital admixture by Solomon et al.[93] As discussed by Parker,[89] the p-orbital contribution is the key element in the reduction of the value of $A_\parallel$ (or A_z).

In principle, in an axially symmetric system, the choice of one of the two sets of in-plane orbitals as $|x^2-y^2>$ and the other $|xy>$ is entirely arbitrary. If, however, the symmetry is lower than axial, e.g., C_{2h} (see Hitchman et al.[76]), the directions of the x and y axes can be chosen to lie along the directions of the out-of-plane orbitals, which become labeled as $|xy>$ and $|yz>$. The ground state orbital then becomes a linear combination of the in-plane orbitals and the out-of-plane orbital along the z direction, $|3z^2-r^2>$. The x and y directions may then be determined by the orientations of the out-of-plane orbitals. By convention, the x and y axes point through the lobes of the $|x^2-y^2>$ orbital (enabling the $|xz>$ and $|yz>$ orbitals to be written as such) and the $|xy>$ orbital lobes are at 45° to these.

In the case of Pc, the symmetry as observed by EPR is axial, so that the labeling of the x and y axes by Solomon et al.[93] and Penfield et al.[94] is correct. However, an examination of the structure around the copper ion indicates that a lower symmetry should be expected, which would enable the $|x^2 - y^2\rangle$ and $|xy\rangle$ orbitals to be distinguished. It is most probable that the ground state does not have a d-orbital component which is purely $|x^2 - y^2\rangle$ or $|xy\rangle$, but rather is a mixture of the two. However, the inclusion of a p_z component, as proposed by Brill and Bryce,[91] does require that there be an $|xy\rangle$ component in the ground state.[101] The question may only be resolved by a determination of the unpaired electron-spin populations of the orbitals, similar to that performed by Williams et al.[79] on a much simpler system. Such an experiment may be feasible in the near future, despite the difficulties of a molecule as large as Pc.

E. Summary

In the previous sections we have developed the spin Hamiltonian for mononuclear copper(II) ions and shown how the experimental spectrum can be used to obtain values of its parameters, namely the components of $\mathbf{g}$, $\mathbf{A}$, $\mathbf{A}_{ligand}$, and $\mathbf{P}$. We have also indicated how these parameters may be related to the electronic structure and symmetry of the copper site, using a simple crystal-field model. At this stage, sophisticated molecular orbital calculations do not appear to give much more satisfactory answers. However, recent developments in the X-ray structure analysis of copper proteins will enable such calculations to be performed with some greater degree of validity.

For the case of the type-1 copper sites, the approach of Brill and Bryce,[91] which uses an $|xy\rangle$ orbital with $|p_z\rangle$ admixture as the ground state, gives excellent agreement with experiment. An alternative approach, using the $|x^2 - y^2\rangle$ orbital as the ground state, does not appear to give as good an agreement,[93] but does not include p-orbital admixture. The problem must therefore be regarded as incompletely resolved at this stage, although one should point out that it is quite probable that the ground state wave function contains a mixture of both $|x^2 - y^2\rangle$ and $|xy\rangle$ character.

The observation of nitrogen superhyperfine structure shows that the simple crystal-field model is not strictly correct, and ligand orbitals need to be taken into account in an exact calculation. Such a calculation is likely to become feasible in the near future, as the structures of the active sites become more clearly defined by X-ray structural analysis. Once the electronic structure of the active site is determined, an understanding of the reaction mechanisms will surely follow.

IV. THE ANALYSIS AND INTERPRETATION OF THE EPR SPECTRA OF BINUCLEAR COPPER(II) IONS, INCLUDING TYPE-3 COPPER IONS

A. Introduction

The active forms of a number of the multicopper proteins, such as laccase, ceruloplasmin, and ascorbate oxidase, give EPR signals whose intensity corresponds to only a fraction of the number of copper ions known to be present.[2-4] Similarly, the active forms of Hc and tyrosinase do not give any EPR signals at all.[1,2,4] This so-called "EPR undetectable copper" or type-3 copper has given rise to considerable speculation regarding the nature of the copper site, even though there is general agreement that it is at least partly in the copper(II) state. A further characteristic of the "EPR undetectable" copper ions is that they seem to occur in pairs, or at least in multiples of two.[1-4] Thus an explanation of the behavior of the copper sites involves a coupling between two copper ions such that no EPR signals are observed. Three possibilities have been suggested,[10] namely:

1.   The EPR signals are broadened to such an extent by magnetic dipole-dipole coupling between copper(II) ions that no resonances can be detected.
2.   Antiferromagnetic exchange interactions between the ions are so large that the system is diamagnetic at the temperature of the measurements.

3.   The copper ions form a mixed valence complex, with the electron transfer between copper(I) and copper(II) being such that no EPR signal can be observed.

Both dipolar and exchange interactions have been observed in a large number of simple copper(II) complexes. The field has been reviewed by Smith and Pilbrow,[103] who give a comprehensive account of the methods used to obtain information from the EPR spectra. Since both types of interaction have been observed in modified forms of copper proteins and have given much information about the metal-binding sites, we will discuss the phenomena in some detail below. It should be emphasized at this point that the dipolar interaction is a classical coupling between two magnetic point dipoles, while exchange is a quantum mechanical phenomenon which arises from the indistinguishability of two electrons when their wave functions overlap. Thus dipolar coupling will always be present when two copper ions are close together, even though exchange, which requires overlap of the wave functions, may not.

Mixed valence complexes have been observed for copper(I)-copper(II) systems,[104] although once more there is no EPR evidence for such interactions in the native proteins. However, modified forms of Hc and tyrosinase have given EPR signals which may be attributed to a mixed valence complex, as we will discuss below.

In this section we will discuss the various possible forms of coupling between pairs of copper ions and the interpretation of the EPR signals attributed to these interactions. We will also mention the unusual behavior of the EPR signals of copper in cytochrome *c* oxidase, which has sometimes been explained on the basis of coupling between copper(II) and other metal ions.

B. Dipolar Coupling between Copper(II) Ions

1. The Dipolar Interaction

We first treat the case where the ions are regarded as point dipoles and the coupling is produced solely by the magnetic field of one dipole acting upon the other. The Hamiltonian of the interaction is written as:

$$\hat{\mathcal{H}}_d = \frac{\hat{\boldsymbol{\mu}}_1 \cdot \hat{\boldsymbol{\mu}}_2}{r^3} - \frac{3(\hat{\boldsymbol{\mu}}_1 \cdot r)(\hat{\boldsymbol{\mu}}_2 \cdot r)}{r^5} \qquad (29)$$

where the ions have magnetic moments $\hat{\boldsymbol{\mu}}_1$ and $\hat{\boldsymbol{\mu}}_2$ and r is the vector joining their nuclei. We can write $\hat{\boldsymbol{\mu}}_1 = -\mathbf{g}_1 \cdot \hat{S}_1$ and $\hat{\boldsymbol{\mu}}_2 = -\mathbf{g}_2 \cdot \hat{S}_2$ where $\mathbf{g}_1$ and $\mathbf{g}_2$ are the $\mathbf{g}$ matrices of the ions and $\hat{S}_1$ and $\hat{S}_2$ are their spins. The dipolar interaction produces a zero-field splitting of the triplet state and separates the singlet state from the center of gravity of the triplet state. When two spins $\hat{S}_1$ and $\hat{S}_2$ are coupled and $S_1 = S_2 = 1/2$ as in the case of copper(II), four possible states arise. Three of these states have a resultant wave function which is antisymmetric, and give rise to a triplet state, with total spin $S = 1$ where the individual states can be labeled by the magnetic spin quantum number as $M = 1, 0,$ or -1. The fourth state is a singlet state, with a symmetric wave function, a total spin $S = 0$, and also $M = 0$. For copper(II) ion pairs, the dipolar splittings are of the order of 0.1 cm^{-1} when r is around 3 Å and of the order of 0.01 cm^{-1} when r is around 7 Å. This may be compared with the magnitudes of typical copper nuclear hyperfine interactions which are of the order of 0.015 cm^{-1}. The spin Hamiltonian of the pair system may be written as:

$$\hat{\mathcal{H}} = \beta(B \cdot \mathbf{g}_1 \cdot \hat{S}_1) + \beta(B \cdot \mathbf{g}_2 \cdot \hat{S}_2) + \hat{I}_1 \cdot \mathbf{A}_1 \cdot \hat{S}_1 + \hat{I}_2 \cdot \mathbf{A}_2 \cdot \hat{S}_2 + \hat{S}_1 \cdot \mathbf{D} \cdot \hat{S}_2 \quad (30)$$

For the copper(II) pairs in proteins it may be assumed that the **g** and **A** matrices of the individual ions are identical, although their principal axes may bear no special relationship to each other or to the principal axes of **D**, the tensor of the dipolar coupling interaction given in Equation (29).

In the simplest case both **g** and **A** are axially symmetric and have parallel axes for both ions, while the internuclear vector is along the z axis of **g**. Thus **D** is also axially symmetric, as is the overall system and there are only five independent parameters, namely $g_\parallel$, $g_\perp$, $A_\parallel$, $A_\perp$, and r, the internuclear distance.

A more complicated case occurs when **g** and **A** have orthorhombic symmetry and the internuclear vector is not along *OZ*, although the principal axes of the **g** matrices of the two ions are still parallel. Here the overall symmetry may be as low as triclinic, and there are nine parameters to be determined, namely g_x, g_y, g_z, A_x, A_y, A_z, r, and the angles ξ and η.

The most complicated case likely to apply to copper proteins occurs when **g** and **A** are orthorhombic and their principal axes are no longer parallel. Three further angles must be introduced to specify the relative orientations of the **g** matrices, although this reduces to a single angle, α, when **g** and **A** are axially symmetric (see Figure 6). In these cases the numbers of parameters are 12 and 8, respectively. Examples of these calculations are given by Smith and Pilbrow.[103]

For completeness, we refer the reader to the case of coupling between a cobalt(II) ion and a radical in some B-12-dependent enzyme reactions. To analyze this system it was necessary to include not only different ions with differently oriented **g** and **A**, but also the effects of small exchange coupling.[105] We will not consider this type of system further here.

2. Solution of the Hamiltonian for Dipolar Coupling

In principle, solution of the Hamiltonian (Equation 30) involves diagonalization of a 64 $\times$ 64 matrix, since copper(II) ions have $S = 1/2$ and $I = 3/2$ and the states may be labeled by a wavefunction $|M_1\,M_2\,m_1\,m_2>$, where M_1 and M_2 can take values of $\pm 1/2$ and m_1 and m_2 can take values $\pm 1/2$ or $\pm 3/2$. The repeated diagonalizations needed to simulate powder or frozen solution spectra would lead to a prohibitive requirement on computer time, so that a perturbation theory approach is used in which the Zeeman interactions (those involving **B**) are considered as the zero-order interactions and the hyperfine and dipolar interactions are considered as perturbations.

The first step in the solution of the Hamiltonian is to write the dipolar interaction in terms of the coordinate system of one of the ions. Next, the total Hamiltonian, including dipolar and hyperfine terms, is transformed into a representation in which the Zeeman interactions of both ions are diagonal.

The transformed Hamiltonian is then applied to the 64 basis states of the uncoupled representation, which is written as:

$$|\pm \pm m_1\,m_2> \equiv |M_1>|M_2>|m_1>|m_2>$$

where:

$$|M_i> = \pm \tfrac{1}{2} \quad \text{and} \quad |m_i> = \pm \tfrac{1}{2} \text{ or } \pm \tfrac{3}{2} \tag{31}$$

If we neglect the dipolar and hyperfine interactions, the states labeled by $|M_1 = 1/2>|M_2 = -1/2>$ and $|M_1 = -1/2>|M_2 = 1/2>$ are degenerate. Thus perturbation theory cannot be applied unless these two states are transformed into a representation in which the degeneracy is removed. This is a familiar problem in perturbation theory and results in four sets of states in the coupled representation, namely:

$$|\Psi_1> = |\tfrac{1}{2}>|\tfrac{1}{2}>|m_1>|m_2>$$

$$|\Psi_2> = |-\tfrac{1}{2}>|-\tfrac{1}{2}>|m_1>|m_2>$$

$$|\Psi_3> = a|\tfrac{1}{2}>|-\tfrac{1}{2}>|m_1>|m_2> \;+\; b|-\tfrac{1}{2}>|\tfrac{1}{2}>|m_1>|m_2>$$

$$|\Psi_4> = c|\tfrac{1}{2}>|-\tfrac{1}{2}>|m_1>|m_2> \;-\; d|-\tfrac{1}{2}>|\tfrac{1}{2}>|m_1>|m_2> \tag{32}$$

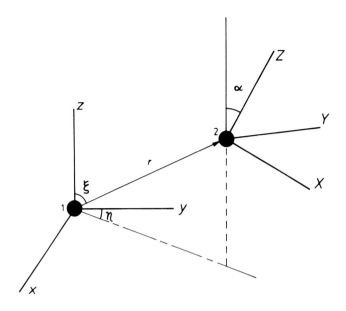

FIGURE 6. Principal axes for coupled copper(II) ions, with the most general orientation of **g** matrices with respect to each other.

where a, b, c, and d are coefficients which can be calculated and which may be complex. Second-order perturbation theory is then used to derive expressions for the energies of the 64 states, and hence for the values of magnetic field at which transitions occur. Transition probabilities are calculated using the first-order wave functions. The expressions obtained are given by Smith and Pilbrow.[103]

In general, 64 transitions of the type $\Delta M = 1$ are allowed. These occur around $g = 2$, i.e., at around 0.3 T for microwave frequencies of 9 GHz. Some of the 64 transitions may overlap. As shown in Figure 7, the main features of a $\Delta M = 1$ spectrum are two strong peaks centered around $g = 2$ whose splitting depends on the magnitude of r, the internuclear distance. Weaker lines due to hyperfine structure are sometimes observed at higher and lower fields than the two main peaks.

Transitions between the states labeled as $|\Psi_1\rangle$ and $|\Psi_2\rangle$ in Equation (32) are also possible in the appropriate circumstances. There are 16 of these so-called forbidden or $\Delta M = 2$ transitions and they occur around $g = 4$, i.e., 0.15 T which is half the field at which the $\Delta M = 1$ transitions occur. Some of the forbidden transitions also overlap. In many cases, a 7-line hyperfine pattern is observed with a splitting between the hyperfine lines of one half that of an individual copper(II) ion. High spectrometer gains are usually required for the observation of the $\Delta M = 2$ spectra.

Examples of computed $\Delta M = 1$ and $\Delta M = 2$ spectra are given in Figure 7. The simulation procedures outlined previously are followed, with the appropriate expressions being used for the field values and transition probabilities.

C. Exchange Coupling between Copper(II) Ions
1. The Exchange Interaction

Exchange interactions arise from quantum mechanical effects as a result of the overlap of the wave functions of the two electrons of the ions. The exchange interaction Hamiltonian can be written as:[106]

$$\hat{\mathcal{H}}_{ex} = -J\hat{S}_1 \cdot \hat{S}_2 + \hat{S}_1 \cdot \mathbf{T} \cdot \hat{S}_2 + J_a \hat{S}_1 \times \hat{S}_2 \tag{33}$$

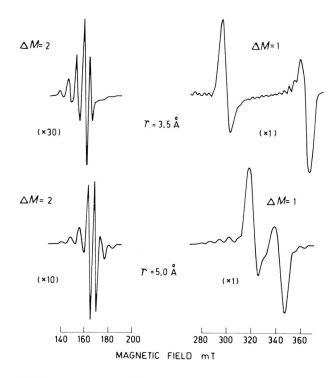

FIGURE 7.   Theoretical EPR spectra due to axial copper(II) binuclear species when the coupling is purely dipolar (i.e., $J = 0$). Other parameters: $g_\parallel = 2.07$, $g_\perp = 2.02$, $A_\parallel = 0.0148\ \mathrm{cm}^{-1}$, $A_\perp = 0.0027\ \mathrm{cm}^{-1}$. Linewidths are 1 mT ($\Delta M = \pm 2$) and 1.8 mT ($\Delta M = \pm 1$). Microwave frequency: 9,450 MHz. The internuclear distances are 3.5 and 5 Å as indicated.

The term $-J\hat{S}_1 \cdot \hat{S}_2$ represents the isotropic exchange interaction, which is much larger than the anisotropic exchange interactions represented by the terms $\hat{S}_1 \cdot \mathbf{T} \cdot \hat{S}_2$ and $J_a \hat{S}_1 \times \hat{S}_2$. If the ions and their sites are identical, $J_a \hat{S}_1 \times \hat{S}_2 = 0$.

The effect of the first term, $-J\hat{S}_1 \cdot \hat{S}_2$ is to separate the antisymmetric triplet state from the symmetric singlet state by an energy J. J is generally negative for copper(II) ions,[107] leaving the singlet state lowest in energy, although some cases are known where J is positive and the triplet state is lowest.[108] The situation for J negative is often called antiferromagnetic coupling, as the electron spins of the two ions of the pair are aligned antiparallel to each other in the singlet state, giving zero net spin. If J is positive, then the triplet is lowest and the coupling may be called ferromagnetic. It is worth noting at this point that the isotropic exchange term may sometimes be written as $+J\hat{S}_1 \cdot \hat{S}_2$, which gives the singlet state lowest for $J > 0$, or as $-2J\hat{S}_1 \cdot \hat{S}_2$, which means that the actual singlet-triplet splitting is $2J$. One must therefore examine the context in which it is used to decide whether $J < 0$ means that the singlet state is lowest (as used in this chapter) or the triplet is lowest, or whether the splitting is J or $2J$.

If $|J| >> h\nu$, where $h\nu$ is the microwave quantum energy, the only possible EPR transitions will be those between the levels of the triplet state. If $J < 0$ and also $|J| >> kT$, where kT is the thermal energy at a temperature T, only the singlet state will be populated and neither a paramagnetic susceptibility nor an EPR signal will be observed.

The anisotropic exchange term, $\hat{S}_1 \cdot \mathbf{T} \cdot \hat{S}_2$, is a traceless tensor quantity and includes the pseudodipolar interaction. This arises from the admixture, via spin-orbit coupling, of some of the higher states into the orbital ground state of the exchange-coupled system. The dipole-dipole interaction, represented by Equation (29) is also a traceless tensor quantity, although it must be emphasized that its origin is entirely different. The effect of the pseudodipolar coupling is to separate further the triplet and singlet states, and to give a zero-field splitting of the triplet state. This zero-field splitting has been expressed as:[109]

$$D = \frac{-J}{8} \left[\frac{1}{4}(g_\parallel - 2)^2 - (g_\perp - 2)^2 \right] \tag{34}$$

Thus for a typical copper(II) system, if $|J| = 4$ cm^{-1}, the magnitude of the zero-field splitting of the triplet state will be similar to that produced by the dipolar interaction of ions 6 Å apart. Measurements of $|J|$ for many copper systems indicate that it is generally much larger than this. However, Bleaney and Bowers[109] also pointed out that the expression for D should involve the value of J for the excited orbital levels and not that for the ground state. The calculations of Ross[110] and Ross and Yates[111] indicate that D, as expressed in Equation (32), may be much too large. From their experimental results, Boyd et al.[55] concluded that this was in fact the case, and that D was negligible even for $|J|$ as large as 30 cm^{-1}. We will return to this point below.

2. Spectral Simulations for Exchange-Coupled Systems

We now consider the method of calculation when the isotropic exchange is not zero, emphasizing that while dipolar coupling must be present in any system involving two (or more) magnetic ions, exchange interactions only occur if the wave functions overlap. We restrict our initial discussion to cases where the axes of the **g** matrices are parallel. Other cases require full diagonalization of the energy matrix due to introduction of large off diagonal terms.

1. If $|J| << h\nu$, the isotropic exchange term is included in the Hamiltonian and it can be treated as a perturbation, along with the effects of dipolar and hyperfine interactions. An example of this (with some additional complications arising from dissimilar ions) is given by Boas et al.[105]
2. If $|J| \approx h\nu$, the zero-order energies arise from the Zeeman and isotropic exchange interactions, and the full 64×64 matrix may have to be diagonalized to obtain the energy levels.
3. If $h\nu << |J| << 30$ cm^{-1}, the singlet state is well separated from the triplet state, and the 32 transitions between the states $|\Psi_1 \rangle \rightarrow |\Psi_3\rangle$ or $|\Psi_2\rangle \rightarrow |\Psi_3\rangle$ are the allowed $\Delta M = 1$ transitions (see Equation 32). The 16 $\Delta M = 2$ transitions will still occur between $|\Psi_1\rangle$ and $|\Psi_2\rangle$. The overall assumption here is that the zero-field splitting within the triplet state is entirely due to dipolar coupling.
4. If $|J| >> 30$ cm^{-1}, it may be expected that the pseudodipolar coupling is large. In this case the hyperfine interaction may be neglected, and the problem treated as one involving a single ion with total spin $S = 1$, using the spin Hamiltonian:

$$\hat{\mathcal{H}} = \beta(B \cdot g \cdot \hat{S}) + \hat{S} \cdot D \cdot \hat{S} \tag{35}$$

which can also be written:

$$\hat{\mathcal{H}} = \beta(B \cdot g \cdot \hat{S}) + D\left[\hat{S}_z^2 - \frac{1}{3}S(S+1)\right] + E(\hat{S}_x^2 - \hat{S}_y^2) \tag{36}$$

In these equations, $\hat{S}$ represents the total spin vector (magnitude $= 1$) and **D** is a true tensor, which may not have the same principal axes as the **g** matrix. D and E are the components of **D** referred to the principal axes. If the symmetry of both **g** and **D** is axial, and the principal axis of **D** is collinear with that of **g**, Equation (36) reduces to:

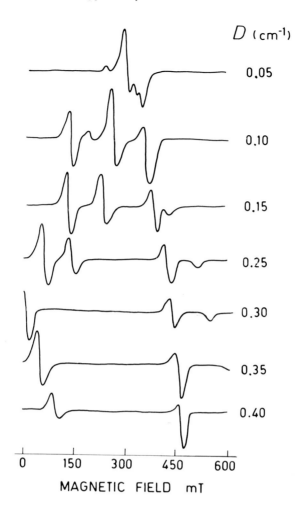

FIGURE 8.   Theoretical EPR spectra due to axial copper(II) binuclear species for various values of the zero-field splitting of the triplet state, *D*. These results include cases where *J* is large and gives a large value of *D*. Other parameters: $g_{\parallel}$ = 2.36, $g_{\perp}$ = 2.09, linewidth = 8 mT. Microwave frequency: 9,160 MHz. The copper hyperfine structure is not included, but is not expected to be resolved, owing to the large linewidth.

$$\hat{\mathcal{H}} = g_{\parallel}\beta B \hat{S}_z \cos\theta \; + \; g_{\perp}\beta B \hat{S}_x \sin\theta \; + \; D\left[\hat{S}_z^2 - \frac{1}{3}S(S+1)\right] \quad (37)$$

Since *D* is large (of the order of *hν*, the microwave quantum energy), the energy levels are found by computer diagonalization of the matrix for each value of θ, and the wave functions are those appropriate to the triplet state.[103,112] Simulated spectra for various values of *D* are shown in Figure 8.

Two major points should be made at this stage. First, the spectra of any triplet state can be fitted to the spin Hamiltonian of Equation (36), whether the zero-field splitting is due to dipolar coupling alone or to the combination of dipolar and exchange coupling. *D* and *E* do not give any information about the internuclear separation of the ions unless the exchange contribution can be evaluated. Even then, the **g** anisotropies and hyperfine interactions need to be taken into account if correct answers are to be obtained. The inclusion of the hyperfine interactions may lead to difficulties as the matrix becomes very large and if perturbation theory is used, a decision must be made as to which interactions are predominant.[113]

Second, there is no a priori reason why **D** and **g** should have the same principal axes. Thus, it is not necessarily justified to neglect the terms in E in Equation (36), as these will make a significant contribution once the internuclear vector is directed more than about 10° from the z axes of **g**. In this context, we should note that the results of Boyd et al.[55] for copper(II) *bis*(pyridine-*N*-oxide) were in excellent agreement with the copper(II) — copper(II) distances and angles deduced from the X-ray analysis. Boyd et al.[55] assumed pure dipolar coupling and simulated their spectra accordingly, without assuming that the dipolar coupling tensor and the **g** matrices possessed coincident axes. In contrast, Lund and Hatfield[114] used the spin Hamiltonian form and had difficulty in relating X-ray and EPR structural data. A more correct form of the calculation, using **D** and **g** has been given by Buluggiu and Vera,[115] although they do not relate their results to the structural data in the manner of Boyd et al.[55]

D. Experimental Results for Coupled Copper(II) Ions in Proteins
1. Magnetic Susceptibility Measurements

Direct evidence of coupling between neighboring paramagnetic ions in proteins can be provided by magnetic susceptibility and EPR measurements. Magnetic susceptibilty measurements have been reported for *Rhus vernicifera* laccase and for OxyHc from *Limulus polyphemus* and from the keyhole limpet, *Megathura crenulata*. In the case of laccase, measurements between 30 and 210 K were interpreted in terms of an antiferromagnetic coupling between the copper ions such that $|J| = 170 \pm 30$ cm^{-1}, i.e., the splitting between singlet (lowest) and triplet states.[116] However, more accurate measurements between 5 and 260 K using a superconducting quantum interference device (SQUID)[117] did not show the complicated temperature dependence expected from a system where the type-3 copper (i.e., the pair) contributed to the susceptibility in addition to the type-1 and type-2 coppers. The latter show the linear dependence on $1/T$ expected from copper(II) ions uncoupled by exchange interactions. The authors deduced that these more recent measurements showed that $|J| > 550$ cm^{-1}, with the singlet state lowest.

For *M. crenulata* Hc, Solomon et al.[116] deduced a lower limit for $|J|$ of 625 cm^{-1}. More accurate measurements were performed on *L. polyphemus* Hc[117] and showed that a lower limit for $|J|$ was 550 cm^{-1}. The authors concluded that for both laccase and OxyHc, the exchange coupling between the copper ions of the type-3 pair was larger than 550 cm^{-1}, once again with the singlet state lowest. This value is considerably larger than that proposed as a result of the earlier magnetic susceptibility measurements of Moss et al.[118] on *Cancer magister* Hc.

Measurements of the magnetic susceptibility of laccases and ceruloplasmin between 50 K and room temperature by Petersson et al.[119] are in agreement with the above, and showed that for ceruloplasmin, $|J| > 400$ cm^{-1}.

The absence of a detectable magnetic susceptibility due to the type-3 copper and the absence of model complexes of copper(II) which exhibited as strong exchange coupling as that proposed for the proteins, called into question the assumption that the copper ions of the pair were copper(II) and strongly antiferromagnetically coupled. However, the recent resonance Raman spectroscopy[12,13] and X-ray fine structure[14-16] measurements have supported the antiferromagnetically coupled copper(II) pair model. In the last few years, binuclear copper(II) complexes have been prepared in the laboratory which have singlet-triplet splittings of between 800 and 1,000 cm^{-1} and copper(II) — copper(II) distances of 3.384[120] and 3.642 Å.[121]

2. EPR Studies
a. Ceruloplasmin

Ceruloplasmin contains seven copper atoms per molecule,[1-7] namely two of type 1 (which exhibit slightly different EPR spectra[122]), one of type 2, and four of type 3. The incubation of reduced ceruloplasmin, which gives no EPR signals, with nitric oxide under anaerobic conditions gave signals at around $g = 2$ and $g = 4$ which could be attributed to the $\Delta M = 1$ and $\Delta M = 2$ transitions from dipolar coupled pairs of copper(II) ions.[123] Since the temperature dependence of the signal intensity did not deviate from normal Curie law behavior (i.e., the intensity increased as $1/T$ down to 10 K), Van Leeuwen et al.[123] concluded that the exchange-coupling constant J was less than 7 cm^{-1}. The authors also pointed out that the $\Delta M = 2$ signals at half field showed some evidence for the presence of two species, but were unable to determine which of the three types of copper was responsible for these signals. Unfortunately, a computer simulation has not be reported, although a visual inspection suggests a copper-copper distance of 5 Å, or slightly less, and a symmetry lower than axial.

More recent experiments showed that although all the forms of copper in ceruloplasmin were affected by nitric oxide, the binuclear signals were most likely to arise from the type-3 copper ions.[124] One may also comment that the signals are most unlikely to be due to any of the copper ions other than the type 3, and that it may well be that the two pairs of these are slightly different, as shown by the $\Delta M = 2$ signals.

b. Hemocyanin

The number of copper atoms in Hc depends on the species, and may vary from around 20 to 100 or more.[125] One oxygen molecule can be bound to each pair of copper ions. Neither OxyHc nor DeoxyHc give EPR signals. However, Schoot Uiterkamp[45] showed that both nitric oxide- and nitrite-treated DeoxyHc gave strong EPR signals, and in the case of the nitric oxide-treated Hc, these could be interpreted as arising from coupled pairs of copper(II) ions. Since the intensity of the signals around $g = 2$ and $g = 4$ attributed to the coupled copper ions did not show any deviation from the Curie law dependence of $1/T$ between 230 and 14 K, the coupling was taken to be purely dipolar.[126]

A detailed study of the spectra of the nitric oxide-treated Hc's of the arthropod *C. magister* and the mollusc *Helix pomatia* was undertaken by Schoot Uiterkamp et al.[61] Systematic simulations of the various possibilities for the internuclear distances and angles led to the conclusion that the overall symmetry was close to axial, with the internuclear vectors being along the z axes of each of the copper complexes. The copper-copper distance was found to be 6.4 ± 0.3 Å using the $\Delta M = 1$ transitions or 6.2 ± 0.5 Å using the $\Delta M = 2$ transitions in the case of *H. pomatia* Hc and 5.7 ± 0.2 Å for *C. magister* Hc. In the latter case, the simulations provided evidence for lower than axial symmetry at each copper site and for the displacement of the internuclear vector away from the z axes of the copper sites by about 10°. This displacement may well reflect the slightly shorter copper(II) - copper(II) distance. Possible errors and the limitations of these calculations are discussed in detail by Schoot Uiterkamp et al.[61]

In a recent series of papers Himmelwright et al.[127] have reported optical and EPR spectroscopic studies of the effects of binding with various ligands to the copper sites of Hc derivatives. Their comparison of the optical spectra and chemical behavior of mollusc and arthropod Hc's indicates that the arthropod active sites were less stable and more distorted than those of the molluscan species, possibly due to a strain induced by the protein.

The studies of Himmelwright et al.[43,47,127-130] have involved ligand binding to the following Hc derivatives:

1. MetHc (EPR-nondetectable MetHc), where both copper ions of the pair are copper(II) but no EPR signal is observed
2. Dimer Hc (EPR-detectable MetHc), where both copper ions are copper(II) and this form is prepared by treatment of DeoxyHc with nitric oxide and trace amounts of oxygen or by the addition of a large excess of sodium nitrite to MetHc at pH <7

3.   Half-MetHc (semi-MetHc), where one copper is copper(II) and the other is copper(I)
4.   Met-ApoHc, in which one of the copper ions of the pair has been removed and the other is copper(II)
5.   Half-ApoHc (semi-ApoHc), in which one of the copper ions of the pair has been removed and the other is copper(I)

The "Dimer" Hc, produced by the action of $NaNO_2$ on DeoxyHc,[127] showed similar EPR spectra to those reported by Schoot Uiterkamp et al.[61] The addition of 100-fold excess of NaN_3 to solutions of "Dimer" Hc resulted in the appearance of new signals at $g = 2$ and $g = 4$ which were attributed to another species of dipolar coupled pairs of copper(II) ions. In the case of the molluscan Hc from *Busycon canaliculatum*,[127] the azide and nitric oxide signals differ, and a visual inspection suggests the possibility of both a larger distance and a different symmetry for the azide form. A detailed comparison of the computer-simulated spectra of both azide and nitric oxide binuclear species of mollusc and arthropod Hc's would be of interest.

From their EPR results and other evidence, Schoot Uiterkamp et al.[61] were able to suggest the following model for the oxygenation reaction in Hc.

In DeoxyHc, the copper ions are in the monovalent state, copper(I), with a distance apart similar to that of the copper(II) ions in the NO-treated Hc's. Considerable movement occurs around the copper ions on oxygenation, allowing both the change to copper(II) and the reduction in the distance between the ions which results in strong antiferromagnetic coupling.

This model has been confirmed in essence by the extended X-ray absorption fine structure (EXAFS) measurements of Hodgson and co-workers,[15] who found a copper(II)-copper-(II) distance of 3.55 Å for OxyHc and a copper(I)-copper(I) distance of 5.6 Å for DeoxyHc.[16] However, one should also note the EXAFS measurements of Brown et al.[14] which could be interpreted on the basis of a copper(I)-copper(I) distance of 3.4 Å in DeoxyHc, although other interpretations were possible. The measurements of Himmelwright and co-workers[43,47,127-130] referred to above have given additional support to this model of the active site and have given detailed information about its chemical and structural properties. Some recent Raman spectroscopic results have ascribed a broad band at 1075 cm^{-1} to the singlet-triplet transition of the coupled Cu^{2+} unit.[13]

c. Tyrosinase

Tyrosinases may contain two or four copper atoms per molecule, depending on their origin.[4] In their natural state, no EPR signals are observed. However, after treatment with NO, the deoxygenated form of H_2O_2-treated *Agaricus bisporus* tyrosinase gave EPR signals around $g = 2$ and $g = 4$ which were interpreted as being due to dipolar coupled copper(II) species.[126] The computer simulations[61] showed an overall symmetry close to axial, and a copper(II)-copper(II) distance of 5.9 ± 0.2 Å.

Himmelwright et al.[130] have carried out studies on *Neurospora crassa* tyrosinase similar to those performed on the Hc derivatives. As for *A. bisporus* tyrosinase, a "Dimer" tyrosinase could be formed by the reaction of nitric oxide and deoxytyrosinase in the presence of trace quantities of oxygen. The EPR signals in the two cases appear to be similar, although Himmelwright et al.[130] do not report the observation of $\Delta M = 2$ transitions near $g = 4$ for *Neurospora* tyrosinase. They were also unable to decide whether an azide-"Dimer" form was produced on addition of excess NaN_3, or whether the small signal which remained after the azide treatment was due to a small amount of half-Met-tyrosinase [i.e., tyrosinase where one copper of the pair is copper(II) and the other is copper(I)]. A computer simulation of the *N. crassa* "Dimer" signals has not been reported as yet.

d. Bovine Superoxide Dismutase

This contains two copper(II) ions and two zinc ions. The copper(II) ions give EPR spectra characteristic of type-2 copper, and are not in close proximity in the natural product.[131] However, substitution of either a cobalt(II) or copper(II) ion for the zinc(II) ion resulted in the observation of magnetic coupling between the paramagnetic species.[132] Substitution of the cobalt(II) resulted in a decrease in the intensity of the copper(II) signal, and the EPR signal due to the cobalt(II) ion was not observed until the copper was reduced with hexa-cyanoferrate(II). The interaction is clearly different than that observed between cobalt(II) and copper(II) tetrasulfonated phthalocyanines in frozen solution, where signals due to dipolar coupling between the paramagnetic ions could be clearly discerned.[37]

Upon substitution of copper(II), a decrease in the EPR spectral intensity with temperature showed that antiferromagnetic coupling occurred, with a singlet-triplet splitting of approximately 52 cm^{-1}. Signals were observed in the $g = 4$ region near 0.15 T and between 0.26 and 0.36 T at 100 K, showing that coupling was occurring between pairs of copper ions.[133] The spectra cannot be interpreted on the basis of pure dipolar coupling alone, as this would lead to a copper-copper distance of around 3.3 Å, which is half the distance of about 6 Å found by Richardson et al.[131] for the separation of the zinc and copper sites. We may note that substitution of the appropriate values of J, $g_\parallel$, and $g_\perp$ into Equation (34) gives a value of the zero-field splitting of the triplet state of $D = 0.15$ cm^{-1}, which is about three times larger than that estimated from the experimental splitting (see Figure 8). This calculation only serves to emphasize that estimates of D from a measured J may be grossly in error, as pointed out earlier.

E. Mixed Valence Complexes in Hemocyanin and Tyrosinase

The EPR spectrum observed by Deinum et al. on the reaction of *Neurospora* tyrosinase with β-mercaptoethanol led to the suggestion that a mixed valence complex was formed.[134] Himmelwright et al.[43,127-130] likewise suggest that mixed valence complexes are formed on the substitution of ligand ions into some Hc and tyrosinase half-Met derivatives. In the case of the half-Met-azide derivative of *B. canaliculatum*,[43] a strong absorption band at around 1,500 nm and a change in the EPR spectrum were observed at temperatures of 77 K and below.[47,128] Both were attributed to the presence of a class II mixed valence complex, with some, but not complete, delocalization of the unpaired electron from one copper site [the copper(II) site] onto the other copper ion. While the changes in the optical and near-IR spectrum may well be due to an intervalence charge transfer, the explanation of the EPR spectral changes is less obvious. In the classification scheme of Robin and Day,[104] three situations may be proposed, namely class I where there is no delocalization, class II where there is an intermediate situation, and class III where there is complete delocalization of the electron. The EPR spectrum of class-I compounds may be expected to show a relatively normal copper(II) EPR spectrum, any effect of the other copper ion only being observed as superhyperfine structure. In class-III compounds, a 7-line hyperfine pattern is expected, as observed by Sigwart et al.[135] for a mixed valence copper(I)-copper(II) acetate complex. The type of signal expected from a class-II complex is not clear,[136,137] but may be of the type observed for the mixed valence complex of copper in frozen aqueous solutions of ($\pm$)-mercaptosuccinic acid at pH ≈ 4. In this system, a single resonance line was observed at 77 K, with a peak-to-peak derivative width of around 35 mT at 9 GHz microwave frequency. No hyperfine structure was observed at this pH value, and this, together with the broad wings observed for the resonance, suggested that the unpaired electron of the copper(II) ion in the center of the complex was delocalized over the other four copper ions, but not to such an extent that individual hyperfine structure could be observed.[10] The similarity between this signal, and those observed by Himmelwright et al. for half-Met-azide Hc[43] and half-Met-2-mercaptoethanol tyrosinase and Hc,[130] may be noted. Himmelwright et al.[47] also observed a complex hyperfine splitting for a number of the half-Met-halide derivatives of Hc, which was attributed to a mixed valence complex. However, it should be noted that the EPR spectra may be capable of other interpretations, such as hyperfine interactions with the halide ions, and in the case of the azide and mercaptoethanol derivatives, as "Dimer" i.e., copper(II)-copper(II) complex formation, with the signal in the $g = 2$ region being due to the $\Delta M = 1$ transitions. A puzzling observation in the case of the half-Met-azide Hc derivative is the apparent change from a more or less normal copper(II) EPR spectrum above 77 K to the "mixed valence" spectrum at 7 K, as the reverse may be expected when the complex changes from a class-II situation to a class-I, the electron becoming less delocalized as the temperature is reduced. Himmelwright et al.[130] propose that the temperature reduction gives a structural change in the azide anion and hence a better pathway for electron delocalization. That they did not observe a signal around $g = 4$ is evidence against the $g = 2$ signal being due to a dipolar coupled species.

It is evident that further experiments are required, and an ENDOR experiment may resolve the hyperfine coupling from the copper ions and the azide nitrogens.

F. Cytochrome *c* Oxidase

Cytochrome *c* oxidase is the terminal oxidase in the respiratory chain and contains two atoms of iron incorporated into type-*a* heme groups and two atoms of copper.[138] The copper may be in one of three forms,[92] the first being a form with well-resolved copper and ligand hyperfine structure, the second exhibiting an unusual EPR signal with some characteristics of type-1 copper, and the third being "EPR nondetectable".[92] It is now thought that the first type is formed in a denaturation process and is biochemically inactive. The second type of copper is called intrinsic copper and is thought to be located between the two heme *a* groups, while the third is close to the a_3 heme and may act as an electron sink.[4]

Since both types of copper are thought to be coupled to the iron atoms of the heme groups, it is convenient to discuss them at this point.

The third type of copper ion, namely the "EPR undetectable copper", is now thought to be copper(II) but strongly coupled to the iron atom of the heme a_3. Magnetic susceptibility measurements[139,140] have indicated that the copper(II) and iron(III) ions are antiferromagnetically coupled to give a ground state with total spin $S = 2$ and a higher state with $S = 3$ at least 200 cm^{-1} above this. Although the ground state of the copper-iron system is paramagnetic, it is by no means obvious what type of EPR signal might be observed. The $S = 2$ state will be split into five states which can be labeled by their magnetic spin quantum number as $M = 0, \pm 1$ or ± 2, each separated by a zero-field splitting. A spin Hamiltonian similar to that of Equation (36) may then be written, and the type of spectrum observed will depend on the relative magnitudes of the Zeeman interaction $\beta(\boldsymbol{B} \cdot \boldsymbol{g} \cdot \hat{\boldsymbol{S}})$ and the terms in D and E. The susceptibility measurements[139,140] have indicated that $D = 9$ cm^{-1} and that E lies between 0.1 and 1 cm^{-1}. The effect of the term in D is to separate the singlet $M = 0$ state and the doublets $M = \pm 1$ and ± 2. Since D is very much larger than the microwave frequency, no transitions will be observed between the doublets or between the $M = \pm 1$ doublet and $M = 0$ singlet. In principle, transitions may be allowed within the doublets because the spin levels are mixed by the magnetic field and the term in E. However, E also has the effect of giving the doublets a zero-field splitting, and since this is of the order of the microwave energy, the situation is far from clear. A detailed discussion is given by Abragam and Bleaney,[141] and all that can be said at this stage is that very careful measurements at a variety of temperatures and microwave frequencies will be required to establish if any of the EPR signals of cytochrome *c* oxidase are related to the copper(II)-iron(III) pair. The possibility of peculiar line shapes, resulting from strain effects, should also be borne in mind.[142]

An alternative model for the copper-iron pair of cytochrome *c* oxidase has been proposed by Seiter and Angelos.[143] They point out that the magnetic susceptibility measurements cannot be decisive, as the splitting between the $S = 2$ and $S = 3$ states has not been measured directly and is assumed to be much larger than the highest temperature at which measurements were made. Seiter and Angelos proposed that a model with an iron(IV) ion (S = 2) and a copper(I) ion would equally well account for the magnetic susceptibility data.[143] This dilemma is similar to that encountered for the type-3 EPR-undetectable copper ions discussed previously and must rely on other techniques for its resolution. At this stage, the bulk of the evidence favors the iron(III)-copper(II) model.

Some recent studies of ligand binding to cytochrome *c* oxidase have given evidence for the close association of a copper ion and the heme a_3 iron atom, e.g., Stevens et al.[144] added azide to an oxidized cytochrome *c* oxidase-NO complex and observed signals near g = 2 and g = 4 which were attributed to coupling between a copper(II) ion and a system with S = 1/2 located near the iron atom. It was suggested that the NO coordinates to the copper atom associated with the a_3 heme and breaks the strong antiferromagnetic coupling between the copper and iron atoms which normally renders them EPR undetectable. Observations of the previously ''EPR-undetectable'' copper of cytochrome *c* oxidase have been made in similar chemical circumstances by Seiter et al.[145] and Reinhammar et al.[146]

To date, the X-ray absorption fine structure measurements have not established the copper-iron distance,[147] although the possibility of a distance of 3 Å has been suggested.[148] This would certainly be close enough for strong exchange coupling to occur.

A possible model system for the copper-iron pair of cytochrome *c* oxidase has recently been investigated by Gunter et al.[149] The compound involved an iron(III) porphyrin, with a copper(II) located some 5.0 Å away and directly connected to the iron atom via a chlorine bridge. The magnetic behavior of this compound turned out to be most complex, and the authors concluded that it did not mimic the magnetic properties of the iron-copper complex of cytochrome oxidase.

The nature of the ''intrinsic'' copper center of cytochrome *c* oxidase is also far from fully clarified. There has been doubt that the signal is due to a copper(II) ion, although recent ENDOR[150] and low frequency EPR[151] experiments discussed later in this chapter seem to have clarified the situation. A current view is that the center is a highly covalent copper(II) ion.[150] The ''intrinsic'' copper signal has *g*-values of 1.99, 2.03, and 2.185, with values of the hyperfine parameters of 0.0020, 0.0025, and 0.0030 cm^{-1} for the *x*, *y*, and *z* axes, respectively.[92,152] These can be explained as arising from an essentially $|xy>$ ground state, as discussed in Section III, and by Greenaway et al.[92] The peculiarity of the intrinsic copper signals lies in their orientation-dependent linewidth and the rapid decrease in spin-lattice relaxation time observed above 40 K, which manifests itself in a further line broadening. Both effects have been proposed as arising from a dipole-dipole interaction between the copper(II) and an iron(III) ion some 7 Å away.[92] It is in this context that the model complex of Gunter et al.[149] is significant, as the EPR spectrum shows signals at 4.2 K characteristic of high spin iron (a signal at g = 5.9) and copper(II). These broaden and disappear above about 40 K, which is somewhat similar to the behavior shown by the copper signal of cytochrome *c* oxidase, although the latter does not disappear. The difference in behavior could be explained as being due to the smaller copper-iron distance in the model complex.

However, a contrary view can also be put forward. While dipolar coupling may well influence the relaxation times, it will also give a coupling of the spins of the individual ions. This is a case of coupling at a distance between two dissimilar ions, only one of which has a nuclear spin. A somewhat analogous case has been documented for some vitamin B-12-dependent enzyme reactions,[105] where both dipolar and exchange coupling was observed between the cobalt(II) ion in the corrin ring and a radical at the active site some distance away. Even at a distance of 10 Å, the dipolar interaction had a significant effect on the appearance of the spectrum, which was unlike that of either the radical or the cobalt(II) ion. In the case of cytochrome *c* oxidase, dipolar coupling between a copper(II) ion and a high spin iron(III) ion (S = 5/2) at a distance of around 7 Å would be expected to produce a quite distinctive spectrum, particularly in view of the $S = 5/2$ nature of the iron(III) ion. If the iron were in the low spin state ($S = 1/2$), the situation would be even more closely analogous to that described by Boas et al.,[105] and once again, a distinctive spectrum might be expected. There must therefore be some uncertainty regarding the explanation of the EPR spectral peculiarities of the intrinsic copper of cytochrome c oxidase in terms of a copper-iron interaction.

G. Summary

The previous sections have described the interpretation of the spectra observed when coupling occurs between the paramagnetic ions of copper proteins. In all cases where such coupling has been observed and interpreted, considerable information about the metal-ion site has been obtained. Only laccase and ascorbate oxidase of the proteins where the copper ions exist in pairs at the active sites have not given spectra attributable to coupled pairs under the appropriate conditions.[146,153] As we have seen, there are still several issues which remain unresolved, not the least of which is the continuing saga of cytochrome c oxidase.

V. PRACTICAL CONSIDERATIONS IN EPR SPECTROSCOPY

A. Introduction

The previous two sections have outlined the means by which EPR spectra can be used to give information about the active sites in copper proteins. However, it is all too easy to take the view that all that needs to be done is to place the appropriate sample in the EPR spectrometer and an intense, well-resolved spectrum is obtained automatically. Unfortunately, this is often not the case, and extracting a useful EPR spectrum may be a difficult task.

Quite apart from the obvious limitation that there may not be enough paramagnetic centers present to enable a resonance signal to be detected, the major limitation of an EPR experiment is the resolution. While this chapter is not the place to engage in a detailed discussion of the factors affecting the performance of EPR spectrometers, there are some experimental aspects which are of particular relevance to recent studies of copper proteins and which will be discussed below. The resolution of an EPR experiment is determined by the widths of the component lines. These component lines may overlap to such an extent that information is lost, and this situation is most likely to occur when powders or frozen solutions are studied. The width of a resonance line is determined by two mechanisms, namely homogeneous and inhomogeneous broadening. Homogeneous broadening mechanisms such as spin-lattice relaxation determine the linewidth by affecting the lifetime of the excited state of the system. An important consequence of a finite spin-lattice relaxation time is that it is possible to saturate the resonance line if enough microwave energy is supplied.

Inhomogeneous broadening arises from nonuniformities in the magnetic field throughout the sample, unresolved fine, hyperfine, or superhyperfine structure, or from dipolar interactions between unlike spins. The result is a spectral distribution of individual resonant lines or spin packets, which merge into a single overall line or envelope. Two magnetic resonance techniques, namely ENDOR and spin-echo spectrometry, can be used to overcome the difficulties caused by inhomogeneous broadening, and these are discussed in the final two sections of this chapter.

B. Some Experimental Techniques in Studies of Copper Proteins

1. The EPR Spectrometer Frequency

The most commonly used microwave frequencies are around 9 and 35 GHz, and require magnetic fields of 0.3 and 1.2 T, respectively, to observe resonances in the $g = 2$ region. The advantages of 9 GHz spectrometers are first that larger sample volumes can be accommodated, second that narrow tailed glass dewars can be readily inserted into the sample cavity, and third that both microwave components and electromagnets are cheaper than for 35 GHz spectrometers. On the other hand, 35 GHz spectrometers have greater sensitivity for small samples such as single crystals.

The choice of microwave frequency may also be influenced by the degree of resolution required and the line broadening mechanisms involved. Ideally, the linewidth is independent of frequency, so that lines with different *g*-values will be better resolved at the higher frequencies. This is useful for studies of frozen solutions, as it may then be possible to resolve the peaks in the resonance due to g_x, g_y, and g_z more clearly than at the lower frequencies. An example of the improved resolution at 35 GHz when compared to that at 9 GHz is given by Schoot Uiterkamp et al.[61]

A situation where a lower frequency may give improved resolution is if the major contribution to the linewidth arises from "*g*-strain" effects.[154] These effects arise from small differences in the local environment of the paramagnetic ion, which give variations in the crystalline electric field at the ion and hence in the splittings between ground and excited states. These splittings are related directly to the *g*-values. Thus random strains in the lattice lead to a line broadening due to the spread in *g*-values, and the effect may be large enough to obscure the splitting of the resonance line due to hyperfine interactions, which are independent of frequency (at least to first order). Measurements at lower frequencies will reduce the line broadening due to *g* strain, and may enable resolution of the hyperfine structure.

An example of a case where low frequency measurements have been advantageous is that of cytochrome *c* oxidase.[151] Studies of the prominent signal around *g* = 2 at 9 and 35 GHz were unable to resolve the copper hyperfine interaction directly, although estimates were made using computer simulations.[92,152] However, measurements made by Froncisz et al. at 4.2 K and microwave frequencies of 3.78 and 2.62 GHz were able to resolve the hyperfine structure due to copper(II), because the *g*-strain effect was much reduced at these lower frequencies.[151] The values obtained are shown in Appendix 1, together with those obtained by the ENDOR experiments of Hoffman et al.[150] and those obtained by computer simulation.[92,152] There is no immediate explanation of the discrepancies between these results. It should also be mentioned that the earlier low frequency measurements, made at 90 K and 3.26 GHz, did not succeed in resolving the hyperfine structure due to copper(II), probably because of lower sensitivity and the line broadening due to relaxation effects.[155]

Low frequency EPR spectrometers operate on similar principles to conventional 9 GHz EPR systems. Sample cavities are of course larger and require larger magnet gaps, although this is offset by the smaller magnetic field required for signals around *g* = 2. The transmission of the microwaves is rather more conveniently performed by coaxial lines than by waveguide, so that the spectrometer may be more compact overall.

2. Temperature

The temperature at which EPR measurements are made affects not only the sensitivity of the measurement, but also the type of information obtained. A major disadvantage of working at room temperature with biological systems is that water has an extremely high dielectric loss at microwave frequencies, so that small samples must be used if overall spectrometer sensitivity is to be maintained. Furthermore, the sample must be placed in a capillary tube or a specially constructed flat solution cell to minimize dielectric losses sufficiently. For these reasons, most studies of copper proteins are made at reduced temperatures, where the solution is frozen.

Nevertheless, there are circumstances in which studies using aqueous solutions can be advantageous. If the copper complex is tumbling in the solution at a rate much greater than the microwave frequency, an averaged spectrum will be observed,[156] where:

$$g_{av} = \frac{1}{3}(g_x + g_y + g_z) \tag{38}$$

and

$$A_{av} = \frac{1}{3}(A_x + A_y + A_z) \tag{39}$$

so that the spectrum will show resonances centered around the average g-value with a splitting corresponding to the average value of the hyperfine interaction. In exceptionally favorable circumstances, superhyperfine structure may also be observed. Equations (38) and (39) involve algebraic sums, and whereas the g-values will all have the same sign, the A-values may not. Hence a tumble-averaged spectrum may be useful in determining the sign of the hyperfine constants.[157]

However, copper proteins are large molecules and therefore cannot tumble rapidly in solution. The result is that the spectrum is similar to that observed for powders or frozen solutions, with the above-mentioned disadvantages of lower sensitivity. However the EPR spectral parameters of the type-1 and type-2 copper ions of *R. vernicifera* laccase were shown to undergo large changes on freezing, indicating a substantial rearrangement of the active sites.[158] Quite apart from eliminating the problems caused by water, low temperatures have the additional advantage of reducing spin-lattice relaxation broadening.

One should note that in general, the broadening due to spin-lattice relaxation in copper complexes is not as serious a problem as it is for some other transition-metal complexes. We should also note that it is difficult to observe spectra from binuclear copper(II) complexes in solution at room temperature, although spectra characteristic of binuclear vanadyl complexes have been observed in the same circumstances.[159,160] Spin-spin interactions are presumably a major factor in the nonobservation of spectra due to binuclear copper complexes tumbling in solution at room temperature, for even small values of isotropic exchange will give a noticeable effect on the spectrum when anisotropic effects (such as those due to dipolar coupling) are averaged out.[160]

For these reasons, most EPR measurements on copper proteins are made at reduced temperatures, using either liquid nitrogen (boiling point 77 K) or liquid helium (boiling point 4.2 K) as refrigerants. The former has the advantage of ready availability, low cost, and ease of handling, while the latter has the advantage of giving much improved sensitivity. Most commercially available spectrometers have facilities for operation at or near liquid nitrogen temperature, either by inserting a narrow tailed dewar into the sample cavity or by forcing cold nitrogen gas through a double-walled glass insert in the cavity. The first method, which introduces liquid nitrogen directly into the cavity, increases the noise level, because of instabilities arising from the bubbling liquid. Poole[19] and Alger[20] suggest how these instabilities can be reduced. The second method is preferable from the noise and instability viewpoints and because temperatures between 77 K and room temperature or higher may be obtained by heating the gas and adjusting the flow rate.

Operation with liquid helium is more difficult, the two most common systems involving either the total immersion of the cavity and its connecting waveguide into liquid helium or the use of a continuous flow system. The first method uses a double dewar assembly in which the liquid helium section is shielded by liquid nitrogen and has the advantage of small liquid helium consumption. In a good dewar, with minimum heat losses, the liquid helium consumption may be 250 mℓ/hr or even less. Such a system is advantageous if repeated measurements on a particular sample are to be made, e.g., measurements of the angular dependence of the resonances of a single crystal, or for ENDOR and spin-echo measurements. A further advantage is that temperatures below 4.2 K can be obtained relatively easily by pumping the liquid helium bath. These lower temperatures will give further increases in sensitivity and will increase spin-lattice relaxation times, which may be advantageous for ENDOR experiments. The disadvantages of a double dewar system are that samples are not readily changed once the dewar contains liquid helium, and that considerable ancillary equipment, such as a vacuum system capable of pumping the cryostat down to better than 10^{-5} Torr, is required. An example of this type of system for single crystal EPR studies is described by Brill and Venable.[161]

The continuous flow method involves pumping liquid helium continuously from a storage dewar through a transfer line and then into an insert in the cavity, so that the liquid evaporates just below the position of the sample. This system has the advantages that samples are readily changed and that the whole system is relatively easily assembled and requires less complex ancillary equipment. The vacuum requirements are generally less stringent and the vacuum system need not be permanently attached to the cryostat system. Furthermore, temperatures above 4.2 K are easily achieved and maintained by heating the gas with a small winding placed just below the cavity and adjusting both the heater current and gas flow. The ability to operate at temperatures other than 4.2 and 77 K is of considerable advantage in studies of exchange interactions between ions and for measurements of linewidth variations and relaxation times. ENDOR experiments on copper proteins have often required operation at temperatures between 4.2 and 77 K for their success, as we shall discuss later in this chapter. An example of a gas flow system for measurements in the temperature range 4.2 to 100 K is given by Lundin and Aasa.[162] Systems of this type are manufactured commercially by such companies as Oxford Instruments, Oxford, and Air Products, Allentown, Pa.

A disadvantage of a gas flow system is that the liquid helium consumption may be as large as 1.5 to 2.0 ℓ/hr if it is desired to maintain temperatures near to 4.2 K. The liquid helium consumption is less at elevated temperatures.

In measurements where either a knowledge of the actual sample temperature or the accurate control of this temperature is important, it should be recognized that the sample temperature may not be accurately measured by a sensor located some millimeters below it. In addition, accurate determinations of temperature around 4 K are not as reliable if a thermocouple is used instead of carbon or germanium resistance thermometers. The sensitivity of a thermocouple is not as great near 4 K as that of a carbon or germanium thermometer, although the opposite applies at temperatures around 77 K. The choice of thermocouple is also important, as, e.g., chromel-gold/iron thermocouples may perform better at low temperatures than the most common copper-copper/constantan type. Another factor to be borne in mind is that the sample may be heated by absorption of microwave or radio frequency power in an ENDOR experiment or by absorption of infra-red radiation from arc lamps or other sources used in photolytic experiments.

3. The Rapid-Freezing Technique

This was first described by Bray[163] and Bray and Pettersson[164] and may be used to monitor the appearance of EPR signals due to intermediate species in biochemical reactions. It has been particularly useful in studies of the kinetics and reaction intermediates in cytochrome *c* oxidase, and at the present time reaction rates with half-lives less than 10 msec can be measured with reasonable accuracy.[165]

In the rapid-freezing technique, reactants from two or three syringes are driven into a mixing tube by pressurized gas or a motor-driven cam system. The reaction time is governed by the length of the tube in which the mixing occurs. The reaction is quenched on leaving the reaction tube by spraying the mixture into cold isopentane at $-145°C$. The ice crystals which form from the reaction mixture are then packed into an EPR sample tube, the isopentane is removed, and the samples are stored in liquid nitrogen before measurement by EPR. The technique and some studies of free-radical intermediates performed by this means have been reviewed by Edmondson.[166]

4. Magnetic Field Modulation Frequency

The magnetic field modulation frequency is generally not a crucial factor in copper protein studies. Most EPR spectrometers use a modulation frequency of around 100 kHz, which allows large modulation amplitudes to be obtained fairly readily and is near the minimum of the noise amplitude-frequency curve of a crystal detector. Broadening of the resonance lines of copper ions due to too high a modulation frequency does not present a problem, as the absorptions are generally around 1 mT in width, much wider than the modulation frequency. Low frequency modulation is sometimes used in spectrometers employing superheterodyne detection. Some commercial spectrometers have provision for a double modulation of the magnetic field, using both 100 kHz and 400 Hz, allowing the second derivative of the absorption to be displayed. This may be advantageous in analysis of complex powder or frozen solution spectra, as it is more sensitive to small changes in the gradient of the absorption than the first derivative presentation.

A novel approach along these lines has been described by Chevion et al.[167] for experiments on nitrosyl hemoglobins which required high resolution.

5. Measurement of Spin Concentrations by EPR

In studies such as those on cytochrome *c* oxidase it is important to determine the relative concentrations of those species giving rise to the EPR signals and it is often equally important to be able to estimate the absolute number of spins in the reacting species. Such measurements have long been regarded as being the most difficult of all EPR measurements, and in particular, absolute measurements require the knowledge of a large number of factors all of which may be considerably in error or vary during the experiment.

As a result, it is more usual to compare the resonance of the unknown system with that of a standard which contains a known number of paramagnetic centers. Since the area under an EPR absorption curve is directly proportional to the number of paramagnetic centers contributing to the resonance, it is relatively simple to compare the areas of the standard and unknown signals and thus determine the spin concentration of the latter.

In practice, the spectrometer output is in the form of a first derivative curve, so that a double integration must be performed. Provided that the line shapes of the standard and unknown spectra are either identical or differ in a mathematically defined way, the comparison can be made using the assumption that the area is proportional to the product of the height and square of the width of the first derivative.[168] In general, the line shapes are not the same, so that double integration is required. One such method, which also can be used to compensate for baseline drift and which requires only a simple hand calculator, is given by Wyard.[169] These approximation methods are discussed by Poole[170] and Alger.[171] The advent of digital computers has made double integration very easy, e.g., Vollmer and Caspary[172] describe an off-line system for full analysis of the experimental data, including the double integration of overlapping systems, while an on-line system is described by Klopfenstein et al.[173]

A review of the factors involved in signal area measurements is given by Eaton and Eaton.[174] Some of the considerations to be borne in mind are that first the linewidths of standard and unknown spectra should be comparable, second the spin concentrations should be comparable, and third the same spectrometer settings should be used. As discussed by Alger,[175] even with these and other precautions, measurements made to better than 10% accuracy are considered to be very good. It should also be remembered that the transition probability factors must be corrected for the *g* anisotropy, as described by Aasa and Vänngård.[56] These authors point out that the total intensity of a powder spectrum is proportional to a factor $g_p{}^{av}$ which may be expressed approximately as:

$$g_p{}^{av} = \left\{ \frac{2}{3} \left[(g_x^2 + g_y^2 + g_z^2)/_3 \right]^{1/2} + \frac{1}{3} (g_x + g_y + g_z) \right\} /_3 \qquad (40)$$

i.e., g_p^{av} is proportional to g and not g^2 as used by some other authors. This point is of considerable importance when comparisons are to be made between signals with different g-values.

In many copper-protein systems, the spectra from different species overlap, or the unknown and standard spectra overlap. Nevertheless, the total integrated intensity can still be found provided that the EPR parameters can be determined experimentally or predicted theoretically. If part of the spectrum can be observed, the rest can be simulated, and the total area found by integration. This method has been applied by Vänngård for ceruloplasmin,[176] although we should note that the transition probability correction factor must be as in the paper by Aasa and Vänngård,[56] and not as shown in Vänngård's earlier work.[176] Examples of the use of the method of Vänngård in the case of cytochrome c oxidase are given by Aasa et al.[152] and in the earlier papers referred to therein.

C. The Limitations of EPR Spectroscopy — Line Broadening Mechanisms
1. Homogeneous Broadening
This occurs for a transition between two levels which are not sharply defined and arises from dipolar interactions between like spins, spin-lattice relaxation, motional narrowing, and other similar effects. The width of the resonance line is determined by the lifetime of the electron in the excited state, as can be shown by the uncertainty principle $\Delta E \, \Delta \tau \approx \hbar$, where ΔE is the width of the state and $\Delta \tau$ its lifetime. In copper proteins, the most important homogeneous broadening mechanism is spin-lattice relaxation, which we discuss briefly below. A full account is given by Abragam and Bleaney.[177]

Spin-lattice relaxation is the mechanism by which the spin system loses energy to the lattice. The interactions arise from the modulation of the orbital motion of the magnetic electron by the crystalline electric field as a result of lattice vibrations. Spin-lattice relaxation times (i.e., the lifetimes of electrons in the excited state) tend to be strongly temperature dependent, with the nature of this dependence being a result of the specific interaction mechanism. The mechanisms to be considered are

1. Direct processes, which involve phonons of the same energy as the microwave quantum and which give a relaxation time proportional to the temperature when $h\nu << kT$ (i.e., in the high temperature limit) and a constant time at low temperatures
2. Raman processes, which involve the entire phonon spectrum and give a temperature dependence of the relaxation time proportional to T^n, where in general n may be between 5 and 9; in copper proteins, $n \approx 9$
3. The so-called Orbach process, involving relaxation via the absorption of a phonon so that the electron reaches a higher state and relaxes back to the ground state by emitting another phonon; this mode of relaxation is also strongly temperature dependent

The first two processes can also be involved in spin-lattice relaxation via the modulation of the interaction between electron spins on neighboring ions. Energy is transferred from one spin to another by the lattice phonons, so that there is a form of spin-spin interaction. In the copper proteins, this may occur between the copper ions of a coupled pair, although the major relaxation process is likely to be a result of the direct dipole-dipole interaction. One should note that the copper pairs are generally well-separated from each other, and that single ions are also likely to be well-separated from other paramagnetic ions.

2. Saturation
Microwave-power saturation effects are a consequence of a spin system having a finite relaxation time. The observation of an EPR line depends on the existence of a population difference between the ground and excited electron spin levels, and the intensity of the

resonance absorption depends on the magnitude of the population difference. In normal circumstances, the population difference is maintained by electrons returning from the excited state to the ground state by losing energy through one of the relaxation processes described above. If a sufficiently intense microwave frequency field is applied, i.e., a high microwave-power level, the number of electrons excited from the ground state may equal the number which return, so that the populations of the two levels tend to equalize and the absorption of microwave energy decreases. This phenomenon is called saturation, and is observed as a decrease in the signal intensity and an increase in the linewidth as the microwave power increases. In the absence of saturation, the signal intensity increases as the square root of the microwave power.

If maximum resolution of the spectra is to be achieved, microwave-power saturation clearly needs to be avoided. In copper proteins, saturation is usually only observed at temperatures below 77 K. Microwave powers as low as 200 μW may cause saturation at 2.0 K.[150]

In some circumstances, the power levels at which copper(II) ions show saturation can be used to differentiate between signals from different sites, e.g., in nitric oxide-treated *H. pomatia* Hc, where the signals from monomeric and dimeric species overlapped, the signal from the monomeric species was saturated out above about 5 mW at 14 K, while that from the dimeric species did not show evidence of saturation even at 100-mW incident microwave power.[61] In the case of *A. bisporus* tyrosinase, at 14 K, the spectra due to monomeric copper(II) showed saturation effects at 1 mW, whereas the spectrum of the dimer species did not show these effects until microwave powers above 100 mW were reached.[61]

Thus although saturation is a phenomenon for the EPR practitioner to be aware of and generally avoid, it may be useful in special circumstances and it is essential for the performance of the ENDOR and electron spin-echo experiments.

3. Inhomogeneous Broadening

This arises from variations in the magnetic field seen by the ion in question as a result of other neighboring paramagnetic ions or the magnetic moments of neighboring nuclei. The interaction is random in direction and is often referred to as the spin-spin interaction. Two particular cases may be mentioned.[178]

First, if the paramagnetic ions are identical, so that their dipole moments precess at the same frequency in the external magnetic field, one induces an alternating field at the other of just the right frequency for magnetic resonance transitions. The effect is that of exchanging energy between the ions, giving broadening and a shortening of the lifetime of the individual ions in a given state.

Second, the effect of nearby nuclei with magnetic moments may be to inhomogeneously broaden the resonance lines. The inherent linewidth of each component may be such that the hyperfine interaction is not resolved. In copper proteins, the nuclei which often produce this inhomogeneous broadening are those due to ^{14}N, ^{1}H, and the copper nuclei since ^{63}Cu and ^{65}Cu have slightly different nuclear magnetic moments.

A third form of inhomogeneous broadening arises from local variations in the position of neighboring ions. These will give rise to variations in the g-values of the individual paramagnetic ions and thus a variation in the magnetic fields at which resonance occurs. This effect is often called "g-strain broadening".[154] A similar effect arises where the total spin S is greater than 1/2, and the spin Hamiltonian contains terms in D and E, as in Equation (36). In these circumstances, D and E can vary from ion to ion, also giving a broadening. This may be an important component in the width of the lines of dipolar or exchange-coupled pairs. The low frequency experiments of Froncisz et al.[151] were devised to overcome the problems caused by the g-strain effect in cytochrome *c* oxidase.

In a randomly oriented sample, the linewidth at a particular orientation will be a function

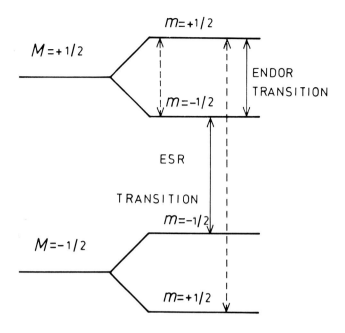

FIGURE 9.   The ENDOR transition in a $S = 1/2$, $I = 1/2$ system.
Possible relaxation paths are shown by the broken lines.

of the linewidths of each of the components arising from the hyperfine or superhyperfine splitting. Each of these components will have its own width, determined by the spin-lattice or spin-spin relaxation and the inhomogeneous broadening of each component. Thus each component, or "spin packet" will have its own characteristic relaxation time, a fact which is used in spin-echo experiments.

VI. ELECTRON NUCLEAR DOUBLE RESONANCE STUDIES OF COPPER PROTEINS

A. The ENDOR Experiment

The ENDOR technique was first used by Feher[179,180] and relies upon the detection of the effect on an electron resonance signal of an intense NMR field. The advantage of ENDOR is that measurements of the energy differences between hyperfine levels, which in general cannot be made directly by a nuclear resonance technique, can be made indirectly and with the much greater sensitivity of an EPR experiment over that of NMR. The principle of ENDOR is shown in Figure 9, and the reader is directed to the numerous extensive discussions of ENDOR techniques and interpretation for a more complete account than that given here.[181,182]

In a conventional EPR experiment, where the microwave power supplied to the sample is much less than that required to saturate the allowed $\Delta M = 1$, $\Delta m = 0$ transitions, the signal intensity is determined primarily by the population difference between the two electron spin levels involved. This difference is proportional to $e^{-h\nu_e/kT}$, i.e., it depends on the magnitude of the microwave quantum, $h\nu_e$ and the temperature T.

If a nuclear resonance frequency is applied to the system, transitions of the type $\Delta M = 0$, $\Delta m = 1$ are induced, and the nuclear resonance signal intensity depends on $e^{-h\nu_n/kT}$, i.e., on the magnitude of the nuclear resonance quantum and the temperature. It is clear that the nuclear resonance absorption will be a factor of approximately 1/1,000 less intense than the electron resonance absorption, i.e., the relative intensities will depend on the ratio of the radiofrequency appropriate for NMR to the microwave frequency appropriate for EPR. Further, provided that both electron and nuclear resonance fields are less intense than those required to cause saturation of the appropriate transition, the relaxation mechanisms are able to maintain the populations of the levels at their initial levels and the two resonance experiments are able to proceed essentially independently of one another.

When the electron resonance transition is saturated, the populations of the two levels are equalized and the EPR absorption signal is reduced in intensity. The levels involved may be labeled as $< -1/2, -1/2|$, and $< +1/2, -1/2|$, as shown in Figure 9. If a nuclear resonance frequency field is applied, with intensity sufficient to saturate the transition $< +1/2, -1/2| \rightarrow < +1/2, +1/2|$, the population of the level $< +1/2, -1/2|$ is reduced, giving now a population difference between the two levels involved in the electron resonance transition. The intensity of the EPR absorption then increases, as the transition is no longer saturated. In effect, the application of the nuclear resonance frequency has provided an additional pathway for the electron in the excited state to relax back to the ground state.

To conduct an ENDOR experiment in practice, the external magnetic field is set at the appropriate position on the EPR line, which is then saturated. The nuclear resonance frequency is then swept through the range of interest, and the EPR signal level monitored as a function of the nuclear frequency. The nuclear resonance frequency range may be from 1 to 100 MHz.

Because of the high accuracy with which the splittings of the nuclear energy levels can be measured using ENDOR, the technique is particularly useful in determining the components of hyperfine and superhyperfine matrices, the components of the nuclear quadrupole coupling tensor and the nuclear *g*-value. The technique can also be used to resolve hyperfine contributions to inhomogeneously broadened lines, in circumstances where the splitting between the hyperfine components is less than the width of the individual spin packets. Furthermore, it can be applied to studies of randomly oriented systems, such as frozen solutions, where an EPR experiment is unable to provide the required resolution.

B. Practical Considerations in ENDOR Experiments

In its simplest form an ENDOR spectrometer is simply an EPR spectrometer with a radiofrequency coil or loop in the sample cavity to enable the application of the nuclear resonance frequency. Details of ENDOR spectrometers may be found in the literature referred to in this section, and we will not discuss ENDOR instrumentation in detail here.

Although in principle ENDOR only requires that the microwave power be sufficient to saturate the electron resonance signal and that there be a means of providing a large radiofrequency field at the sample, the practice of ENDOR is anything but straightforward. The discussion above indicates that the observation of ENDOR is critically dependent on the relaxation times of the various levels and the amount of microwave and radiofrequency power applied. Considerable effort is usually required to find the conditions for obtaining any ENDOR resonance at all and then to optimize the signals obtained.

Low temperatures are nearly always required to reduce the spin-lattice relaxation times sufficiently for saturation to occur. Since the electron resonance must remain stable for the time taken to sweep the radiofrequency field through the region of interest, good thermal stability is essential. This may not be a trivial problem if the high radiofrequency currents are required to produce the necessary nuclear resonance field at the sample, e.g., Brown and Hoffman[41] required 80 W to produce a 0.1 mT radiofrequency at the sample. Fortunately, for the experimenter, most of this power is dissipated externally and not in the cavity and low temperature system. A further consideration is that the optimum conditions for ENDOR may not be obtained at the boiling points of the two most common refrigerants, namely liquid helium (4.2 K) or liquid nitrogen (77 K), but either in between, e.g., between 20 and 30 K,[41] or at temperatures lower than 4.2 K, e.g., 2.0 K.[150] The dependence on the temperature arises not only because of the effect of the relaxation times on the ENDOR conditions, but also because the best ENDOR signals are not necessarily obtained at the microwave-power level at which the EPR transition is completely saturated. It is more usual for the optimum ENDOR conditions to be obtained at a power level between this and the level at which saturation first occurs, e.g., Brown and Hoffman[41] used a power level 3 db above that at which the maximum intensity of their EPR signals occurred.

A number of additional factors need to be borne in mind. First, the conventional magnetic field modulation frequency of 100 kHz may be of the same order as the relaxation frequency of the electron spin levels. This will affect the saturation conditions, and a lower frequency must be used, such as 1 kHz. This requirement, together with the fact that low microwave powers are generally sufficient to saturate the EPR transitions, make superheterodyne detection schemes a possibility.

Second, as shown by Roberts et al.,[35] the rate at which the nuclear frequency is swept through the resonance may be decisive in determining whether or not an ENDOR signal is observed.

Third, the ENDOR effect may only be about 10% or less of the EPR signal, so that EPR signal-to-noise ratios of 100:1 or better are required before ENDOR is likely to be successful.

From the above discussion, it can be seen that the implementation of ENDOR has several pitfalls for the unwary.

C. Some Examples of ENDOR Studies of Copper Proteins
1. Stellacyanin

The ENDOR study of stellacyanin (St) by Rist et al.[183] showed that at least one of the ligands was nitrogen, that there was at least one proton strongly coupled to the copper, and that there were other less strongly coupled protons on nearby amino acids. Copper ENDOR signals were not observed.

The studies of Roberts et al.[35] confirmed these results and showed that the copper in St had at least two nitrogen ligands. They were also able to determine the components of the copper hyperfine interaction and the quadrupole interaction tensor. The copper resonances were weaker and broader than those of ^1H or ^{14}N, and occurred primarily at higher radiofrequencies. Their appearance depended on the rate at which the radiofrequency field was swept through the resonance, with sweep rates greater than 10 MHz/sec giving enhanced EPR absorptions. Decreased EPR absorptions were observed under other conditions.

2. Cytochrome c Oxidase

The original EPR studies attributed the intense signal near $g = 2$ to copper(II) ions.[184] However, several features of this signal led to alternative explanations, namely the absence at conventional microwave frequencies of any resolved hyperfine structure, the g-values of $g_x = 1.99$, $g_y = 2.02$, and $g_z = 2.18$, and the temperature dependence of the linewidth referred to earlier. One suggestion was that the signal was due to a thiyl π radical, R-S[.] rather than to copper(II).[9] The early ENDOR experiments[185] showed that nitrogen and proton ENDOR could be detected, but not the ENDOR from a copper nucleus.

More recent ENDOR experiments[150] at a temperature of 2.0 K, a microwave power of 20 μW, and a radiofrequency sweep rate of 7 MHz/sec confirmed the earlier observations of ^1H and ^{14}N coupling. However, if the microwave power was increased to 200 μW and the sweep rate to 160 MHz/sec, the ^1H and ^{14}N resonances became smeared out and broader resonances of the opposite phase were observed. These were attributed to copper(II) ions, and the unusual g-values ($g_x < 2.0$) and very low value of A_z attributed to the ion being highly covalent, with most of the spin density being on the coordinating sulfur ions. We should note here that the values of the copper hyperfine interaction obtained by ENDOR differ from those obtained by low frequency EPR. There is no explanation for this discrepancy.

VII. ELECTRON SPIN-ECHO SPECTROMETRY

A. Principles of Electron Spin-Echo Spectrometry

Electron spin-echo spectrometry can be applied to the same situations as ENDOR, namely where the hyperfine and superhyperfine interactions give an inhomogeneously broadened line. However, the spin-echo technique is a useful alternative to ENDOR when the splittings are small and conventional ENDOR difficult to perform, although both the experimental apparatus and the analysis of the results are more complex than for ENDOR.

The principles of an electron spin-echo experiment are most simply explained using a macroscopic classical model,[186-188] although Rowan et al.[187] also present the quantum mechanical approach using the density matrix formalism to represent the macroscopic properties of the spin system. We will content ourselves with a brief discussion of the classical model and its application to the determination of the surrounding ligands and the symmetry of the copper-binding site.

In thermal equilibrium, a spin system in a strong stationary magnetic field of magnitude and direction B has its magnetic moment vector M aligned along B which defines the z axis. When an alternating field of magnitude B_1 and frequency ν such that the resonance condition $h\nu = g\beta B$ is satisfied is applied perpendicular to B, i.e., in the x-y plane, the vector M will nutate through an angle θ while still precessing about z. The angle of nutation, θ, is given by:

$$\theta = \tfrac{1}{2} \nu B_1 \tau \qquad\qquad (41)$$

where τ is the duration of the pulse, i.e., the time for which the alternating field is applied. If the static magnetic field B is large, all the components of M will appear to rotate together, so that if τ is chosen so that $\theta = 90°$, the vector M will precess in the x-y plane, and a so-called free induction signal will be induced in the resonant cavity of the spectrometer. As transverse relaxation occurs, the components of M will fan out in the x-y plane because of the variations in B over the sample, and the free induction signal will decay.

If after a time t a further pulse is applied for a length of time 2τ such that $\theta = 180°$ in Equation (41), the components of M will be rotated through $180°$ in the x-y plane, i.e., the components will point in the opposite direction. Once again, the components will precess in this plane, but this time, instead of dispersing, they will recluster because of the inversion of their relative positions. After a time $2t$, there will be perfect reclustering, so that the magnetization vector M will have its maximum value in the x-y plane before dispersion and decrease in M occurs again. These variations in the magnetization induce the echo signal, which will show an increase and then a decay. The process of producing an echo signal can be repeated by application of another $180°$ pulse. The two most generally used pulse sequences for generating echoes are the two-pulse (denoted for obvious reasons as the $90°$, $180°$ pulse sequence) or the three-pulse (stimulated echo) sequence, where the pulses are $90°$, $180°$, $180°$. The shape of the echo signal will depend on the precession time of the spins in each of the spin packets making up the inhomogeneously broadened line. Hence a variation of the time between pulses will give a variation in the overall shape (and hence amplitude) of the echo signal, which will be reflected in a modulation of the echo amplitude as a function of the time between pulses. This modulation can be related back to the precession frequencies of the individual spin packets,[189] and hence to the magnitudes of the hyperfine and superhyperfine interactions.

The function obtained by plotting the amplitude of the echo signals against the time between the microwave pulses which generate them is called the envelope of electron spin echoes.[190,191] As we have seen, the envelope is modulated by the superposition of the echoes from each of the spin packets. Mims and co-workers[190,191] have used this envelope to obtain information about both the superhyperfine coupling at the copper(II) site and the symmetry at the copper(II) site. Information about the superhyperfine coupling is obtained because of the modulation of the amplitude of the echoes as described above, while information about the site symmetry is obtained using the linear electric field effect (LEFE), in which the envelope of echoes is modified by the application of an electric field at the time of the second pulse. We discuss each of these below, although the reader is referred to the book by Mims[192] for a full discussion of the LEFE.

B. Experimental Apparatus for Spin-Echo Measurements

Similar spectrometers are used for both types of experiment, and detailed descriptions are given by Blumberg et al.[191] and Mims.[192,193] The spectrometer has a similar structure to a conventional EPR spectrometer, with a microwave-power generator, a resonant cavity in a magnetic field, and a detection system. In the system described by Blumberg et al.[191] the microwave pulses are provided by a 1-kW traveling wave tube (TWT). This is driven by the 2-W output from a second TWT which acts as a microwave amplifier and is driven by a conventional klystron. Pulses as short as 10 nsec can be delivered by the main TWT, which is controlled by a modulator unit. The pulses are fed via a circulator to the sample cavity, which is specially designed to have a low Q and a high filling factor. The requirement of low Q arises from the need to avoid ringing of the cavity on application of the pulse.

Echo signals returning from the cavity via a circulator are amplified by a low noise TWT microwave amplifier and fed into a homodyne mixer, which also receives power from the signal klystron via an attenuator and phase shifter. A homodyne rather than a superheterodyne mixer is employed because the former is simpler and more suitable for short pulses. The mixer output is fed to a video amplifier and then to a boxcar integrator. The output of the boxcar integrator is then fed to the data accumulation and presentation circuits, which can be either a multichannel analyzer or a chart recorder. Details of the rather specialized electronics, such as the voltage-step generator for the LEFE experiments and the various pulse timing and data averaging and collection circuits are given by Blumberg et al.[191] and Mims.[193]

C. Determination of the Coordinating Ligands by Spin-Echo Spectrometry

As discussed previously, the envelope of the spin echoes is made up from the superposition of the echoes from each of the spin packets comprising the inhomogeneously broadened line, and the presence of these packets may give rise to a modulation of the envelope. It can be shown that the amplitude of this modulation is determined by the relative magnitudes of the nuclear Zeeman, nuclear quadrupole, and electron-nuclear coupling terms in the spin Hamiltonian.[187,190,191] Thus, the echo envelope is capable of yielding similar information to that obtained from an ENDOR experiment.

In principle, all that is required is to find the Fourier cosine transforms of the echo-envelope function, so that each superhyperfine (or hyperfine) line appears to be centered on the frequency corresponding to the superhyperfine splitting and have a line shape corresponding to the Fourier transform of a function which describes the decay of the spin packet corresponding to that superhyperfine component.[191] Fourier transform spectroscopy is well-known in NMR, so that the analysis presents no great difficulty, at least in principle. Some of the problems which may arise are due to the dead time of the apparatus,[191] which may give a decrease in the signal-to-noise ratio, an alteration of the relative amplitudes of the lines, and the disappearance of some of the superhyperfine components. The dead time may also give rise to overlapping lines with partially absorptive and partially dispersive line shapes and the presence of a d.c.-like component. It should also be noted that the two-pulse echo envelope contains sidebands in addition to the ENDOR frequencies. The sidebands do not appear in a three-pulse, or stimulated-echo experiment.[190]

Although the same information can be obtained from both echo and ENDOR experiments, the spin-echo method is to a certain extent limited by the requirement that there be mixing of the spin states, which can occur either by nuclear quadrupole interactions or electron-nuclear dipolar coupling. If one of the interactions is very much stronger than the other, sufficient state mixing will not occur, with the nuclear spin magnetic quantum number, m, being a "good" quantum number. The same applies in a large magnetic field. These limitations do not apply to ENDOR. Against this must be balanced the fact that relaxation effects within the spin packet are less important for spin-echo than for ENDOR experiments, so that the spin-echo experiment may be easier to perform. Both experiments have much the same sensitivity under optimum conditions, and are capable of comparable frequency resolution since the inhomogeneous broadening of the spin packet itself is the limiting factor.[194] The choice between ENDOR or spin-echo techniques is best left to the means and inclination of the user, who may well have to consider the inherent behavior of the copper protein system under study.

D. Representative Spin-Echo Results

Mims and Peisach[195] examined the echo spectra of a number of simple copper complexes and of copper in bovine serum albumin and St in a qualitative fashion. The echo spectra for the copper(II) aquo complex and for copper(II) glycylglycine showed only the effects of coupling to protons, namely a high frequency modulation of the echo, while those of copper(II) hexaglycine showed the effects of proton coupling and only a weak feature attributable to nuclei other than protons. No evidence of nitrogen coupling was observed in the case of the copper(II) glycylglycine, presumably because the nitrogen ligand coordinates directly to the copper(II) ion and thus gives rise to an $\hat{I} \cdot \mathbf{A} \cdot \hat{S}$ term which is too large to give a modulation effect.

The echo spectra of copper(II) (imidazole)$_4$ and copper(II) (guanidine)$_4$ showed strong modulation from nitrogen atoms not coordinated directly to the copper ion. The echo envelopes were different in each case. In the case of copper(II) bovine serum albumin, where the copper is known to be coordinated to an amino nitrogen, an imidazole nitrogen, and two peptide nitrogens, the echo spectra are interpreted in terms of the N-1 nitrogen of the imidazole group. This is not coordinated directly to the copper, and it was proposed by Mims and Peisach[195] that the nitrogens directly coordinated to the copper give too large a coupling for modulation of the echo to be observed.

In the case of St, the low frequency modulation pattern showed a strong similarity to that of bovine serum albumin, leading to the assignment of imidazole as one of the copper ligands. The proton-induced modulation was less marked than for bovine serum albumin, indicating that fewer protons are close to the copper in St. Mims and Peisach[195] did not rule out the possibility of a second imidazole ligand. These results are consistent with the ENDOR studies of Roberts et al.,[35] who showed that there were at least two nitrogen atoms coordinated directly to the copper ion, as well as a strongly coupled proton.

Studies of galactose oxidase[196] showed that there was at least one imidazole ligand, although it was not clear whether a second imidazole was bound to the copper site. The effects of addition of the ligands imidazole, fluoride, and cyanide were examined using a two-pulse echo sequence.[196] The changes observed in the echo spectrum on the addition of imidazole and cyanide, but not on the addition of fluoride, led the authors to suggest that the copper site of galactose oxidase contains a ligand which is easily displaced by any of the above three. We note here that the fluoride anion is expected to coordinate directly to the copper, giving a strong interaction and hence no modulation effect. We may also note that the possibility of the copper in galactose oxidase being copper(III) is now discredited.[5]

Spin-echo experiments on *R. vernicifera* laccase indicated that both the type-1 and type-2 copper ions were bound to one imidazole ligand each.[197] Similarly, the results demonstrated the coordination of a single imidazole to the type-2 copper of ceruloplasmin. The similarity of the symmetry of the type-1 site in laccase to that of the single copper site in St and Az lends support to the belief that the type-1 sites in the copper oxidases are similar to those in the dark blue single copper proteins.[197]

In the case of cytochrome *c* oxidase, the spin-echo results were quite unlike those for either simple copper complexes or for copper(II) in an essentially tetrahedral site.[198] The nuclear modulation pattern was different than that obtained for either type-1 or type-2 copper sites, and when considered in conjunction with the LEFE measurements discussed below, it cast doubt on the model of Greenaway et al.[92]

E. The Linear Electric Field Effect and the Study of Low Symmetry Copper Sites

Conventional EPR techniques are unable to distinguish between paramagnetic sites which are related by an inversion symmetry operation, e.g., one is unable to distinguish whether the resonance condition for the magnetic field acting in the direction **AB** is different from that for the field acting in the direction **BA**. However, the application of an electric field to the sample gives a shift of the resonance lines from such centers in opposite directions. This effect was first noticed by Ludwig and Woodbury[199] for Fe° centers in silicon. In principle, very strong electric fields, with intensities of 100 kV/cm or greater, are required to produce *g* shifts which exceed the width of inhomogeneously broadened lines in single crystals. Even greater electric fields would be required for the powder or frozen solution samples encountered in biological systems. This difficulty may be overcome by a spin-echo experiment as described above, where a voltage step is applied to the sample in coincidence with the second pulse (i.e., the 180° pulse) of a two-pulse echo sequence. The effect of the voltage step is to change the precession frequencies, as a result of the LEFE, and thus alter the convergence of the magnetic moments of the spin packets which make up the inhomogeneously broadened line. The electric field effect is observed as a reduction in amplitude or as an inversion of the echo. The Fourier transform of the function obtained by taking the ratio of the echo signals observed with and without the applied electric field for a series of times between pulses (i.e., the echo envelope difference function) gives the distribution of frequencies and hence the *g* shifts on application of the electric field.[193]

In a powder or frozen solution system, the sites of different paramagnetic centers are not related by an inversion symmetry operation. However, the sites themselves may possess inversion symmetry if, e.g., two of the ligands are different or if the electron distribution around the ion is different in opposite directions. To perform the experiment, the spectrometer is set at the value of the magnetic field corresponding to the direction along which the asymmetry is to be measured, e.g., at the field value corresponding to the peak associated with g_z. Descriptions of the apparatus and details of the analysis of the results are given in the book by Mims.[192]

F. Representative Linear Electric Field Effect Results

Peisach and Mims[200,201] have studied the LEFE on the copper sites of several proteins. In the studies of the blue copper proteins St and Az, a large shift was observed, as well as an unusual dependence on the magnetic field. These phenomena were not observed for simple copper complexes which involved oxygen or nitrogen coordination to the copper site. The explanation of the behavior in the proteins was that it was due to an inequivalence in the bonding to the copper site, possibly arising from a ligand-to-metal charge transfer.[202] In galactose oxidase,[196] the effect was smaller and more akin to that observed for the simple copper complexes.

A study of the LEFE in cytochrome *c* oxidase, made in conjunction with the spin-echo measurements described above,[198] showed that the magnetic field dependence of the electric field-induced *g* shifts was quite unlike those for simple copper(II) complexes or for tetrahedrally coordinated copper(II). Neither the small *g* shift nor the unusual symmetry could be reconciled with the tetrahedral model of the intrinsic copper(II) proposed by Greenaway et al.[92] Mims et al.[198] proposed that this typical behavior could be explained if an otherwise typical copper(II) center were to receive the donation of an electron from an RS⁻ ligand, thus giving a center in which the unpaired electron spin resides mainly on the radical and only to a lesser extent on the copper. This suggestion is in accord with the earlier proposal of Peisach and Blumberg[9] and the ENDOR measurements of Hoffman et al.[150] However, the degree of delocalization of the unpaired electron from the copper ion onto the ligand must still be regarded as an unresolved question at this stage.

The results indicate that measurements of the LEFE are possible for copper proteins, and may give information about the charge distribution and local symmetry at the copper site. Such information is of value in discussions of the biochemical and electronic mechanisms of the copper site, but a full explanation of the effects requires a detailed structural knowledge which does not exist at present.

VIII. CONCLUDING REMARKS

Since EPR was first applied to studies of the copper proteins some 20 years ago, it has provided some spectacular successes, but has also raised many unanswered questions. The most spectacular success has probably been the demonstration that three types of binding site exist in copper(II) proteins, that only two of these are detectable by EPR, and that one of these is in a most unusual site. However, EPR alone has been unable to provide any conclusive and direct evidence as to the nature of this unusual binding site, nor has it been able to provide any evidence as to the nature of the ''EPR undetectable'' copper site. In these respects, EPR studies of copper proteins have failed to meet many of the objectives suggested by Brill and Venable.[17]

However indirect evidence using chemically modified copper proteins and direct evidence from EPR-related techniques such as ENDOR and spin-echo spectrometry are now starting to fill the gaps left by conventional EPR. When these results are combined with those obtained using resonance Raman spectroscopy, EXAFS, and other spectroscopic techniques, a detailed picture of the structure and function of the active sites of copper proteins is expected to emerge. The recent advances in X-ray crystallographic studies of copper proteins have given the first detailed structural model of an active site, and it now remains to relate the structure to the function on an electronic level. As will be seen in other chapters in these volumes, many issues in the copper proteins are still both unresolved and contentious.

ACKNOWLEDGMENTS

The author wishes to thank Dr. R. L. Calvert of the Australian Radiation Laboratory and Dr. J. R. Pilbrow of the Department of Physics, Monash University, for their helpful and constructive comments. Grateful acknowledgment is also made to Mrs. Chris Telfer for the typing of this chapter.

Appendix 1

REPRESENTATIVE g- AND A^a-VALUES OF SOME COPPER PROTEINS DISCUSSED IN THIS CHAPTER

Protein	g_x	$g_\perp$	g_y	$g_z, g_\parallel$	A_x	$A_\perp$	A_y	$A_z, A_\parallel$	Notes	Ref.
Single copper "blue" proteins										
Stellacyanin	2.025		2.077	2.287	0.0057		0.0029	0.0035	EPR	203
					0.0056		0.0029	0.0032	ENDOR	35
Plastocyanin		2.053		2.226		<0.0017		0.0063	EPR	3
Azurin		2.052		2.260		≈0.0000		0.0060	EPR	204
Multicopper oxidases										
Ceruloplasmin (human)										
Type 1a		2.06		2.215		0.0010		0.0095	EPR	122
Type 1b		2.05		2.206		0.0010		0.0074		
Type 2		2.06		2.247		0.0024		0.0189		
Laccase (*Rhus vernicifera*)										
Type 1		2.047		2.289		0.0017		0.0043	EPR	203
Type 2		2.053		2.237		0.0018		0.0200		
Ascorbate oxidase (*Cucurbita pepo medullosa*)										
Type 1	2.036		2.058	2.227		0.0005		0.0052	EPR	205
Type 2		2.053		2.242		0.0005		0.0200		
Cytochrome *c* oxidase (bovine heart)										
	1.99		2.03	2.185	0.0020		0.0025	0.0030	9 GHz EPR	92
	1.990		2.109	2.182	No hyperfine structure resolved				35 GHz EPR	152
	1.99		2.02	2.18	0.0009		0.0042	≤0.0040	2 and 4 GHz EPR	151
					0.0023		0.0033	0.0030	9 GHz ENDOR	150
"Nonblue" copper proteins										
Galactose oxidase (*Polyporus circinatus*)		2.055		2.277		<0.0001		0.0186	EPR	44
Superoxide dismutase (bovine)		2.087	2.28				—	0.014	EPR	133
Plasma amine oxidase (porcine)										
Axial component		2.078		2.286		0.0010		0.0165	EPR	206
Rhombic component		2.039	2.069	2.294		—		0.0150		

^{a} All values of A are in cm^{-1}.

REFERENCES

1. **Peisach, J., Aisen, P., and Blumberg, W. E., Eds.,** *The Biochemistry of Copper,* Academic Press, New York, 1966.
2. **Malkin, R. and Malmström, B. G.,** The state and function of copper in biological systems, *Adv. Enzymol.,* 33, 177, 1970.
3. **Fee, J. A.,** Copper proteins. Systems containing the "blue" copper center, *Struct. Bonding (Berlin),* 23, 1, 1975.
4. **Boas, J. F., Pilbrow, J. R., and Smith, T. D.,** ESR of copper in biological systems, in *Magnetic Resonance in Biological Systems,* Berliner, L. J. and Reuben, J., Eds., Plenum Press, New York, 1978, chap. 7.
5. **Beinert, H.,** Structure and function of copper proteins, *Coord. Chem. Rev.,* 33, 55, 1980.
6. **Malmström, B. G. and Vänngård, T.,** Electron spin resonance of copper proteins and some model complexes, *J. Mol. Biol.,* 2, 118, 1960.
7. **Broman, L., Malmström, B. G., Aasa, R., and Vänngård, T.,** Quantitative electron spin resonance studies on native and denatured ceruloplasmin and laccase, *J. Mol. Biol.,* 5, 301, 1962.
8. **Blumberg, W. E., Levine, W. G., Margolis, S., and Peisach, J.,** On the nature of copper in two proteins obtained from *Rhus vernicifera* latex, *Biochem. Biophys. Res. Commun.,* 15, 277, 1964.
9. **Peisach, J. and Blumberg, W. E.,** Structural implications derived from the analysis of electron paramagnetic resonance spectra of natural and artificial copper proteins, *Arch. Biochem. Biophys.,* 165, 691, 1974.
10. **Boas, J. F., Pilbrow, J. R., Troup, G. J., Moore, C., and Smith, T. D.,** Electron spin resonance studies of copper(II) in haemocyanin, its chelates with ($\pm$)-mercaptosuccinic acid and some related systems, *J. Chem. Soc. Abstr.,* 965, 1969.
11. **Freedman, T. B., Loehr, J. S., and Loehr, Th. M.,** A resonance Raman study of the copper protein, hemocyanin. New evidence for the structure of the oxygen-binding site, *J. Am. Chem. Soc.,* 98, 2809, 1976.
12. **Thamann, T. J., Loehr, J. S., and Loehr, Th. M.,** Resonance Raman study of oxyhemocyanin with unsymmetrically labeled oxygen, *J. Am. Chem. Soc.,* 99, 4187, 1977.
13. **Larrabee, J. A. and Spiro, Th. G.,** Structural studies of the hemocyanin active site. II. Resonance Raman spectroscopy, *J. Am. Chem. Soc.,* 102, 4217, 1980.
14. **Brown, J. M., Powers, L., Kincaid, B., Larrabee, J. A., and Spiro, Th. G.,** Structural studies of the hemocyanin active site. I. Extended X-ray absorption fine structure (EXAFS) analysis, *J. Am. Chem. Soc.,* 102, 4210, 1980.
15. **Co, M. S., Hodgson, K. O., Eccles, Th. K., and Lontie, R.,** Copper site of molluscan oxyhemocyanins. Structural evidence from X-ray absorption spectroscopy, *J. Am. Chem. Soc.,* 103, 984, 1981.
16. **Co, M. S. and Hodgson, K. O.,** Copper sites of deoxyhemocyanin. Structural evidence from X-ray absorption spectroscopy, *J. Am. Chem. Soc.,* 103, 3200, 1981.
17. **Brill, A. S. and Venable, J. H., Jr.,** Electron paramagnetic resonance in single crystals of cupric insulin, *Nature (London),* 203, 752, 1964.
18. **Abragam, A. and Bleaney, B.,** *Electron Paramagnetic Resonance of Transition Ions,* Clarendon Press, Oxford, 1970.
19. **Poole, C. P.,** *Electron Spin Resonance. A Comprehensive Treatise on Experimental Techniques,* Wiley-Interscience, New York, 1967.
20. **Alger, R. S.,** *Electron Paramagnetic Resonance: Techniques and Applications,* Wiley-Interscience, New York, 1968.
21. **Ingram, D. J. E.,** *Biological and Biochemical Applications of Electron Spin Resonance,* Hilger, London, 1969.
22. **Pilbrow, J. R. and Lowrey, M. R.,** Low symmetry effects in electron paramagnetic resonance, *Rep. Prog. Phys.,* 43, 433, 1980.
23. **Abragam, A. and Bleaney, B.,** *Electron Paramagnetic Resonance of Transition Ions,* Clarendon Press, Oxford, 1970, 134.
24. **Abragam, A. and Bleaney, B.,** *Electron Paramagnetic Resonance of Transition Ions,* Clarendon Press, Oxford, 1970, 135.
25. **Abragam, A. and Bleaney, B.,** *Electron Paramagnetic Resonance of Transition Ions,* Clarendon Press, Oxford, 1970, 406 and chap. 17.
26. **Abragam, A. and Bleaney, B.,** *Electron Paramagnetic Resonance of Transition Ions,* Clarendon Press, Oxford, 1970, 166.
27. **Pilbrow, J. R. and Winfield, M. E.,** Computer simulation of low symmetry ESR spectra due to vitamin B_{12r} and model systems, *Mol. Phys.,* 25, 1073, 1973.
28. **Abragam, A. and Bleaney, B.,** *Electron Paramagnetic Resonance of Transition Ions,* Clarendon Press, Oxford, 1970, 168.

29. **Rotilio, G., Morpurgo, L., Giovagnoli, C., Calabrese, L., and Mondovì, B.,** Studies of the metal sites of copper proteins. Symmetry of copper in bovine superoxide dismutase and its functional significance, *Biochemistry,* 11, 2187, 1972.

30. **Bereman, R. D. and Kosman, D. J.,** Stereoelectronic properties of metalloenzymes. V. Identification and assignment of ligand hyperfine splittings in the electron spin resonance spectrum of galactose oxidase, *J. Am. Chem. Soc.,* 99, 7322, 1977.

31. **Abragam, A. and Bleaney, B.,** *Electron Paramagnetic Resonance of Transition Ions,* Clarendon Press, Oxford, 1970, 166.

32. **Bleaney, B., Bowers, K. D., and Ingram, D. J. E.,** Paramagnetic resonance in diluted copper salts. I. Hyperfine structure in diluted copper Tutton salts, *Proc. R. Soc. (London),* A228, 147, 1955.

33. **Maki, A. H. and McGarvey, B. R.,** Electron spin resonance in transition metal chelates. I. Copper(II) bis-acetylacetonate, *J. Chem. Phys.,* 29, 31, 1958.

34. **White, L. K. and Belford, R. L.,** Quadrupole coupling constants of square-planar copper(II)-sulfur complexes from single-crystal electron paramagnetic resonance spectroscopy, *J. Am. Chem. Soc.,* 98, 4428, 1976.

35. **Roberts, J. E., Brown, T. G., Hoffman, B. M., and Peisach, J.,** Electron nuclear double resonance spectra of stellacyanin, a blue copper protein, *J. Am. Chem. Soc.,* 102, 825, 1980.

36. **Rollmann, L. D. and Chan, S. I.,** Quadrupole effects in electron paramagnetic resonance spectra of polycrystalline copper and cobalt complexes, *J. Chem. Phys.,* 50, 3416, 1969.

37. **De Bolfo, J. A., Smith, T. D., Boas, J. F., and Pilbrow, J. R.,** Electron spin resonance study of solute-solute interactions in aqueous solutions containing transition metal ion chelates of 4,4′,4″,4‴-tetrasulpho-phthalocyanine, *J. Chem. Soc. Faraday II,* 72, 481, 1976.

38. **Abragam, A. and Bleaney, B.,** *Electron Paramagnetic Resonance of Transition Ions,* Clarendon Press, Oxford, 1970, 167.

39. **Abragam, A. and Bleaney, B.,** *Electron Paramagnetic Resonance of Transition Ions,* Clarendon Press, Oxford, 1970, 192.

40. **Guzy, C. M., Raynor, J. B., and Symons, M. C. R.,** Electron spin resonance spectrum of copper-63 phthalocyanin. A reassessment of the bonding parameters, *J. Chem. Soc. Abstr.,* 2299, 1969.

41. **Brown, T. G. and Hoffman, B. M.,** ^{14}N, ^{1}H and metal ENDOR of single crystal Ag(II)(TPP) and Cu(II)(TPP), *Mol. Phys.,* 39, 1073, 1980.

42. **Harrison, S. E. and Assour, J. M.,** Relationship of electron spin resonance and semiconduction in phthalocyanines, *J. Chem. Phys.,* 40, 365, 1964.

43. **Himmelwright, R. S., Eickman, N. C., and Solomon, E. I.,** Chemical and spectroscopic conformation of an exogenous ligand bridge in half met hemocyanin, *Biochem. Biophys. Res. Commun.,* 84, 300, 1978.

44. **Marwedel, B. J., Kosman, D. J., Bereman, R. D., and Kurland, R. J.,** Magnetic resonance studies of cyanide and fluoride binding to galactose oxidase copper(II): evidence for two exogenous ligand sites, *J. Am. Chem. Soc.,* 103, 2842, 1981.

45. **Schoot Uiterkamp, A. J. M.,** Monomer and magnetic dipole-coupled Cu^{2+} EPR signals in nitrosylhemocyanin, *FEBS Lett.,* 20, 93, 1972.

46. **Malkin, R., Malmström, B. G., and Vänngård, T.,** The requirement of the ''non-blue'' copper(II) for the activity of fungal laccase, *FEBS Lett.,* 1, 50, 1968.

47. **Himmelwright, R. S., Eickman, N. C., and Solomon, E. I.,** Comparison of half-met and met-apo hemocyanin. Ligand bridging at the binuclear copper active site, *J. Am. Chem. Soc.,* 101, 1576, 1979.

48. **Aasa, R., Brändén, R., Deinum, J., Malmström, B. G., Reinhammar, B., and Vänngård, T.,** A ^{17}O-effect on the EPR spectrum of the intermediate in the dioxygen-laccase reaction, *Biochem. Biophys. Res. Commun.,* 70, 1204, 1976.

49. **Abragam, A. and Bleaney, B.,** *Electron Paramagnetic Resonance of Transition Ions,* Clarendon Press, Oxford, 1970, chap. 3.

50. **Waller, W. G. and Rogers, M. T.,** A generalization for methods of determining *g* tensors, *J. Magn. Reson.,* 9, 92, 1973.

51. **Brill, A. S. and Venable, J. H., Jr.,** The binding of transition metal ions in insulin crystals, *J. Mol. Biol.,* 36, 343, 1968.

52. **Sands, R. H.,** Paramagnetic resonance absorption in glass, *Phys. Rev.,* 99, 1222, 1955.

53. **Venable, J. H., Jr.,** Electron paramagnetic resonance spectroscopy of protein single crystals. II. Computational methods, in *Magnetic Resonance in Biological Systems,* Ehrenberg, A., Malmström, B. G., and Vänngård, T., Eds., Pergamon Press, Oxford, 1967, 373.

54. **Boas, J. F., Dunhill, R. H., Pilbrow, J. R., Srivastava, R. C., and Smith, T. D.,** Electron spin resonance studies of copper(II) hydroxy-carboxylic acid chelates in aqueous and non-aqueous solutions, *J. Chem. Soc. Abstr.,* 94, 1969.

55. **Boyd, P. D. W., Toy, A. D., Smith, T. D., and Pilbrow, J. R.,** A theoretical and experimental study of the electron spin resonance of a number of low symmetry copper(II) dimers, *J. Chem. Soc. Dalton Trans.,* 1549, 1973.

56. **Aasa, R. and Vänngård, T.,** EPR signal intensity and powder shapes: a reexamination, *J. Magn. Reson.,* 19, 308, 1975.
57. **Bleaney, B.,** Electron spin resonance intensity in anisotropic substances, *Proc. Phys. Soc. (London),* A75, 621, 1960.
58. **Kneubühl, F. K. and Natterer, B.,** Paramagnetic resonance intensity of anisotropic substances and its influence on line shapes, *Helv. Phys. Acta,* 34, 710, 1961.
59. **Pilbrow, J. R.,** Anisotropic transition probability factor in ESR, *Mol. Phys.,* 16, 307, 1969.
60. **Bonamo, R. P. and Pilbrow, J. R.,** An EPR single crystal study of Cu^{2+} in cadmium acetate: a seven-coordinated low symmetry site, *J. Magn. Reson.,* 45, 404, 1981.
61. **Schoot Uiterkamp, A. J. M., Van der Deen, H., Berendsen, H. C. J., and Boas, J. F.,** Computer simulation of the EPR spectra of mononuclear and dipolar coupled Cu(II) ions in nitric oxide- and nitrite-treated hemocyanins and tyrosinase, *Biochim. Biophys. Acta,* 372, 407, 1974.
62. **Pryce, M. H. L.,** A modified perturbation procedure for a problem in paramagnetism, *Proc. Phys. Soc. (London),* A63, 25, 1950.
63. **Abragam, A. and Pryce, M. H. L.,** Theory of the nuclear hyperfine structure of paramagnetic resonance spectra in crystals, *Proc. R. Soc. (London),* A205, 135, 1951.
64. **Abragam, A. and Pryce, M. H. L.,** The theory of the nuclear hyperfine structure of paramagnetic resonance spectra in the copper Tutton salts, *Proc. R. Soc. (London),* A206, 164, 1951.
65. **Abragam, A. and Bleaney, B.,** *Electron Paramagnetic Resonance of Transition Ions,* Clarendon Press, Oxford, 1970, chap. 19.
66. **Keijzers, C. P. and de Boer, E.,** ESR study of copper and silver, N,N-dialkyldiselenocarbamates. II. Interpretation of spectra measured in host lattices with monomeric structures, *Mol. Phys.,* 29, 1007, 1975.
67. **Nickerson, K. W. and Phelan, N. F.,** A molecular orbital treatment of Cu(II) proteins, *Bioinorg. Chem.,* 4, 79, 1974.
68. **Buluggiu, E., Dascola, G., Giori, D. C., and Vera, A.,** ESR studies of covalent copper complexes with a rhombic arrangement, *J. Chem. Phys.,* 54, 2191, 1971.
69. **Swalen, J. D., Johnson, B., and Gladney, H. M.,** Covalency and electronic structure of Cu^{2+} in ZnF_2 by EPR, *J. Chem. Phys.,* 52, 4078, 1970.
70. **Weeks, M. J. and Fackler, J. P.,** Single-crystal electron paramagnetic resonance studies of copper diethyldithiocarbamate, *Inorg. Chem.,* 7, 2548, 1968.
71. **Kivelson, D. and Neiman, R.,** ESR studies on the bonding in copper complexes, *J. Chem. Phys.,* 35, 149, 1961.
72. **Abragam, A. and Bleaney, B.,** *Electron Paramagnetic Resonance of Transition Ions,* Clarendon Press, Oxford, 1970, 455.
73. **Cotton, F. A.,** *Chemical Applications of Group Theory,* Wiley-Interscience, New York, 1963, 196.
74. **Carrington, A. and McLachlan, A. D.,** *Introduction to Magnetic Resonance,* Harper & Row, New York, 1967, chap. 10.
75. **Abragam, A. and Bleaney, B.,** *Electron Paramagnetic Resonance of Transition Ions,* Clarendon Press, Oxford, 1970, 371 and chap. 21.
76. **Hitchman, M. A., Olson, C. D., and Belford, R. L.,** Behavior of the in-plane *g* tensor in low-symmetry d^1 and d^9 systems with application to copper and vanadyl chelates, *J. Chem. Phys.,* 50, 1195, 1969.
77. **Cotton, F. A.,** *Chemical Applications of Group Theory,* Wiley-Interscience, New York, 1963, 209.
78. **Figgis, B. N., Mason, R., Smith, A. R. P., and Williams, G. A.,** Valence electrons in open-shell molecules: experimental studies using polarized neutron scattering, *J. Am. Chem. Soc.,* 101, 3673, 1979.
79. **Williams, G. A., Figgis, B. N., and Mason, R.,** Spin density and cobalt electronic structure in phthalocyaninatocobalt(II): a polarised neutron diffraction study, *J. Chem. Soc. Dalton Trans.,* 734, 1981.
80. **Abragam, A. and Bleaney, B.,** *Electron Paramagnetic Resonance of Transition Ions,* Clarendon Press, Oxford, 1970, 456.
81. **Pilbrow, J. R. and Spaeth, J. M.,** ESR studies of Cu^{2+} in NH_4Cl single crystals between 4.2 and 453°K. I. Spin-Hamiltonian parameters and lattice constant for NH_4Cl, *Phys. Stat. Sol.,* 20, 225, 1967; II. Theoretical analysis: Vibrational admixtures, spin polarization and $<r^{-3}>$ for Cu^{2+} ions, *Phys. Stat. Sol.,* 20, 237, 1967.
82. **Abragam, A. and Bleaney, B.,** *Electron Paramagnetic Resonance of Transition Ions,* Clarendon Press, Oxford, 1970, 45 and chap. 17.
83. **Bleaney, B., Bowers, K. D., and Pryce, M. H. L.,** Paramagnetic resonance in diluted copper salts. III. Theory and evaluation of the nuclear electric quadrupole moments of ^{63}Cu and ^{65}Cu, *Proc. R. Soc. (London),* A228, 166, 1955.
84. **Schatz, G. C. and McMillan, J. A.,** Electron paramagnetic resonance of magnetically dilute cupric $(3d^9, ^2D)$ ion in single crystals of zinc 3-pyridine sulfonate, *J. Chem. Phys.,* 55, 2342, 1971.
85. **De Bolfo, J. A., Smith, T. D., Boas, J. F., and Pilbrow, J. R.,** ESR studies of the solute-solute interactions between copper(II) and nickel(II) thiosemicarbazone chelates in non-aqueous solutions, *Aust. J. Chem.,* 29, 2583, 1976.

86. **Bates, C. A., Moore, W. S., Standley, K. J., and Stevens, K. W. H.,** Paramagnetic resonance of a Cu^{2+} ion in a tetrahedral crystal field, *Proc. Phys. Soc. (London),* 79, 73, 1962.

87. **Bates, C. A.,** The effects of distortion on the spectra of a Cu^{2+} ion in a tetrahedral crystal field, *Proc. Phys. Soc. (London),* 83, 465, 1964.

88. **Hausmann, A. and Schreiber, P.,** Electron spin resonance of divalent copper in zinc oxide, *Solid State Commun.,* 7, 631, 1969.

89. **Parker, I. H.,** Cu^{2+} in ammonium fluoride — a tetrahedral site, *J. Phys. C. Solid State Phys.,* 4, 2967, 1971.

90. **Sharnoff, M.,** Electron paramagnetic resonance and the primarily $3d$ wavefunctions of the tetrachlorocuprate ion, *J. Chem. Phys.,* 42, 3383, 1965.

91. **Brill, A. S. and Bryce, G. F.,** Cupric ion in blue proteins, *J. Chem. Phys.,* 48, 4398, 1968.

92. **Greenaway, F. T., Chan, S. H. P., and Vincow, G.,** An EPR study of the lineshape of copper in cytochrome c oxidase, *Biochim. Biophys. Acta,* 490, 62, 1977.

93. **Solomon, E. I., Hare, J. W., Dooley, D. M., Dawson, J. H., Stephens, P. J., and Gray, H. B.,** Spectroscopic studies of stellacyanin, plastocyanin, and azurin. Electronic structure of the blue copper sites, *J. Am. Chem. Soc.,* 102, 168, 1980.

94. **Penfield, K. W., Gay, P. R., Himmelwright, R. S., Eickman, N. C., Norris, V. A., Freeman, H. C., and Solomon, E. I.,** Spectroscopic studies on plastocyanin single crystals: a detailed electronic structure determination of the blue copper active site, *J. Am. Chem. Soc.,* 103, 4382, 1981.

95. **Colman, P. M., Freeman, H. C., Guss, J. M., Murata, M., Norris, V. A., Ramshaw, J. A. M., and Venkatappa, M. P.,** X-ray crystal structure analysis of plastocyanin at 2.7 Å resolution. *Nature (London),* 272, 319, 1978.

96. **Adman, E. T., Stenkamp, R. E., Sieker, L. C., and Jensen, L. H.,** A crystallographic model for azurin at 3 Å resolution, *J. Mol. Biol.,* 123, 35, 1978.

97. **Dooley, D. M., Rawlings, J., Dawson, J. H., Stephens, P. J., Andréasson, L.-E., Malmström, B. G., and Gray, H. B.,** Spectroscopic studies of *Rhus vernicifera* and *Polyporus versicolor* laccase. Electronic structures of the copper sites, *J. Am. Chem. Soc.,* 101, 5038, 1979.

98. **Dawson, J. H., Dooley, D. M., Clark, R., Stephens, P. J., and Gray, H. B.,** Spectroscopic studies of ceruloplasmin. Electronic structures of the copper sites, *J. Am. Chem. Soc.,* 101, 5046, 1979.

99. **Krishnan, V. G.,** Electron spin resonance studies on Cu^{2+} doped at tetrahedrally coordinated Zn^{2+} sites in the single crystals of guanadinium zinc sulfate, *J. Chem. Phys.,* 68, 660, 1978.

100. **Hathaway, B. J. and Hodgson, P. G.,** The single-crystal electronic and electron-spin resonance spectra of catena-μ-*bis* (1,2-diphenylphosphinyl)ethane dichlorocopper(II), *Spectrochim. Acta,* 30A, 1465, 1974.

101. **Cotton, F. A.,** *Chemical Applications of Group Theory,* Wiley-Interscience, New York, 1963, 209, and appendix II.

102. **Bacci, M.,** Static distortions induced by vibronic interactions in copper(II) complexes. An approach to the interpretation of spectroscopic and structural properties of blue copper proteins, *J. Inorg. Biochem.,* 13, 49, 1980.

103. **Smith, T. D. and Pilbrow, J. R.,** The determination of structural properties of dimeric transition metal ion complexes from EPR spectra, *Coord. Chem. Revs.,* 13, 173, 1974.

104. **Robin, M. B. and Day, P.,** Mixed valence chemistry — a survey and classification, *Adv. Inorg. Chem. Radiochem.,* 10, 247, 1967.

105. **Boas, J. F., Hicks, P. R., Pilbrow, J. R., and Smith, T. D.,** Interpretation of electron spin resonance spectra due to some B_{12}-dependent enzyme reactions, *J. Chem. Soc. Faraday II,* 74, 417, 1978.

106. **Erdos, P.,** Theory of ion pairs coupled by exchange interaction, *J. Phys. Chem. Solids,* 27, 1705, 1966.

107. **Kato, M., Jonassen, H. B., and Fanning, J. C.,** Copper(II) complexes with subnormal magnetic moments, *Chem. Rev.,* 64, 99, 1964.

108. **Hatfield, W. E.,** Spin-spin coupling in parallel planar copper(II) dimers, *Inorg. Chem.,* 11, 216, 1972.

109. **Bleaney, B. and Bowers, K. D.,** Anomalous paramagnetism of copper acetate, *Proc. R. Soc. (London),* A214, 451, 1952.

110. **Ross, I. G.,** The metal-metal bond in binuclear copper acetate. I. Confirmation of the δ-bond, *Trans. Faraday Soc.,* 55, 1057, 1959.

111. **Ross, I. G. and Yates, J.,** The metal-metal bond in binuclear copper acetate. II. Non-empirical calculation of the singlet-triplet separation for δ- and σ-bonds, *Trans. Faraday Soc.,* 55, 1064, 1959.

112. **Price, J. H., Pilbrow, J. R., Murray, K. S., and Smith, T. D.,** Electron spin resonance study of exchange and dipole-dipole coupling in copper(II) chelates of cyclopentanetetracarboxylic acid and related systems, *J. Chem. Soc. Abstr.,* 968, 1970.

113. **Claesson, O. and Lund, A.,** Calculation of EPR spectra of triplet-state molecules with hyperfine and nuclear quadrupole interactions, *J. Magn. Reson.,* 41, 106, 1980.

114. **Lund, T. and Hatfield, W. E.,** Simulation of triplet state EPR spectra from dimers: g and D tensors not coaxial, *J. Chem. Phys.,* 59, 885, 1973.

115. **Buluggiu, E. and Vera, A.,** Computation of principal values and directions of tensors in the spin Hamiltonian, *J. Magn. Reson.,* 41, 195, 1980.
116. **Solomon, E. I., Dooley, D. M., Wang, R.-H., Gray, H. B., Cerdonio, M., Mogno, F., and Romani, G. L.,** Susceptibility studies of laccase and oxyhemocyanin using an ultrasensitive magnetometer. Antiferromagnetic behavior of the type 3 copper in *Rhus* laccase, *J. Am. Chem. Soc.,* 98, 1029, 1976.
117. **Dooley, D. M., Scott, R. A., Ellinghaus, J., Solomon, E. I., and Gray, H. B.,** Magnetic susceptibility studies of laccase and oxyhemocyanin, *Proc. Natl. Acad. Sci. U.S.A.,* 75, 3019, 1978.
118. **Moss, T. H., Gould, D. C., Ehrenberg, A., Loehr, J. S., and Mason, H. S.,** Magnetic properties of *Cancer magister* hemocyanin, *Biochemistry,* 12, 2444, 1973.
119. **Petersson, L., Ångström, J., and Ehrenberg, A.,** Magnetic susceptibility of laccases and ceruloplasmin, *Biochim. Biophys. Acta,* 526, 311, 1978.
120. **Burk, P. L., Osborn, J. A., Youinou, M.-T., Agnus, Y., Louis, R., and Weiss, R.,** Binuclear copper complexes: an open and shut case. A strong antiferromagnetically coupled μ-monohydroxo bridged complex, *J. Am. Chem. Soc.,* 103, 1273, 1981.
121. **Coughlin, P. K. and Lippard, S. J.,** A monohydroxo bridged, strongly antiferromagnetically coupled dicopper(II) center in a binucleating macrocycle. Comparisons with binuclear copper sites in biology, *J. Am. Chem. Soc.,* 103, 3228, 1981.
122. **Gunnarsson, P.-O., Nylén, Y., and Pettersson, G.,** Effect of pH on electron-paramagnetic-resonance spectra of ceruloplasmin, *Eur. J. Biochem.,* 37, 47, 1973.
123. **Van Leeuwen, F. X. R., Wever, R., and Van Gelder, B. F.,** EPR study of nitric oxide-treated reduced ceruloplasmin, *Biochim. Biophys. Acta,* 315, 200, 1973.
124. **Van Leeuwen, F. X. R. and Van Gelder, B. F.,** A spectroscopic study of nitric-oxide-treated ceruloplasmin, *Eur. J. Biochem.,* 87, 305, 1978.
125. **Ghiretti, F.,** Hemerythrin and hemocyanin, in *Oxygenases,* Hayaishi, O., Ed., Academic Press, New York, 1962, chap. 12.
126. **Schoot Uiterkamp, A. J. M. and Mason, H. S.,** Magnetic dipole-dipole coupled Cu(II) pairs in nitric oxide-treated tyrosinase: a structural relationship between the active sites of hemocyanin and tyrosinase, *Proc. Natl. Acad. Sci. U.S.A.,* 70, 993, 1973.
127. **Himmelwright, R. S., Eickman, N. C., and Solomon, E. I.,** Reactions and interconversion of met and dimer hemocyanin, *Biochem. Biophys. Res. Commun.,* 86, 628, 1979.
128. **Hepp, A. F., Himmelwright, R. S., Eickman, N. C., and Solomon, E. I.,** Ligand displacement reactions of oxyhemocyanin: comparison of reactivities of arthropods and molluscs, *Biochem. Biophys. Res. Commun.,* 89, 1050, 1979.
129. **Himmelwright, R. S., Eickman, N. C., LuBien, C. D., and Solomon, E. I.,** Chemical and spectroscopic comparison of the binuclear copper active site of mollusc and arthropod hemocyanins, *J. Am. Chem. Soc.,* 102, 5378, 1980.
130. **Himmelwright, R. S., Eickman, N. C., LuBien, C. D., Lerch, K., and Solomon, E. I.,** Chemical and spectroscopic studies of the binuclear copper active site of *Neurospora* tyrosinase: comparison to hemocyanins, *J. Am. Chem. Soc.,* 102, 7339, 1980.
131. **Richardson, J. S., Thomas, K. A., Rubin, B. H., and Richardson, D. C.,** Crystal structure of bovine Cu,Zn superoxide dismutase at 3 Å resolution: chain tracing and metal ligands, *Proc. Natl. Acad. Sci. U.S.A.,* 72, 1349, 1975.
132. **Weser, U.,** Structural aspects and biochemical function of erythrocuprein, *Struct. Bonding (Berlin),* 17, 1, 1973.
133. **Fee, J. A. and Briggs, R. G.,** Studies on the reconstitution of bovine erythrocyte superoxide dismutase. V. Preparation and properties of derivatives in which both zinc and copper sites contain copper, *Biochim. Biophys. Acta,* 400, 439, 1975.
134. **Deinum, J., Lerch, K., and Reinhammar, B.,** An EPR study of *Neurospora* tyrosinase, *FEBS Lett.,* 69, 161, 1976.
135. **Sigwart, C., Hemmerich, P., and Spence, J. T.,** A binuclear mixed-valence copper acetate complex as a model for copper-copper interaction in enzymes, *Inorg. Chem.,* 7, 2545, 1968.
136. **Cooper, S. R., Dismukes, G. C., Klein, M. P., and Calvin, M.,** Mixed valence interactions in di-μ-oxo bridged manganese complexes. Electron paramagnetic resonance and magnetic susceptibility studies, *J. Am. Chem. Soc.,* 100, 7248, 1978.
137. **Wong, K. Y. and Schatz, P. N.,** A dynamic model for mixed-valence compounds, *Prog. Inorg. Chem.,* 28, 369, 1981.
138. **Malmström, B. G.,** Cytochrome *c* oxidase. Structure and catalytic activity, *Biochim. Biophys. Acta,* 549, 281, 1979.
139. **Tweedle, M. F., Wilson, L. J., García-Iñiguez, L., Babcock, G. T., and Palmer, G.,** Electronic state of heme in cytochrome oxidase. III, *J. Biol. Chem.,* 253, 8065, 1978.
140. **Moss, T. H., Shapiro, E., King, T. E., Beinert, H., and Hartzell, C.,** The magnetic susceptibility of cytochrome oxidase in the 4.2-1.5 K range, *J. Biol. Chem.,* 253, 8072, 1978.

141. **Abragam, A. and Bleaney, B.,** *Electron Paramagnetic Resonance of Transition Ions,* Clarendon Press, Oxford, 1970, 209.

142. **Abragam, A. and Bleaney, B.,** *Electron Paramagnetic Resonance of Transition Ions,* Clarendon Press, Oxford, 1970, 213.

143. **Seiter, C. H. A. and Angelos, S. G.,** Cytochrome oxidase: an alternative model, *Proc. Natl. Acad. Sci. U.S.A.,* 77, 1806, 1980.

144. **Stevens, T. H., Brudvig, G. W., Bocian, D. F., and Chan, S. I.,** Structure of cytochrome a_3-Cu_{a_3} couple in cytochrome c oxidase as revealed by nitric oxide binding studies, *Proc. Natl. Acad. Sci. U.S.A.,* 76, 3320, 1979.

145. **Seiter, C. H. A., Angelos, S. G., and Perreault, R. A.,** An EPR signal from the invisible copper of cytochrome oxidase, *Biochem. Biophys. Res. Commun.,* 78, 761, 1977.

146. **Reinhammar, B., Malkin, R., Jensen, P., Karlsson, B., Andréasson, L.-E., Aasa, R., Vänngård, T., and Malmström, B. G.,** A new copper(II) electron paramagnetic resonance signal in two laccases and in cytochrome c oxidase, *J. Biol. Chem.,* 255, 5000, 1980.

147. **Powers, L., Blumberg, W. E., Chance, B., Barlow, C. H., Leigh, J. S., Jr., Smith, J., Yonetani, T., Vik, S., and Peisach, J.,** The nature of the copper atoms of cytochrome c oxidase as studied by optical and X-ray absorption edge spectroscopy, *Biochim. Biophys. Acta,* 546, 520, 1979.

148. **Blumberg, W. E. and Peisach, J.,** Possibility of a μ-oxo bridge between iron and copper in cytochrome c oxidase, *Biophys. J.,* 25, 37a, 1979.

149. **Gunter, M. J., Mander, L. N., McLaughlin, G. M., Murray, K. S., Berry, K. J., Clark, P. E., and Buckingham, D. A.,** Towards synthetic models for cytochrome oxidase: a binuclear iron(III) porphyrin-copper(II) complex, *J. Am. Chem. Soc.,* 102, 1470, 1980.

150. **Hoffman, B. M., Roberts, J. E., Swanson, M., Speck, S. H., and Margoliash, E.,** Copper electron-nuclear double resonance of cytochrome c oxidase, *Proc. Natl. Acad. Sci. U.S.A.,* 77, 1452, 1980.

151. **Froncisz, W., Scholes, C. P., Hyde, J. S., Wei, Y.-H., King, T. E., Shaw, R. W., and Beinert, H.,** Hyperfine structure resolved by 2 to 4 GHz EPR of cytochrome c oxidase, *J. Biol. Chem.,* 242, 7482, 1979.

152. **Aasa, R., Albracht, S. P. J., Falk, K.-E., Lanne, B., and Vänngård, T.,** EPR signals from cytochrome c oxidase, *Biochim. Biophys. Acta,* 422, 260, 1976.

153. **Van Leeuwen, F. X. R., Wever, R., Van Gelder, B. F., Avigliano, L., and Mondovì, B.,** The interaction of nitric oxide with ascorbate oxidase, *Biochim. Biophys. Acta,* 403, 285, 1975.

154. **Abragam, A. and Bleaney, B.,** *Electron Paramagnetic Resonance of Transition Ions,* Clarendon Press, Oxford, 1970, 205.

155. **Hagen, W. R. and Albracht, S. P. J.,** EPR studies on cytochrome c oxidase, in *Structure and Function of Energy-Transducing Membranes,* Van Dam, K. and Van Gelder, B. F., Eds., Elsevier/North-Holland Biomedical Press, Amsterdam, 1977, 23.

156. **Carrington, A. and McLachlan, A. D.,** *Introduction to Magnetic Resonance,* Harper & Row, New York, 1967, 26.

157. **Belford, R. L. and Pilbrow, J. R.,** Motionally averaged hyperfine structure and asymmetries in low symmetry EPR, *J. Magn. Reson.,* 11, 381, 1973.

158. **Morpurgo, L., Calabrese, L., Desideri, A., and Rotilio, G.,** Dependence on freezing of the geometry and redox potential of type 1 and type 2 copper sites of Japanese-lacquer-tree *(Rhus vernicifera)* laccase, *Biochem. J.,* 193, 639, 1981.

159. **Dunhill, R. H. and Symons, M. C. R.,** Triplet state ESR spectra of vanadyl tartrate chelates in aqueous solution, *Mol. Phys.,* 15, 105, 1968.

160. **Smith, T. D., Boas, J. F., and Pilbrow, J. R.,** An electron spin resonance study of certain vanadyl polyaminocarboxylate chelates formed in aqueous and frozen aqueous solutions, *Aust. J. Chem.,* 27, 2535, 1974.

161. **Brill, A. S. and Venable, J. H., Jr.,** Electron paramagnetic resonance spectroscopy of protein single crystals. I. Experimental methods, in *Magnetic Resonance in Biological Systems,* Ehrenberg, A., Malmström, B. G., and Vänngård, T., Eds., Pergamon Press, Oxford, 1967, 365.

162. **Lundin, A. and Aasa, R.,** A simple device to maintain temperatures in the range 4.2-100 K for EPR measurements, *J. Magn. Reson.,* 8, 70, 1972.

163. **Bray, R. C.,** Sudden freezing as a technique for the study of rapid reactions, *Biochem. J.,* 81, 189, 1961.

164. **Bray, R. C. and Pettersson, R.,** Electron-spin-resonance measurements, *Biochem. J.,* 81, 194, 1961.

165. **Beinert, H., Hansen, R. E., and Hartzell, C. R.,** Kinetic studies on cytochrome c oxidase by combined ESR and reflectance spectroscopy after rapid freezing, *Biochim. Biophys. Acta,* 423, 339, 1976.

166. **Edmondson, D. E.,** ESR of free radicals in enzymatic systems, in *Biological Magnetic Resonance,* Vol. 1, Berliner, L. J. and Reuben, J., Eds., Plenum Press, New York, 1978, chap. 5.

167. **Chevion, M., Traum, M. M., Blumberg, W. E., and Peisach, J.,** High resolution EPR studies of the fine structure of heme proteins. Third harmonic detection approach, *Biochim. Biophys. Acta,* 490, 272, 1977.

168. **Chesnut, D. B.,** On the use of the AW^2 method for integrated line intensities from first-derivative presentations, *J. Magn. Reson.,* 25, 373, 1977.
169. **Wyard, S. J.,** Double integration of electron spin resonance spectra, *J. Sci. Instrum.,* 42, 769, 1965.
170. **Poole, C. P.,** *Electron Spin Resonance. A Comprehensive Treatise on Experimental Techniques,* Wiley-Interscience, New York, 1967, 589 and chap. 20.
171. **Alger, R. S.,** *Electron Paramagnetic Resonance: Techniques and Applications,* Wiley-Interscience, New York, 1968, 213.
172. **Vollmer, R. T. and Caspary, W. J.,** Computer analysis of ESR data. Integration for concentration determination and resolution of multiple-component spectra, *J. Magn. Reson.,* 27, 181, 1977.
173. **Klopfenstein, C., Jost, P., and Griffith, O. H.,** The dedicated computer in electron spin resonance spectroscopy, in *Computers in Chemical and Biochemical Research,* Klopfenstein, C. E. and Wilkins, C. L., Eds., Academic Press, New York, 1972, 175.
174. **Eaton, S. S. and Eaton, G. R.,** Signals area measurements in EPR, *Bull. Magn. Reson.,* 1, 130, 1980.
175. **Alger, R. S.,** *Electron Paramagnetic Resonance: Techniques and Applications,* Wiley-Interscience, New York, 1968, 200.
176. **Vänngård, T.,** Some properties of ceruloplasmin copper as studied by ESR spectroscopy, in *Magnetic Resonance in Biological Systems,* Ehrenberg, A., Malmström, B. G., and Vänngård, T., Eds., Pergamon Press, Oxford, 1967, 213.
177. **Abragam, A. and Bleaney, B.,** *Electron Paramagnetic Resonance of Transition Ions,* Clarendon Press, Oxford, 1970, 60 and chap. 10.
178. **Abragam, A. and Bleaney, B.,** *Electron Paramagnetic Resonance of Transition Ions,* Claredon Press, Oxford, 1970, 52 and chap. 9.
179. **Feher, G.,** Method of polarizing nuclei in paramagnetic substances, *Phys. Rev.,* 103, 500, 1956.
180. **Feher, G.,** Observation of nuclear magnetic resonances via the electron spin resonance line, *Phys. Rev.,* 103, 834, 1956.
181. **Abragam, A. and Bleaney, B.,** *Electron Paramagnetic Resonance of Transition Ions,* Clarendon Press, Oxford, 1970, chap. 4.
182. **Kevan, L. and Kispert, L. D.,** *Electron Spin Double Resonance Spectroscopy,* John Wiley & Sons, New York, 1976.
183. **Rist, G. H., Hyde, J. S., and Vänngård, T.,** Electron-nuclear double resonance of a protein that contains copper: evidence for nitrogen coordination to Cu(II) in stellacyanin, *Proc. Natl. Acad. Sci. U.S.A.,* 67, 79, 1970.
184. **Beinert, H., Griffiths, D. E., Wharton, D. C., and Sands, R. H.,** Properties of the copper associated with cytochrome oxidase as studied by paramagnetic resonance spectroscopy, *J. Biol. Chem.,* 237, 2337, 1962.
185. **Van Camp, H. L., Wei, Y.-H., Scholes, C. P., and King, T. E.,** Electron nuclear double resonance of cytochrome oxidase: nitrogen and proton ENDOR from the ''copper'' EPR signal, *Biochim. Biophys. Acta,* 537, 238, 1978.
186. **Abragam, A. and Bleaney, B.,** *Electron Paramagnetic Resonance of Transition Ions,* Clarendon Press, Oxford, 1970, 113.
187. **Rowan, L. G., Hahn, E. L., and Mims, W. B.,** Electron-spin-echo envelope modulation, *Phys. Rev.,* 137, A61, 1965.
188. **Alger, R. S.,** *Electron Paramagnetic Resonance: Techniques and Applications,* Wiley-Interscience, New York, 1968, 230.
189. **Mims, W. B., Nassau, K., and McGee, J. D.,** Spectral diffusion in electron resonance lines, *Phys. Rev.,* 123, 2059, 1961.
190. **Mims, W. B.,** Envelope modulation in spin echo experiments, *Phys. Rev. B,* 5, 2409, 1972.
191. **Blumberg, W. E., Mims, W. B., and Zuckerman, D.,** Electron spin echo envelope spectrometry, *Rev. Sci. Instrum.,* 44, 546, 1973.
192. **Mims, W. B.,** *The Linear Electric Field Effect in Paramagnetic Resonance,* Clarendon Press, Oxford, 1976.
193. **Mims, W. B.,** Measurement of the linear electric field effect in EPR using the spin echo method, *Rev. Sci. Instrum.,* 45, 1583, 1974.
194. **Mims, W. B.,** Electron echo methods in spin resonance spectrometry, *Rev. Sci. Instrum.,* 36, 1472, 1965.
195. **Mims, W. B. and Peisach, J.,** Assignment of a ligand in stellacyanin by a pulsed electron paramagnetic resonance method, *Biochemistry,* 15, 3863, 1976.
196. **Kosman, D. J., Peisach, J., and Mims, W. B.,** Pulsed electron paramagnetic resonance studies of the copper(II) site in galactose oxidase, *Biochemistry,* 19, 1304, 1980.
197. **Mondovì, B., Graziani, M. T., Mims, W. B., Oltzik, R., and Peisach, J.,** Pulsed electron paramagnetic resonance studies of types I and II copper of *Rhus vernicifera* laccase and porcine ceruloplasmin, *Biochemistry,* 16, 4198, 1977.

198. **Mims, W. B., Peisach, J., Shaw, R. W., and Beinert, H.,** Electron spin echo studies of cytochrome *c* oxidase, *J. Biol. Chem.,* 255, 6843, 1980.
199. **Ludwig, G. W. and Woodbury, H. H.,** Splitting of electron spin resonance lines by an applied electric field, *Phys. Rev. Lett.,* 7, 240, 1961.
200. **Peisach, J. and Mims, W. B.,** The linear electric field effect in stellacyanin, azurin and in some simple model compounds, *Eur. J. Biochem.,* 84, 207, 1978.
201. **Peisach, J. and Mims, W. B.,** Deviations from centrosymmetry in some simple Cu^{2+} complexes, *Chem. Phys. Lett.,* 37, 307, 1976.
202. **McMillin, D. R., Rosenberg, R. C., and Gray, H. B.,** Preparation and spectroscopic studies of cobalt(II) derivatives of blue copper proteins, *Proc. Natl. Acad. Sci. U.S.A.,* 71, 4760, 1974.
203. **Malmström, B. G., Reinhammar, B., and Vänngård, T.,** The state of copper in stellacyanin and laccase from the lacquer tree *Rhus vernicifera, Biochim. Biophys. Acta,* 205, 48, 1970.
204. **Brill, A. S., Bryce, G. F., and Maria, H. J.,** Optical and magnetic properties of *Pseudomonas* azurins, *Biochim. Biophys. Acta,* 154, 342, 1968.
205. **Marchesini, A. and Kroneck, P. M. H.,** Ascorbate oxidase from *Cucurbita pepo medullosa.* New method of purification and reinvestigation of properties, *Eur. J. Biochem.,* 101, 65, 1979.
206. **Barker, R., Boden, N., Cayley, G., Charlton, S. C., Henson, R., Holmes, M. C., Kelly, I. D., and Knowles, P. F.,** Properties of cupric ions in benzylamine oxidase from pig plasma as studied by magnetic-resonance and kinetic methods, *Biochem. J.,* 177, 289, 1979.

Chapter 3

NUCLEAR MAGNETIC RESONANCE SPECTROSCOPY OF COPPER PROTEINS

Anthony E. G. Cass and H. Allen O. Hill

TABLE OF CONTENTS

I. INTRODUCTION

Of all the techniques currently used to study macromolecules in solution, only nuclear magnetic resonance (NMR) spectroscopy has the power to reveal details of molecular structure and motion at atomic resolution.[1-5] Though the complexity of the spectra of most molecules of interest makes it difficult to express this power and capitalize on the information provided, recent advances in spectrometer design and data manipulation have considerably extended its range of fruitful applications. Compare, e.g., an early (about 1975) ^{1}H NMR spectrum of azurin with one obtained recently (Figure 1) in which the increased resolution and signal-to-noise ratio are only two of the improved features apparent. Many other advances will be revealed in examples considered in detail later in this chapter. First, we provide a brief description of the object of our spectroscopic attentions and the salient features of the technique for those unfortunate to have not yet made its acquaintance.

II. COPPER PROTEINS

The diverse roles played by copper proteins are described in Table 1. They provide an interesting parallel to the functions of iron-containing proteins. Perhaps only the relatively low abundance of copper prevents it from occupying the paramount position among redox-active metals in biology. Copper centers in proteins have previously been classified as type 1, type 2, or type 3. This classification was orignally used to describe the composition of proteins containing *more than one* copper center. It was based on the optical spectra and on the parameters derived from the EPR spectra.[6] We have recently extended it to direct attention to the details of molecular structure now provided by crystallography and to highlight the function of the copper ions in the proteins.[7] Group-I proteins are those that contain a single copper ion having four ligands in a distorted tetrahedral geometry. The ligands are coordinated to the copper via sulfur and nitrogen donor atoms, with two nitrogens and two sulfur atoms being the most common arrangement. Important physical properties such as the redox potential are modulated by the details of the coordination environment, particularly the number and nature of the sulfur ligands. Group-II copper proteins also contain a single copper ion at the catalytic center though in this case the ligands are coordinated via nitrogen and oxygen donor atoms with the geometry dominated by a square-planar arrangement around the copper. An important feature of this arrangement is that there is at least one readily accessible coordination site, either in the plane or along the tetragonal axis, that binds substrate or other ligands. In these proteins, the Lewis acid properties of copper(II) are employed with, or without, concomitant redox activity. Group-III proteins contain one or more binuclear copper centers with a coordination environment consisting mainly of nitrogen donor atoms. In the copper(I) form of the protein, two ligand atoms may dominate the coordination environment thereby allowing the protein to fulfill its function — the coordination of dioxygen. Group-IV proteins contain two or more of the centers that characterize groups I to III. Thus we can characterize laccase as group IV$_{I,II,III}$, and the classification of a number of copper proteins is given in Table 1.

III. NUCLEAR MAGNETIC RESONANCE SPECTROSCOPY

The intrinsic angular momentum of a nucleus is described in terms of a quantum number, I. Since nuclei are also charged, those which have $I > 0$ give rise to observable magnetic properties when placed in a magnetic field. Most applications of NMR spectroscopy to biological problems make use of nuclei with $I = 1/2$. The magnetic moment of these nuclei is given by $\mu = \gamma \hbar I$. Interaction with the applied, static magnetic field gives rise to two energy levels as shown in Figure 2. Absorption of incident electromagnetic radiation causes transitions between these two states. The energy difference between the two states is so

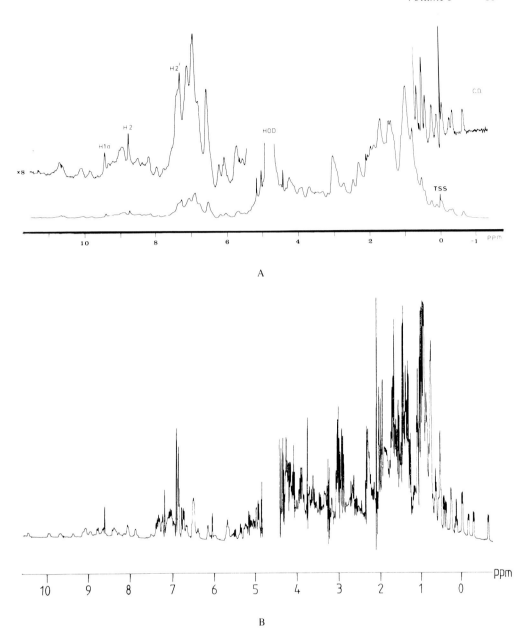

FIGURE 1. (A) An early (1975) ¹H NMR spectrum of *Pseudomonas aeruginosa* azurin obtained at 270 MHz. H1a, H2, and H2′ are resonances from histidine residues and are among the sharpest lines in the spectrum. Inset in the top right hand corner is an example of an early resolution enhancement method known as convolution difference (C.D.). TSS is the internal chemical shift standard. (From Hill, H. A. O., Leer, J. C., Smith, B. E., Storm, C. B., and Ambler, R. P., *Biochem. Biophys. Res.Commun.*, 70, 331, 1976. With permission.) (B) A recent spectrum of *P. aeruginosa* azurin obtained at 300 MHz. In this spectrum resolution is of the whole spectrum and has been performed by using the Gaussian multiplication routine.

small that the population difference at room temperature is slight ($\approx$ 1 in 10^6). Since the net absorption of energy from the electromagnetic radiation is proportional to this difference, NMR spectroscopy is a rather insensitive technique. Nevertheless, technical advances have allowed instruments to be developed that bring many biological problems within the reach of this informative method for probing molecular structure. The power of the method resides

Table 1
CLASSIFICATION OF COPPER PROTEINS

Role	Classification	Examples
Electron transfer	Group I	Azurin, plastocyanin, rusticyanin, stellacyanin
Substrate oxidation with hydrogen peroxide formation	Group II	Galactose oxidase
		Benzylamine oxidase
Superoxide dismutation	Group II	Superoxide dismutase
Monooxygenation	Group II	Dopamine β-monooxygenase
	Group III	Tyrosinase
Oxygen transport	Group III	Hemocyanin
Terminal oxidase	Group IV	Laccase
		Ceruloplasmin
		Ascorbate oxidase

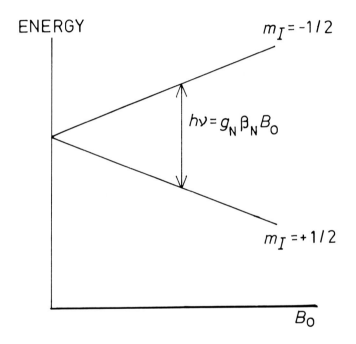

FIGURE 2.   Energy levels for an $I = 1/2$ nucleus in a magnetic field.

in the interpretation of four parameters which serve to describe each and every magnetic nucleus in a molecule.

The chemical shift — The relationship between the applied field, B_o, and the frequency of the incident radiation, ν_o is

$$\nu_o = \gamma B_o \qquad (1)$$

where γ is the magnetogyric ratio, which is characteristic of the type of nucleus undergoing resonance. If NMR spectroscopy was described only by Equation (1), it would be of little value to the chemist or biochemist. Fortunately this is not the case; the field that the nucleus experiences is not the applied field, B_o, but one that is modified by the local magnetic fields. The resonance condition then becomes

$$\nu_{oi} = \gamma B_o (1 - \sigma_i) \qquad (2)$$

where σ_i is the shielding constant for nucleus i. The frequencies at which resonance occurs are usually measured with respect to a standard, and division by the frequency of the spectrometer facilitates comparison between spectra measured at different frequencies. The resulting parameter is the *chemical shift, δ_i*. It can be considered as the sum of two contributions: a primary shift determined by the immediate electronic environment associated with a given chemical environment (e.g., aliphatic, aromatic) and a secondary shift due to nearby groups that give rise to more long-range effects (e.g., proximity to aromatic residues, charged groups, peptide bonds, etc.).

Spin-spin coupling constants — The energy of a nucleus with $I \neq 0$ in an applied magnetic field will be sensitive to the presence in the same molecule of other magnetic nuclei. The coupling between nuclei is described in terms of a field-independent constant, J, and takes place via the electrons in the intervening bonds. The magnitude of the effect is quite small for protons, <15 Hz, and falls off very rapidly with the number of bonds between the two nuclei in question.

Spin-lattice relaxation time — The energy that is absorbed from the applied electromagnetic radiation by a system of spins must be dissipated if saturation is not to ensue since the energy absorbed is proportional to the difference in population between the ground and excited nuclear spin states. One way this might be accomplished is by the conversion of the nuclear spin energy into thermal motion of the surrounding atoms and molecules. The time constant for the rate of this conversion is the *spin-lattice relaxation time, T_1*, so-called because the effect was first described in solids. The mechanism by which this conversion is accomplished is through the effect of magnetic motions that give rise to local magnetic field fluctuating at the NMR frequency ($\nu_o \approx 10^8$ Hz for protons). The T_1 values give information of the rate of such molecular motions.

Spin-spin relaxation time — In addition to exchange of energy between the spin system and the surrounding medium, there can also be interchange of energy *within* the spin system itself. The time constant for this process is known as the spin-spin relaxation time, T_2. Molecular motions that contribute to spin-lattice relaxation may also affect the interchange of energy within the spin system. However, the latter is also affected by processes occurring at much lower frequencies, 100 Hz or less. An important example of this is the exchange of a nucleus between two chemical environments. The spin-spin relaxation time is related to the linewidth of a resonance by:

$$\nu_{1/2} = 1/(\pi T_2) \tag{3}$$

Relaxation properties of copper(II) — Although many different types of motion may contribute to the relaxation of a nucleus, the large fluctuating magnetic moment associated with unpaired electrons will be particularly effective. Thus the copper(II) ion, with its single unpaired electron, will dominate the relaxation processes of nearby magnetic nuclei. A quantitative description of the contribution of copper(II) ions to the relaxation times of nuclei with $I = 1/2$ is provided by the Solomon-Bloembergen equations:

$$\frac{1}{T_{1M}} = \frac{K}{r^{-6}} \frac{3\tau_c}{1 + \omega_o^2 \tau_c^2} \tag{4}$$

$$\frac{1}{T_{2M}} = \frac{K}{2r^{-6}} \left[4\tau_c + \frac{3\tau_c}{1 + \omega_o^2 \tau_c^2} \right] \tag{5}$$

where K is $2\gamma^2 g_{eff}^2/15$, τ_c is the correlation time or inverse rate constant for the motion modulating the electron-nuclear interaction, r is the electron-to-nucleus distance, and $\mathbf{g}_{eff}$ is the effective magnetic moment of the electron. These equations only describe those interactions that take place "through space", the so-called dipolar term. They neglect the "through-bond'' effects, the contact terms. The correlation time, τ_c can be considered as incorporating three separate terms:

$$\tau_c^{-1} = \tau_R^{-1} + \tau_S^{-1} + \tau_M^{-1} \qquad (6)$$

where τ_R is the rotational correlation time, τ_S is the electron-spin relaxation time, and τ_M the lifetime of the nucleus in the vicinity of the copper(II) ion.

IV. NMR STUDIES OF GROUP-I PROTEINS

There are two, inherently complementary, approaches to the study of the NMR properties of copper proteins. The first exploits the enhancement of the relaxation processes of freely exchanging nuclei by the copper(II) form of the protein. It provides information about the copper site and the exchanging ligand. To be reliable, it needs to be carried out at different applied radio frequencies and preferably at different temperatures. The second approach is one whereby the high resolution NMR spectrum of the copper protein is measured. Much effort is required to arrive at unambiguous assignments of the resonances. The effect of the magnetic properties of the copper(II) ion provides an *intrinsic* paramagnetic perturbation of the spectrum which might, in favorable cases, lead to structural information.

A. Proton NMR

The group-I proteins, azurin (Az) and plastocyanin (Pc), have been intensively studied by every available spectroscopic method. The interpretation is more secure, though less challenging, since the X-ray crystal structures became available.[8,9]

Early measurements of the water relaxation times of aqueous solutions of Pc[10] and Az and umecyanin[11] led to the conclusion that the copper(II) center in these molecules is inaccessible to the solvent. An analysis of the frequency dependence of the water relaxation times of the solutions of Az at a number of applied magnetic fields was carried out by Koenig and Brown.[12] They concluded that the small effect observed could be accounted for by outer-sphere relaxation and suggested that the copper(II) ion was about 5 Å from the surface of the molecule, a conclusion since confirmed by the crystal structure.

Az and Pc are relatively small molecules that can be isolated in good yield. As such they are good candidates for high resolution NMR studies. Nevertheless they still present formidable problems in assignment. Though recent advances in instrumentation have somewhat alleviated the problem of assignment to *type* of amino-acid residue, second stage assignments to *specific* amino acids in the sequence of the protein are still laborious and fraught with uncertainty. It goes without saying that the sequence is required; the availability of the crystal structure, even at low resolution, is of enormous help. It may aid the reader if we consider the NMR spectra of these group-I proteins in terms of the information revealed by the study of the various types of amino-acid residues in the proteins.

1. Investigations of Histidine Residues

The resonances associated with the C2 and C4 protons of histidine residues are the most readily identified peaks in the ^1H NMR spectra as they usually give rise to a large upfield shift as the pH of the solution is increased. In their study of French bean Pc, Beattie et al.[13] observed two resonances that showed a pH-dependent chemical shift in the reduced form of the protein and were absent in the oxidized form. They assigned these to one or both of the histidine residues. Markley et al.[14,15] studied the pH-dependence of the resonances in the spectrum of spinach apoPc and found two freely titrating resonances with pK_a values of 6.5 to 6.8, typical of the imidazole group of histidine. By contrast, the pK_a-values associated with these resonances were decreased to 4.9 and 4.5 in the copper(I) form. The authors argued that this shift in pK_a reflected coordination of the metal ion to two histidine residues, the copper(I) competing with protons for the histidines. The resonances associated with the two histidine residues are broadened beyond observation in the copper(II) form consistent with their known role as ligands.

In *Pseudomonas aeruginosa* Az, there are four histidine residues and the X-ray diffraction map can be interpreted in terms of two, 46 and 117, acting as ligands, one, 35, as close, and the remaining one, 83, further away. The last is clearly responsible for the freely titrating C2-proton resonance in both copper(I) and copper(II) forms of the protein. It has a pK_a of 7.6. In the reduced form of the protein, there is an additional resonance at 9.4 ppm which loses intensity as the pH is increased and concomitantly a new resonance at 8.1 ppm gains intensity. In their first paper, Hill et al.[16] assigned these resonances to the C2 proton of a histidine residue whose acid and base forms were in slow exchange on the ^{1}H NMR timescale. The same phenomenon was observed by Ugurbil and Bersohn[17] and similarly interpreted.[17] Although this is a plausible explanation, it does not *prove* that these two resonances arise from the same proton in different environments. Hill and Smith,[18] using saturation-transfer methods, were able to show that the two resonances are so related. If a resonance is irradiated by a strong pulse, the nuclei cannot dissipate the energy rapidly enough and therefore the associated resonance loses intensity and it is said to be saturated. When the nuclei move to a different environment, they are still partially saturated and thus give rise to a resonance with reduced intensity. The results of this type of experiment are shown in Figure 3, which demonstrates that the two resonances arise from the same proton. In their first paper, Hill et al.[16] were able to place an upper limit on the exchange rate by using the criterion for slow exchange, viz.:

$$\tau_M \Delta\omega \gg 1 \qquad (7)$$

where τ_M is the reciprocal of the rate constant and $\Delta\omega$ the frequency difference of the two resonances. A value of $k \ll 2{,}300$ sec^{-1} was derived. However, in the saturation-transfer experiment, the following relationship holds:

$$I/I_o = (1/T_1)(1/T_1 + k)^{-1} \qquad (8)$$

where I and I_o are the peak intensities in the presence and absence of the saturating radiation. Applying this equation yields a value for $k > 1$ sec^{-1}. Furthermore, the exchange contribution to the linewidth, ν_{ex}, in slow exchange is given by:

$$\pi \nu_{ex}^{1/2} = k \qquad (9)$$

Using an estimated ν_{ex} of 10 Hz gives $k < 35$ sec^{-1}. Thus we have a value for the rate of exchange of this histidine of between 1 and 35 sec^{-1} which can be contrasted with the corresponding value for the freely titrating His-83, which is greater than 14,000 sec^{-1}. From the crystal structure, it is obvious that the residue with the anomalous proton-exchange properties is His-35. Its properties including the pK_a (6.9 to 7.3) are presumably influenced by the nearby ligand His-46.

His-35 may influence the rates of electron transfer between Az and inorganic complexes and other redox proteins, cytochrome c_{551}, and *Pseudomonas* nitrite reductase.[19] The rates have a pH-dependence that has been interpreted as reflecting a conformational change between redox-active and inactive forms of Az. Although NMR shows no evidence of a major conformational change, both the pK_a and the rate constant are consistent with the protonation/deprotonation of His-35 being involved.

Resonances from coordinated histidine residues may, in favorable circumstances, be iden-

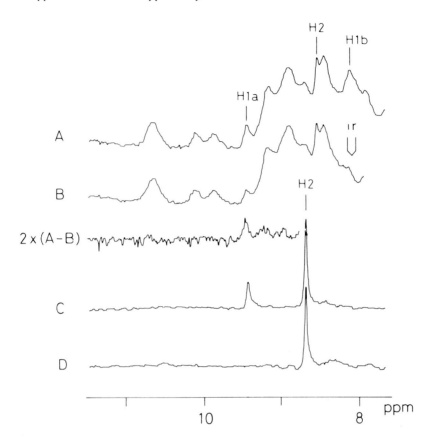

FIGURE 3.   Cross saturation of resonances from His-35 in *Pseudomonas aeruginosa* azurin. (A) Low field region of the spectrum of copper(I) azurin. (B) Spectrum of the same region after irradiation (ir) at the frequency of the C2 proton in the imidazole form of His-35 (H1b) showing the loss of intensity of the peak due to the same proton in the imidazolium form (H1a). Spin-echo spectra at 28 (C) and 47°C (D). The increasing temperature causes a change in exchange rate from slow to intermediate for His-35. This results in a decrease in the relaxation time and loss of the resonance from the (relaxation time sensitive) spectrum. His-118 (H2) is in fast exchange at both temperatures and is hardly affected. (From Hill, H. A. O. and Smith, B. E., *J. Inorg. Biochem.*, 11, 79, 1979. With permission.)

tified by their absence of multiplet structure, narrow linewidths and chemical shifts. In copper(I) Az one has been identified by these criteria.[20] The peak is shifted upfield and an examination of the crystal structure reveals His-117 lies above a phenylalanine residue that forms part of the "aromatic box" close to the copper center, and therefore is ring-current shifted.

ApoAz has an additional titrating C2 proton as well as those described above and this obviously arises from one of those histidine residues that act as a ligand to the copper. Interestingly, removal of the metal ion from Az does not appear to lead to an unfolding of protein, the ring current-shifted methyl resonances R1-R7, which owe their high-field positions to their orientation above nearby aromatic rings, are hardly affected. This implies that those parts of the protein in which these groups are found are virtually the same in holo- and apoproteins. However, there are changes elsewhere, e.g., His-35 shows a distinctly lower pK_a in apo- and mercury(II) Az's (approximately 6) than in the copper(I) form.[18] Furthermore, the exchange of those peptide-backbone protons that are relatively buried occurs very much faster in the apoprotein as illustrated in Figure 4. This particular behavior seems to be a general one for apo- vs. holoproteins and may reflect a greater time-averaged accessibility in the former without substantial changes in the backbone structure.

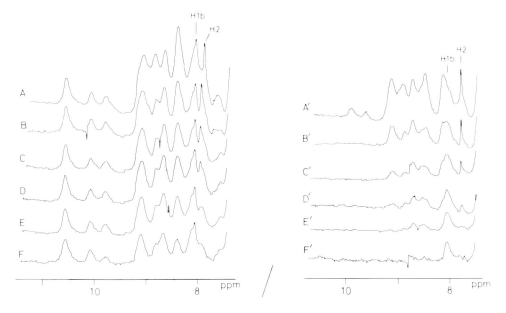

FIGURE 4. The exchange of peptide NH protons in apo (A′ to F′) and copper(I) (A to F) azurin as a function of time from 1.5 hr (A,A′), 12.5 hr (B,B′), 47.5 hr (C,C′), 4 days (D,D′), 9 days (E,E′), to 25 days (F,F′) after dissolving the freeze-dried protein in D$_2$O. (From Hill, H. A. O. and Smith, B. E., *J. Inorg. Biochem.*, 11, 79, 1979. With permission.)

Table 2
ASSIGNMENTS IN *PSEUDOMONAS AERUGINOSA* AZURIN[16-18,20]

Residues	Assigned protons
His-35, -46, -83, -117	C2(H), C4(H)
Tyr-72, -108	All the ring protons
Phe-15, -29, -97, -110, -111/114	All the ring protons
Trp-48	C2(H), C4(H), C5(H), C6(H), C7(H)
Met-13/44, -56/109, -64, -121	ε-Methyl group
Ile-7	δ-Methyl and γ-methylene groups
Ile-20	δ-Methyl group
Val-5	γ-Methyl group
Val-22, -31	γ-Methyl and α- and β-methine groups
Leu-33	δ-Methyl group

2. Studies of Other Residues

A number of other first-stage assignments have been made in both Pc's and Az's. In reduced Pc the resonances from the three tyrosine residues, at positions 70, 80, and 83, have been identified by their coupling patterns. The six doublet resonances arising from the three tyrosine residues have all been identified in French bean Pc.[21] However, the assignment to specific residues required both a knowledge of the X-ray crystal structure and comparison of spectra from several species. Tyr-70 is not conserved in all Pcs and thus is easily identified. Tyr-80 and Tyr-83 are, respectively, far from and close to the copper center and thus only the former is observed in the oxidized protein. Decoupling methods have also been employed to assign some of the aromatic resonances in Az,[20] and the assignments for both these proteins are collected in Tables 2 and 3.

Another easily identified resonance is that from the S-methyl group of a methionine residue

Table 3
ASSIGNMENTS IN
PLASTOCYANIN[13,15,21]

Residues	Assigned protons
His-37/87, -87/37	C2(H) and C4(H)
Tyr-80/83, -83/80, -70	*Ortho* and *meta*
Phe	All the ring protons
Met-57, -92	S-methyl

as this usually gives rise to a sharp singlet resonance in the aliphatic region. Additional interest attaches to this residue as it provides one of the ligands to the copper ion in both Az and Pc. The S-methyl resonance from the ligand methionine in Pc is readily assigned as this methionine is the only one conserved. An unusual feature of this resonance is its substantial upfield shift ($\geq$ 1.3 ppm) from the usual position. In some species of Pc, an additional methionine yields a resonance of about 1.88 ppm.[21]

Az from *P. aeruginosa* is again more complicated as it contains six methionine residues; the ligand one has now been assigned,[20] and some distinction can be made among the five identifiable resonances by a consideration of the effects of differential broadening in mixtures of the oxidized and reduced proteins *(vide intra)*.

Tryptophan and histidine residues also have NH protons which resonate at low field (11 to 15 ppm) in solutions of the protein in H_2O and are exchanged in D_2O. Such protons are readily observed in *P. aeruginosa* and tentative assignments have been proposed.[17] The authors used a combination of real-time deuterium-exchange kinetics, pH dependence, and comparison with the tryptophan-free Az from *P. fluorescens* to assign the resonances.

By comparing the spectra of oxidized, reduced, and apostellacyanin and by a consideration of sequence homologies, Hill and Lee[22,23] have suggested the presence of two histidine and two cysteine residues in the coordination sphere of the copper ion in this protein.

3. Metal-Substituted Azurins

The copper-binding site in Az is able to accommodate a variety of other metal ions, and NMR spectra of the mercury(II),[18] cobalt(II),[24] and nickel(II)[25] Az's have been reported. Mercury(II) Az shows many similarities to the native protein. In the other derivatives the presence of dipolar and contact paramagnetic shifts leads to substantial changes in the spectrum. Cobalt(II) Az gives rise to resonances in the chemical shift range -32 to 60 ppm (Figure 5), all of which show a marked temperature dependence. In the cobalt Az, the C2 protons of histidine residues 35 and 83 can be identified from their characteristic pH dependencies and their extrapolated diamagnetic shifts (from their temperature dependence) agree well with those measured for the copper(I) protein. Of the remaining resonances that are shifted out of the normal diamagnetic region, those that show the largest shifts (and the greatest linewidths) probably arise from the metal ligands and suffer both contact and dipolar interactions. Apart from these resonances, the temperature dependence of the majority of the others extrapolate to diamagnetic chemical shift in the aliphatic region of the spectrum. Nickel(II) Az shows even larger downfield shifts, to 110 ppm, although no analysis of the spectrum has been reported.

B. Carbon-13 NMR

Often the application of ^{13}C NMR spectroscopy provides information complementary to that available from proton NMR studies. The first ^{13}C NMR spectra of Pc were presented by Markley et al.[26] and were natural-abundance spectra obtained at 67.9 MHz. The authors proposed a number of tentative assignments based upon a comparison of oxidized and reduced

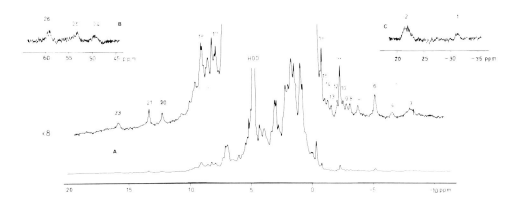

FIGURE 5. The ¹H NMR spectrum of cobalt(II) azurin. The central region (A) shows the unshifted resonances and those shifted by a pseudocontact mechanism; (B) and (C) show the more extensively shifted peaks to low and high field, respectively, probably by a contact mechanism. (From Hill, H. A. O., Smith, B. E., Storm, C. B., and Ambler, R. P., *Biochem. Biophys. Res. Commun.*, 70, 783, 1976. With permission.)

samples and sequence homologies between proteins from different species. Later this work was repeated and extended by obtaining spectra at 90.5 MHz[14] and using the known crystal structure to aid in interpretation. One interesting result of this subsequent investigation was the observation of additional resonances in the copper(II) form of the protein not present in the copper(I) form; these were attributed to a conformational change between the two oxidation states. This result, if the interpretation is correct, implies that a change in redox state can lead to structural perturbations many angstroms from the copper site.

One particular piece of information provided by ¹³C NMR is the site of deprotonation of histidine residues. The imidazolium ring can be deprotonated at either N1 or N3 and, in both cases, the C2 proton-resonance shifts upfield. Deprotonation of N3 shifts the C2 resonance upfield while deprotonation at N1 shifts it downfield. Applying these criteria, the authors concluded that both ligand histidine residues are coordinated through N1 to the copper ion, a result also arrived at by crystallographic studies.

Az has also been the subject of a ¹³C NMR investigation, the main focus being the nonprotonated carbon atoms (Figure 6).[27] His-83, previously assigned in ¹H NMR spectra, is also readily identified in the ¹³C NMR spectra and appears to undergo deprotonation at N3 which is relatively unusual. His-35 shows the same slow exchange behavior on the ¹³C NMR time scale as it did for ¹H NMR. For the ligand histidine residues, coordination to the copper again appears to be through N1 as it was for Pc. Assignments have also been made to carbon atoms from tryptophan, phenylalanine, arginine, and tyrosine residues as well as an unusual carbonyl group that shows a pH dependence probably arising from the titration of a nearby group. Both of the tyrosine resonances show rather high pK_a values, consistent with an environment buried in the protein matrix.

From a consideration of the paramagnetic effect of the copper(II) center, Ugurbil et al.[27] conclude that Trp-48, one of the two tyrosines, His-35, and three or four of the six phenylalanines as well as the unusually shifted carbonyl group are close to the copper ion.

C. Mechanisms of Electron Transfer
1. NMR Studies of Electron Exchange

If a nucleus is exchanged between a paramagnetic and a diamagnetic environment, the spin-spin relaxation time of the resonances of the diamagnetic form will have a contribution from the paramagnetic complex:[5]

$$\frac{1}{T_{2p}} = \frac{P_M}{T_{2M} + \tau_M}$$

(10)

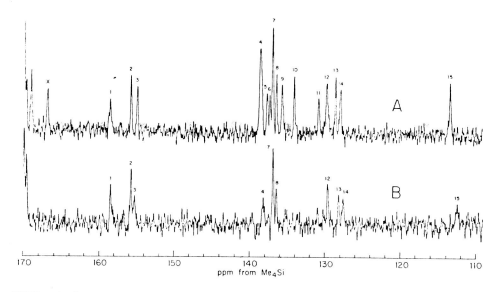

FIGURE 6. ^{13}C NMR spectrum of reduced (A) and oxidized (B) *Pseudomonas aeruginosa* azurin showing resonances from carbon nuclei in aromatic and arginine residues. In addition, an upfield shifted peptide carbonyl resonance (x) can be observed in the reduced protein. (From Ugurbil, K., Norton, R. S., Allerhand, A., and Bersohn, R., *Biochemistry*, 16, 886, 1977. With permission.)

and

$$\frac{1}{T_2} = \frac{1}{T_{2p}} + \frac{1}{T_{2A}} \qquad (11)$$

where T_2 is the observed relaxation time, T_{2A} is the relaxation time in the absence of exchange, and T_{2M} is the relaxation time of the resonance of the paramagnetic form. P_M is the mole fraction of the paramagnetic form and τ_M is the exchange lifetime. Equation (10) is only valid if there is a negligible chemical shift difference between the two forms and if the value of P_M is relatively small ($\lesssim 0.2$). There are two limiting forms of Equation (10), if $\tau_M \ll T_{2M}$, i.e., exchange is fast:

$$\frac{1}{T_{2p}} = \frac{P_M}{T_{2M}} \qquad (12)$$

From this equation we can see that the linewidth will increase in proportion to the amount of the copper(II) form present and, furthermore the r^{-6} dependence of T_{2M} (Equation 5) means that the broadening will depend upon the distance of the nucleus from the copper. In the slow exchange limit, $\tau_M \gg T_{2M}$ and

$$\frac{1}{T_{2p}} = \frac{P_M}{\tau_M} \qquad (13)$$

In this case there is no distance dependence and so to derive relative distances we must be sure of being in the fast exchange limit for all of the resonances of interest. Notice in particular that for resonances that arise from nuclei close to the copper, T_{2M} is very short and the fast exchange condition correspondingly hard to fulfill. By using the hexacyano-ferrates, Hill and Smith[18] were able to achieve fast electron exchange in Az and so obtain a relative ranking of distances of some of the protons from the copper(II) center. These authors were also able to estimate the absolute distances by assuming a value for τ_c although they caution that such a calculation is obviously dependent on the value of τ_c that is chosen. Ugurbil et al.[27] applied the same method in their ^{13}C NMR investigations of Az to distinguish between those resonances that arose from carbon atoms close enough to be in the ligand residues and those that, although not detected in the spectrum of the oxidized protein, are too far away to be potential ligands.

Information on rates of exchange between copper(II) and copper(I) forms of the protein may also be derived if a particular nucleus gives rise to a resonance that has a different chemical shift in the two oxidation states. If the rate of exchange between the two forms is fast compared to the chemical shift difference ($\Delta\omega$):

$$\tau_M \Delta\omega \ll 1 \qquad\qquad (14)$$

then the resonance will show a continuous shift with increasing proportion of one form. Conversely, if exchange is slow then resonances will be observed from both forms with their intensities in proportion to the concentrations of each form present. In French bean Pc the former condition holds in the presence of hexacyanoferrates and the latter in their absence.[13] Analysis of the observed chemical shift differences yielded exchange rates of much less than $2 \cdot 10^4 \ M^{-1} \ \text{sec}^{-1}$ for the self-exchange rate and much greater than $10^5 \ M^{-1} \ \text{sec}^{-1}$ for the catalysed exchange rate.

In *Pseudomonas* Az, small but well-defined shifts in two of the upfield shifted methyl resonances and in two of the methionine resonances are observed in the course of an oxidative titration. These results and those obtained on French bean Pc as well as on spinach Pc clearly show that a change in the redox state of the copper can lead to conformational changes distant (>10 Å) from the metal site.

2. Binding of Inorganic Complexes

In an attempt to elucidate the mechanisms of electron transfer in the group-I copper proteins, extensive kinetic investigations of their reactions with inorganic complexes of iron, cobalt, and ruthenium have been carried out.[28] Some of the rate laws that result involve an association of the protein and complexes before electron transfer.[29] To try and determine the sites of interaction, the effect of chromium(III) complexes on the spectra have been investigated. Chromium(III) has a number of advantages as a perturbation probe; it has a sufficiently low redox potential so there are no complicating electron-transfer processes. Its long τ_s results in it being an effective broadening agent and its complexes are inert to ligand-substitution reactions.

Both Cookson et al.[30,31] and Handford et al.[32] have observed the broadening of specific resonances in the spectrum of French bean and parsley Pc, respectively, by both positively charged [$Cr(NH_3)_6^{3+}$] $Cr(phen)_3^{3+}$] and negatively charged [$Cr(CN)_6^{3-}$] complexes. The former are thought to bind some 15 Å from the copper center, close to Tyr-83, while the latter binds over the copper center at the "top" of the molecule.

V. NMR STUDIES OF GROUP-II COPPER PROTEINS

A. Relaxation Enhancement Studies

Unlike the group-I copper proteins whose metal centers are inaccessible to water, those copper proteins that comprise group II have, as a common feature, one or more coordination positions that can bind water molecules or other exogenous ligands. If the ligand is in rapid exchange between the copper ion and bulk solution, then the observed relaxation times of nuclei in the latter environment can be directly related to their properties in the former. Analysis, therefore, of the relaxation enhancement in solutions of the copper(II) protein can, in principle, yield the number of ligands bound, their distance(s) from the copper ion, and their dissociation rates.

1. Galactose Oxidase

Only a very weak water proton relaxation enhancement occurs in solutions of galactose oxidase, and much of this is due to nonspecific binding of water to the protein. In contrast to this, fluoride ion binds specifically to copper ion and shows large relaxation enhancement of the $^{19}F^-$ NMR signal, only 10% or less of which is due to nonspecific effects of the protein.[33] There are two binding sites for fluoride at the copper center of galactose oxidase. That associated with EPR superhyperfine structure is equatorially coordinated, has a tighter binding constant (≈ 1 M) and does not contribute to the observed relaxation as it is in slow exchange. The second fluoride ion binds much more weakly (≈ 50 M) at an axial position and is responsible for the observed relaxation enhancement.[34]

By analysing the temperature and frequency dependence of both T_1 and T_2 values, Kurland and Marwedel[35] concluded that the dominant dipolar relaxation mechanism appears to arise from unpaired spin density in fluorine-centered p orbitals. This implies that changes in ^{19}F relaxation times can be used to probe the electronic structure of the copper(II) center and two examples of the use of this method have been described.

Cyanide has been demonstrated to bind to an equatorial position in galactose oxidase and will displace the equatorially coordinated fluoride ion. Despite the fact that this fluoride ion makes no contribution to the observed relaxation enhancement, cyanide binding still leads to a decrease in the relaxation rate. The mechanism of the de-enhancement probably involves a change in the superhyperfine interaction and/or binding constant of the axially coordinated fluoride ion.[34] Addition of substrates or products to the enzyme-fluoride complex also results in a de-enhancement of ^{19}F relaxation times, although in this case the mechanism probably is one of direct replacement of the axial ligand.[36] From an investigation of the concentration dependence of the de-enhancement, dissociation constants for the various substrates and products can be derived.[37] Substrate binding at an axial, or outer-sphere, site is further confirmed by measurement of the ^{13}C relaxation times of substrate analogs in solutions of the enzyme. No decrease in either T_1 or T_2 is observed, a result that is also consistent with kinetic and EPR competition experiments against fluoride and cyanide.

2. Amine Oxidases

These proteins contain an additional prosthetic group closely related to pyridoxal and are involved in oxidative deamination reactions. Kluetz and Schmidt[38] have investigated the effect of a diamine oxidase from pig kidney on the proton relaxation times of water and substrate and product analogs as well as ^{15}N relaxation rates of ammonia. They used the Solomon equations (Equations 5 and 6) to derive copper(II) to proton distances for the various substrates and products, and from their results they concluded that neither the amino portion of the substrate nor the product bind to the copper(II) center. Similarly the lack of effect of the enzyme upon ^{15}N relaxation times in ammonia shows that this product too does not bind to the metal ion. Although there is an enhancement of water relaxation rates by the copper(II), there is no competition by oxygen so this substrate also apparently does not interact with the copper. All of these results are interpreted by the authors as implying that the copper(II) ion is not involved in the catalysis. Instead they suggest that the binding of a second substrate molecule occurs at the metal site and that this modulates the activity of the catalytic (pyridoxal) center and accounts for the observed substrate inhibition.

Pig plasma benzylamine oxidase is another enzyme of this type and this also exhibits substantial water proton relaxation enhancement. In a study of the temperature and frequency dependence of the water relaxation times, Barker et al.[39] were able to derive a rate of $5 \cdot 10^5$ sec^{-1} for the exchange process. They also noted that the relaxation properties of the copper(II) center were substantially affected by thermal denaturation. The water relaxation rates were also very pH dependent with an observed pK_a of 8. As the pH was raised the relaxation rate fell to half its low-pH value, a result consistent with deprotonation of a

Table 4
NMR PARAMETERS FOR EXCHANGEABLE WATER
PROTONS IN BOVINE SUPEROXIDE DISMUTASE[44]

Parameter	Site I (axial)	Site II (equatorial)
Exchange lifetime/sec	$1.6 \cdot 10^{-7}$	$1 \cdot 10^{-5}$
Cu-^{1}H distance/nm	0.34	0.27
Number of water molecules coordinated	1	$^1/_2$
Correlation time/sec (at 60 MHz)	$2.7 \cdot 10^{-9}$	$2.7 \cdot 10^{-9}$

copper-bound water molecule. In a similar manner cyanide or azide, which are competitive inhibitors for oxygen, also halved the observed relaxation rates. Furthermore, binding constants for these anions derived from the water relaxation data agree well with those obtained from enzymatic inhibition studies. In order to account for the halving of relaxation rate the authors suggest that two water molecules bind to the copper center, axially and equatorially and it is only the latter that is displaced by inhibitory anions.

In contrast, amphetamine, which is competitive with benzylamine as an inhibitor, has no effect on the water relaxation so the authors conclude that the copper is not involved in the oxidative deamination, but takes part in the reoxidation of the enzyme by oxygen.

3. Superoxide Dismutase

The bovine enzyme has been the subject of a number of relaxation enhancement studies employing both proton and fluorine nuclei. Koenig and Brown[12] and Gaber et al.[40] measured water T_1 values over a range of temperatures and fields and concluded that the copper(II) was coordinated to at least one water molecule with a copper-oxygen bond length of 2 Å and with an exchange rate of $2.5 \cdot 10^5$ to 10^8 sec^{-1}. Addition of inhibitory anions, such as cyanide or azide, lowered the water relaxivity to a value typical of the copper(I) protein. Later Fee and Gaber[41] and Rigo et al.[42] used water relaxation measurements to measure anion binding to the enzyme. The dependence of water proton T_1 values on pH and temperature was investigated by Terenzi et al.,[43] who observed two increases in T_1 at pH values of 10 and 12.5; the latter is certainly due to denaturation and is not reversible when the pH is lowered. However, the nature of the first transition is more contentious. These authors claim that it is due to deprotonation of the copper-bound water molecule with a subsequent reduction in the copper-oxygen bond length and increase in the proton relaxation time. Boden et al.[44] have challenged this interpretation as they observed that in benzylamine oxidase, deprotonation lowers the relaxation rate rather than increasing it. By measuring both T_1 and T_2 values as a function of both field and temperature and by fitting their results to the Solomon-Bloembergen equations (including a contact interaction), they were able to obtain estimates for the various parameters in these equations which are shown in Table 4 (site I). Notice in particular that τ_M is within the limits derived from the results of Gaber et al.,[40] although the value for τ_c is a factor of ten lower than that previously assumed by taking τ_c equal to τ_R and calculating the latter from the Stokes-Einstein equation.

Although the pH dependence for the water proton relaxation times obtained by Boden et al.[44] is very similar to that obtained by Terenzi et al.,[43] the former authors considered a decrease in the copper-proton distance of 0.8 Å upon deprotonation unreasonably large. Instead they suggested that at high pH one of the histidine ligands to the copper is displaced by a hydroxide ion while the water molecule is retained. Starting from this premise they derive the exchange parameters for this hydroxide ion and these are also tabulated in Table 4 (site II). Comparison of the properties of water and hydroxide ligands to copper(II) with those in superoxide dismutase (SOD) suggests that the initial water molecule is coordinated in an axial position while the hydroxide ion binds equatorially. Support for the interpretation of Boden et al.[44] comes from ^{17}O water relaxation measurements by Bertini et al.[45] The dominant relaxation mechanism for oxygen in this case is expected to be a contact one and thus T_2 is much more affected than T_1. At low pH (<9) there is little paramagnetic relaxation enhancement (T_2) of the oxygen nuclei and this is because most of the unpaired spin is in the copper(II) $d_{x^2-y^2}$ orbital and thus in the equatorial plane while the water is coordinated axially. As the pH is increased there is a tenfold enhancement in the T_2 relaxation rate, due, perhaps, to equatorial coordination of hydroxide.

Bovine SOD also effectively relaxes fluoride nuclei and the association constant for fluoride binding has been determined by this technique to be $\approx 2\ M^{-1}$, depending upon temperature, pH, and ionic strength; an off-rate for fluoride of $2.5 \cdot 10^7\ sec^{-1}$ has also been measured.[46] As the degree of relaxation enhancement of fluoride ions is linear in SOD concentration up to about 10 μM, measurement of fluoride T_1 values can be used to determine the oxidized enzyme concentration in both purified and crude preparations,[47] as well as during catalysis.[48] The concentrations so determined correlate well with those from activity measurements, and the interference from other metalloproteins appears to be negligible.

Determination of ligand nuclear relaxation times can also yield information on exchange in the copper protein even if the nucleus is quadrupolar $(I \geq 1)$.[49] For a diamagnetic protein the dominant relaxation process is due to the modulation of the nuclear quadrupole moment by electric-field gradients. ^{35}Cl is such a quadrupolar nucleus $(I = 3/2)$ and the electric field at the chloride nucleus is much more symmetric when the ion is hydrated in solution than when it is bound to the copper(I) ion. Therefore chloride binding will decrease the $^{35}Cl\ T_2$ value and this method has been employed by Fee and Ward[50] to show that anions will bind to reduced bovine SOD.

B. High Resolution Studies

SOD is the only group-II copper protein that has been investigated by the high resolution NMR methods whose application to group-I proteins we have described. The first spectra of both bovine and human enzymes were published by Stokes et al.[51,52] and clearly showed that histidine residues were involved in copper coordination, a result later confirmed by the X-ray crystal structure.[53] Improvements in the performance of NMR spectrometers led to a later study of the bovine enzyme and enabled a number of assignments to be proposed for resonances in the apo-, reduced, and oxidized forms of the enzyme.

1. Histidine Residues

In the apoprotein, C2 proton resonances from all eight histidine residues can be detected and identified by their chemical shifts and pH dependence (Figure 7); however the titration curves for six of the resonances (c to h in Figure 8A) are not as expected for a simple deprotonation. Deviations from the normal Henderson-Hasselbalch type curve had been observed previously in proteins and can be interpreted in terms of interactions between titrating groups where the extent of ionization and/or the chemical shift of one group is affected by the degree of ionization of a nearby group. The copper-zinc binding site in this enzyme has six histidine residues that act as ligands to the two metal ions and we might expect that removal of the metal ions would leave the ligands in sufficiently close proximity to show the type of interactions observed. Further support for this interpretation comes from the reduced protein where none of the resonances show distorted pH titration curves (Figure 8B), a result consistent with resonances c to h in the apoprotein arising from metal ligands.[54]

Two resonances show similar behavior in the apo (a and b) and copper(I) (1 and 5) spectra; resonances a and 5 exhibit a normal pH titration curve and have the same pK_a value (6.7) and may thus be tentatively assigned to the same nonligand residue. Resonances b and 1 exhibit an unusual pH dependence, both resonances only start to shift upfield at pH values

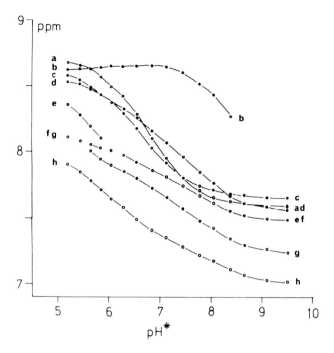

FIGURE 7. NMR pH titration curves of histidine resonances in bovine aposuperoxide dismutase.

greater than 7.5, and both rapidly lose intensity, lowering the pH does not restore the intensity. Irrespective of the reason for this behavior the similarity in both apo- and reduced bovine SOD is strong evidence for the assignment to the same nonligand histidine residue. The intensity of these resonances can be restored by prolonged incubation ($\approx$ 15 hr) in H_2O rather than D_2O solutions, and the loss of intensity is explained by exchange of the C2 proton by deuterium, a reaction known to occur in proteins. This ready exchange ($t_{1/2}$ = 9.2 hr at pH 8.1 and 40°C) provides a method of assigning this resonance by tritium labeling and peptide mapping. Exchange of the C2 proton by tritium and subsequent identification of the radioactively labeled residue showed that resonance 1 (and b) arose from His-41;[55] the crystal structure reveals that this residue is not a ligand. This means that resonances 5 and a must arise from the other nonligand residue, His-19. Further support for these assignments comes from considering the spectrum of the oxidized protein (Figure 8C). As expected, the resonances assigned to the ligands have been broadened out by the paramagnetic effect of the copper(II). In addition, the resonance corresponding to 1 in the reduced protein is also missing, while that corresponding to 5 is still present and shows the same pH titration behavior. Examination of the crystal structure shows that His-41 is 12 Å and His-19 some 20 Å from the copper ion.

After distinguishing ligand from nonligand histidine residues, the next step is to try and assign the ligand histidine resonances to either the copper or the zinc site. A common approach to aid the interpretation of NMR spectra is to use the effect perturbing the protein and analysing the resulting spectroscopic changes. Such perturbations can be temperature changes, chemical modifications of selected residues, or binding of ligands, and it is the latter two methods that have been used with SOD. The $^{35}Cl^-$ relaxation data of Fee and Ward[50] showed that chloride would bind to the copper(I) form of the enzyme, and we therefore anticipated that it would selectively alter the chemical shifts of those histidine resonances from the ligands to the copper. Such was indeed the case, and those resonances labeled 2 and 4 in Figure 8B showed a continuous downfield shift as the chloride concentration

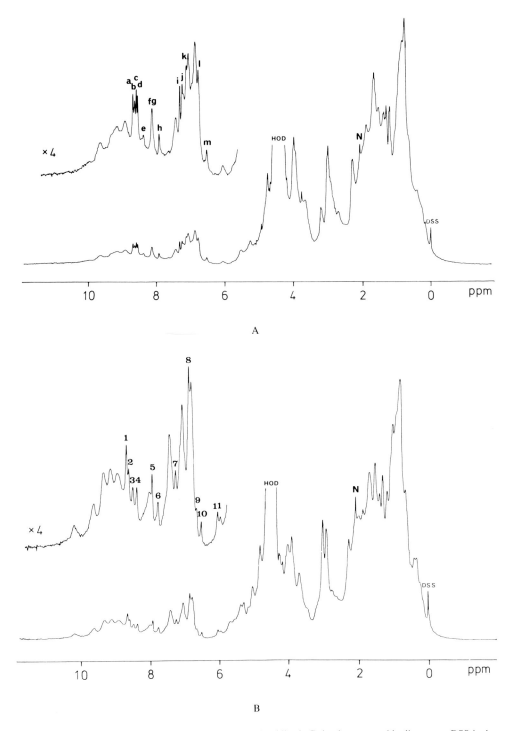

FIGURE 8.   NMR Spectra of apo (A), reduced (B), and oxidized (C) bovine superoxide dismutase. DSS is the internal chemical shift standard. (From Cass, A. E. G., Hill, H. A. O., Smith, B. E., Bannister, J. V., and Bannister, W. H., *Biochemistry*, 16, 3061, 1977. With permission.)

was increased from zero to 1 *M* at fixed ionic strength. Neither resonances 1 and 5 (from the nonligands) nor resonances 3 and 6, which were also assigned to histidine C2 protons by virtue of their linewidth and chemical shift, were affected. Similar changes in chemical

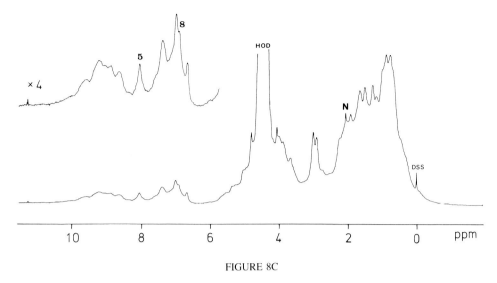

FIGURE 8C

shift were observed for the other halides: a downfield shift, the magnitude of which depended upon the halide.[56]

An alternative and much more generally applicable method for assigning histidine residues that are metal ligands is based upon the observation that the rate of C2 proton exchange is substantially decreased when the imidazole ring is coordinated. This effect has been reported in a variety of heterocyclic bases with exchangeable CH groups,[57] in cobalt(III) imidazole complexes,[58] and in the zinc protein β-lactamase II.[59] Incubation of holo-SOD in D_2O solutions at alkaline pH results in the loss of resonances 1 and 5 already assigned to nonligands. When the apoprotein is similarly treated and reconstituted, resonances 1 to 6 and also resonance 10 are missing. Although the latter peak had originally been assigned to a C4 proton on the basis of its chemical shift, the much slower exchange rate of C4 protons (never observed in proteins) suggests that it should be reassigned to a C2 proton.

These deuterium-exchange studies confirmed the previous assignnments. However, a distinction may be made by subjecting the protein containing only Zn^{2+} to the same exchange conditions. In this case the NMR spectrum, after reconstitution, showed that resonances 2 and 4 had lost intensity and these can therefore be assigned to copper ligands, consistent with the anion binding studies.[60] If the copper-only protein is treated in the same manner then, after reconstitution, all the ligand resonances show a partial loss of intensity, consistent with the migration of the copper ion between the copper and zinc binding sites.[61] Exactly similar results were obtained in parallel studies of the enzyme from *Saccharomyces cerevisiae*, although here further confirmation of the reliability of the method was provided by carrying out the tritium-exchange experiments. In this case the apo-, holo-, and zinc proteins were incubated in tritiated water and the incorporation of tritium into the individual residues determined. The results clearly demonstrated that the extent of tritium exchange is dependent upon whether or not the histidine is coordinated to a metal ion (Table 5).[62] However, in addition a number of interesting and intriguing results were revealed by the tritium exchange. Briefly these were that the incorporation into His-63 was indicative of a unique coordination environment of this residue. Also the incorporation of tritium into the reduced, but not the oxidized, enzyme is dependent upon phosphate, being increased by a factor of two in its presence. Finally there is also appreciable uptake of tritium into a nonhistidine residue in the holoprotein; although the residue was not identified it was in a heptapeptide that contains the essential arginine residue.[63]

2. Other Residues

In bovine SOD the resonances of the single tyrosine residue, at position 108, are readily

Table 5
³H INCORPORATIONª INTO HISTIDINE
RESIDUES IN YEAST SUPEROXIDE
DISMUTASE[62]

Histidine residue	Apoenzyme	Zn enzyme	Cu/Zn enzyme
46, 48	1.00	0.22	0.03
63	1.00	0.30	0.16
71	1.00	0.06	0.04
80	1.00	0.09	0.06
120	1.00	0.94	0.05

ª Incorporation given relative to apoprotein = 1.00. Actual incorporation of ³H in residues of apoprotein = 599 ± 12 cpm nmol⁻¹ histidine.

identified by their multiplet structure and also by their pH titration properties (pK_a = 10.5). The only other assigned resonance is the sharp singlet at 2.01 ppm, N in Figure 8; the possibilities for this are either the S-methyl group of the sole methionine or the N-terminal acetyl group. From an approximate calculation of the extent of "broadening beyond detection" of the copper(II) ion, the methionine resonance should be absent from the copper(II) spectrum and the N-acetyl resonance present. As Figure 8C reveals the latter is the case and the resonance is thus assigned to the N-acetyl group. In general, all of those resonances expected to be severely broadened by the copper(II) ion on the basis of the assignment and the information derived from the crystal structure are, in fact, so effected, showing that, at least as judged by these resonances, the crystal structure is an accurate reflection of the solution structure.[53]

3. Exchangeable Protons

The exchangeable protons in bovine SOD, arising from histidine residues, have been observed in aqueous solutions. Apo-SOD shows only a few broad resonances due to the rapid rate of exchange of imidazole NH protons with solvent. However, coordination of one nitrogen of the imidazole ring to a metal ion appreciably slows the rate of proton exchange of the other nitrogen. In the holo-reduced protein some of these resonances have been assigned by a combination of measuring solvent-exchange rates, nuclear Overhauser effects,[64] and by comparison with the oxidized protein.[65] Histidine N1 and N3 protons from residue 41 were assigned by the observation of a nuclear Overhauser effect from these resonances to the previously assigned C2 proton of this residue. This technique should prove very powerful in cross-correlating resonances when combined with the selective C2 proton exchange by deuterium. Two histidine NH protons show very slow solvent-exchange rates and these have been assigned to the N3 protons of the histidine ligands 44 (copper) and 69 (zinc) by a comparison with the crystal structure.[64] Burger et al.[66] have suggested an alternative assignment for the resonance assigned to His-44 and propose instead the assignment to a zinc ligand based upon the effects of zinc binding and the fact that it shows a nuclear Overhauser effect to a C2 proton assigned to a zinc ligand.

Stoesz et al.[64] have also determined the exchange rates for the histidine NH protons that give rise to observable resonances and these are collected in Table 6. Note that two of these are hydroxide ion catalysed and this has been confirmed by measuring the transfer of saturation from water. Previously these authors had shown that the dominant mechanism for this process was protein-proton mediated cross relaxation.[67] However, the strong pH dependence for two of the resonances is consistent with chemical exchange being important in these cases.

Table 6
ASSIGNMENTS IN BOVINE
SUPEROXIDE DISMUTASE[54,55,60,64,65]

Residue	Assigned protons
His-19	C2(H), C4(H)
His-41	C2(H), C4(H), N1(H), N3(H)
His	
Cu ligands	2 of the C2(H)
	1 of the NH
Zn ligands	3 of C2(H)
	2 or 3 of the NH
Tyr-108	*Ortho* and *meta* protons
N-terminal acetyl	Methyl group

There is a complete summary of the assignment of resonances in the spectrum of bovine Cu(I)/Zn(II)-SOD in Table 6.

4. Metal Binding

Once a number of assignments have been made in a NMR spectrum it becomes possible to use the changes in the properties of the assigned resonances to follow the behavior of individual atoms in the protein. An interesting problem that has been approached in this fashion is to monitor changes in the enzyme when the zinc ion binds to the apoprotein. Although it is clear that the catalytic center in SOD is the copper ion, the role of the nearby zinc ion is also of interest. In many ways the two metals are so closely associated that they could be considered to form a single, contiguous, active site.

Lippard et al.[65] in their study of the exchangeable protons in bovine SOD found that when the apoprotein was reconstituted with 1 mol of zinc(II) per subunit, the chemical shifts of the NH protons were similar to those in the reduced holoprotein. If a second mole of zinc(II) was added to the protein there were further changes in the pattern of NH protons. However, addition of substantial further quantities of zinc(II) resulted in no other changes. A metal ion-dependent organization of the active site was proposed to explain these results; the first zinc(II) ion added binds at its native site and second zinc(II) ion binds at the site normally occupied by copper. Further support for this interpretation was provided by the results of chemical modification of the histidine residues with diethylpyrocarbonate. Stokes et al.[52] had shown that more histidine residues can be modified by this reagent in the apoprotein(4) than in the holoprotein(1). Lippard et al.[65] found conditions where all eight histidine residues in the apoprotein were modified and showed that the degree of modification was proportional to the amount of added zinc(II).

Cass et al.[68] followed the zinc ion-dependent structural change by changes in the chemical shifts of the nonexchangeable protons. They concluded that two zinc(II) ions per subunit bind to the apoenzyme, that both ions are in slow exchange on the ¹H NMR timescale, and that the binding of the first zinc ion (at its native site) is at least an order of magnitude tighter than the binding of the second zinc ion. An interesting observation made by these authors was that the proton-exchange rate of the noncoordinated histidine residues in the mono-zinc protein was in the intermediate exchange region and that there was a pH-dependent, intermediate exchange conformational change in the copper binding site. This result plus the fact that changes in the aliphatic region of the spectrum were only observed during the binding of the first zinc ion led these authors to propose that when the first zinc ion bound it preformed the copper binding site.

Many divalent metal ions can replace the zinc(II) ion in bovine SOD with retention of

Table 7
CHEMICAL SHIFTS OF HISTIDINE C2
PROTONS IN SUPEROXIDE DISMUTASE
FROM VARIOUS SPECIES[73-75]

Species	His-19	His-41	Cu ligands	Zn ligands
Cow	Titrates	8.62	8.58, 8.33	8.47, 7.72, 6.48
Yeast	—[a]	—	8.54, 8.30	8.59, 7.76, 6.50
Swordfish	—	8.61	8.53, 8.27	8.50, 7.73, 6.91
Man	—	8.63[b]	8.53, 8.35	8.48, 7.76, 6.48

[a] — Indicates that this residue has been substituted by another amino acid.

[b] His-43.

activity[69] and the NMR properties of these nuclei may shed additional light on this region of the protein. The ^{113}Cd NMR of the cadmium only and copper(I)-cadmium protein has been investigated by two groups although their results are quite different. While Bailey et al.[70] claim there is only a small change in chemical shift of the cadmium resonance between the two forms (310 and 320 ppm, respectively), Armitage et al.[71] find a substantial change (170 and 8.6 ppm). In addition, the former authors claim that there is essentially no change in linewidth between the two forms and the latter observe a substantial change; it is not clear what these discrepancies are due to. Cass et al.[72] have made indirect use of the properties of the ^{199}Hg nucleus when they observed spin-spin coupling between the mercury ($I = 1/2$) and the C2 proton of a histidine residue ($J = 27$ Hz), thus providing direct evidence for histidine coordination of this metal ion.

5. Species Comparisons

One advantageous feature of NMR spectroscopy is that, even without complete assignment, the spectrum is still a sensitive indicator of the overall structure. Thus a comparison of the NMR spectra of the same protein isolated from several species provides a most convenient way of judging the differences in structures. The first detailed comparison of SOD was of the bovine enzyme with the enzyme from yeast.[73] Resonances in the latter protein were assigned by the same procedures as used for the bovine enzyme and it soon became apparent that there were considerable "active-site homologies" between the two proteins, at least as judged by the chemical shifts of the ligand histidine C2 proton resonances. This comparison also clearly showed that His-19 and His-41 were absent in the yeast enzyme, a result confirmed by the sequence determination that was being carried out at the same time. A later NMR study of the human enzyme reinforced this idea of a strongly conserved structure of the active site.[74] In addition His-41 was found in the human enzyme and the rapid deuterium-exchange properties of this residue testified to the conservation of its whole microenvironment. The chemical shifts of the histidine C-2 proton resonances of the three enzymes, along with those of the enzyme from swordfish liver, are collected in Table 7.[75]

Exchangeable protons may also be used to study structural similarities. Burger et al.[66] have compared the chemical shifts of histidine NH protons of the bovine and two wheatgerm SOD's. The chemical shifts and deuterium-exchange properties as well as the effect of reconstitution in all three enzymes are rather similar. In particular it appears that His-41 is present in both wheatgerm enzymes. All of these results combine to suggest that although

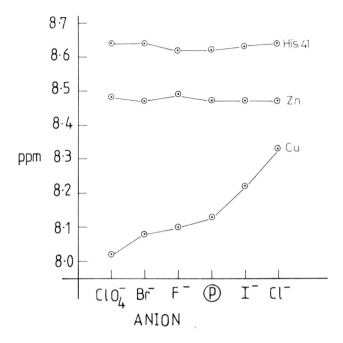

FIGURE 9.   The effect of anions on the chemical shifts of different histidine residues in reduced bovine superoxide dismutase. (From Cass, A. E. G. and Hill, H. A. O., *Chemical and Biochemical Aspects of Superoxide and Superoxide Dismutase*, Bannister, J. V. and Hill, H. A. O., Eds., Elsevier/North-Holland, New York, 1980, 290. With permission.)

the enzyme carries out an apparently simple reaction, the structural constraints at the active site are such as to result in a high degree of conservation from species to species.

6. Anion Binding

The substrate for SOD is a monovalent anion and several other anions bind at the active site and act as competitive inhibitors. Fluoride binding and the perturbation of the ^{1}H NMR spectrum by halides have been described in previous sections of this chapter; however, the question remains: can we derive any additional insight into the mechanism of the enzyme from anion binding studies? Cyanide and azide, though they lead to large changes in EPR and optical spectra of the copper(II) enzyme and are good inhibitors, may not be suitable substrate analogs for the following reason. Pulse-radiolytic studies show the rate of reaction of the enzyme with superoxide has a rate constant of $\approx 10^9 \, M^{-1} \, \text{sec}^{-1}$,[76] while polarographic results suggest a Michaelis constant which we can roughly associate with a dissociation constant of ≈ 0.4 mM,[77] thus implying an off-rate for superoxide of $\approx 10^6 \, \text{sec}^{-1}$, i.e., a rate comparable with fast exchange on the ^{1}H NMR timescale. Both azide and cyanide are in slow exchange on the ^{1}H NMR timescale. On the other hand, the halides are in fast exchange,[56] and consequently they may prove to be more suitable substrate analogs.

Figure 9 shows the change in the chemical shifts on halide binding for a copper ligand, a zinc ligand, and His-41, and as expected only the former is affected. However, the effects shown in this figure are not straightforward as the following two points illustrate. Iodide has been claimed not to inhibit the enzyme because, the authors suggest, it is too large to fit in the active site,[42] yet its effect on the copper ligand resonances is second only to that of chloride. Second, the shifts at low ionic strength fall in the middle of those observed for the different anions (all at 1 M concentration). These observations lead us to consider the role of the active site water molecules and the other nearby residues in the enzyme mechanism.

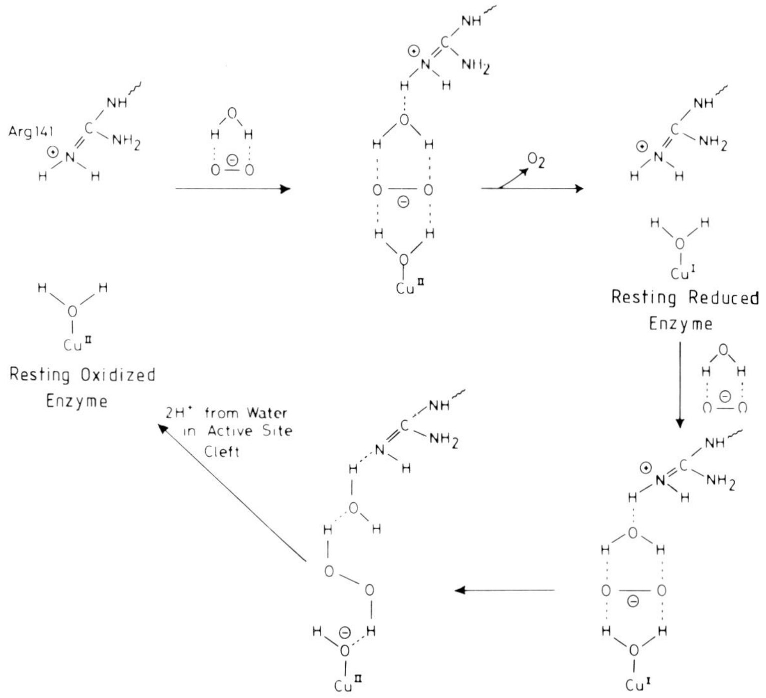

FIGURE 10.   A possible mechanism for Cu/Zn-superoxide dismutase. (From Cass, A. E.
G. and Hill, H. A. O., *Chemical and Biochemical Aspects of Superoxide and Superoxide
Dismutase*, Bannister, J. V. and Hill, H. A. O., Eds., Elsevier/North-Holland, New York,
1980, 290. With permission.)

Chemical modification studies have revealed the importance of Arg-141 in the enzyme
turnover:[63] this arginine is approximately 5 Å from the copper. We have incorporated this
in our mechanism (Figure 10) where we propose that the superoxide ion, as a hydrate,
hydrogen bonds between the arginine and the axial water molecule. All electron transfers
are outer-sphere and in the reduction of the superoxide, the arginine serves to neutralize the
charge on the substrate and to protonate the developing peroxide dianion: the second proton
comes via the copper-bound water molecule so that proton and electron transfer occur in a
concerted and rapid process. Reprotonation of the metal bound water and the arginine then
occurs from water molecules in the active site and the resulting hydroxide ions are released
into the bulk medium by a series of proton transfers by the ordered water in the protein in
a mechanism analogous to proton/hydroxide transport in ice (Grotthus mechanism). Anions
such as iodide and perchlorate (the so-called chaotropic ions) could therefore also affect the
copper site via changes in the water structure.

VI. NMR STUDIES OF GROUP-IV PROTEINS

An early report by Blumberg et al.[78] on the water proton T_1 values of solutions of
ceruloplasmin measured the relaxation times in the oxidized enzyme at various stages of
partial reduction. The nonequivalence of the copper ions was shown by the nonlinear dependence
of the water relaxation rate on the extent of reduction. It was suggested that those copper
ions that were not accessible to water molecules were also those most easily reduced. Later
Koenig and Brown[12] presented data on the frequency dependence of T_1 and they concluded
that a water molecule is bound some 3 to 7 Å from at least one of the copper centers. They
also determined that the residence time of the water molecule on the protein was relatively
long (tens of microseconds).

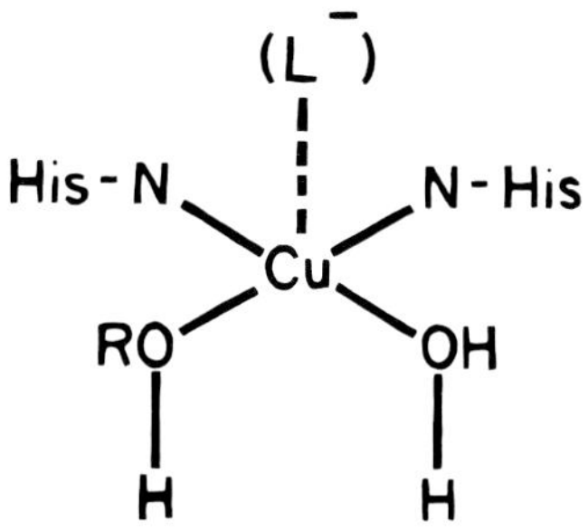

$$(\overline{L})$$

$$\text{His-N} \quad \text{N-His}$$
$$\text{Cu}$$
$$\text{RO} \quad \text{OH}$$
$$\text{H} \quad \text{H}$$

FIGURE 11. Proposed model for the coordination sphere around the nonblue center in laccase. (From Goldberg, M., Vuk-Pavlovié, S., and Pecht, I., *Bio-chemistry*, 19, 5181, 1980. With permission.)

An interesting paper by Goldberg et al.[79] has appeared on the ^1H and ^{17}O nuclear relaxation enhancements of water molecules in solutions of laccase. These authors attribute virtually all (>80%) of the enhancement to the "nonblue" copper center. Although the protons are in rapid exchange on the NMR timescale no relaxation enhancement is seen for the ^{17}O nucleus, and they interpret their results in terms of proton transfer from the equatorially coordinated water molecule, via a protein residue, to bulk solution. The pH dependence of the ^1H relaxation rates shows two pK_as at 6.2 and 8.6 attributed to the coordinated water molecule and a hydroxyl group from a ligand serine or threonine. Fluoride, which is known to bind to the same copper center as does the water molecule, does not affect the proton relaxation times and the authors suggest that it coordinates to the normally vacant axial position. This interpretation is in distinct contrast to the galactose oxidase results where the equatorially bound fluoride gives rise to superhyperfine splitting in the EPR spectrum and axially bound fluoride does not. If the water molecule were axially coordinated then no enhancement of the oxygen relaxation would be expected due to the very small contact interaction exactly analogous to the situation with SOD *(vide infra)*.

Notwithstanding these comments the authors propose a model for the coordination sphere of copper that is illustrated in Figure 11, and they explain the water proton, but not oxygen, relaxation by assuming that the cavity in which the copper ion is located can only be penetrated by protons from the bulk solution.

VII. CONCLUSIONS

We have reached the end of the beginning of the application of NMR spectroscopy to the study of copper proteins. For those of low relative molecular mass, sufficient assignments can be made, especially if the crystal structure is available, to allow both structural and functional features to be revealed. Larger proteins still present some problems of resolution, but the use of selected pulse sequences, specific isotopic substitution, or subtle applications of difference spectroscopy can be rewarding. NMR spectroscopy should come into its own when details of local structural changes are required. This is particularly true when the dynamics of such changes are of interest. The application to copper proteins always will have one singular inherent advantage: they contain, in one oxidation state, an inherent, intrinsic paramagnetic perturbant which provides a central reference point from which distance to other residues can be assessed. No extrinsic perturbation need be applied, no replacement of active metal ion by a more informative but nonactive metal ion need be contemplated. With the ever-improving techniques available, the subtlety of the insight into the structures of these ubiquitous proteins should continue increasing.

ACKNOWLEDGMENTS

Dr. Cass acknowledges the award of a British Petroleum Research Fellowship at St. Hugh's College, Oxford. This is a contribution from the Oxford Enzyme Group of which Dr. Hill is a member.

We thank Mr. N. A. Kitchen and Dr. G. W. Canters for providing unpublished results and for fruitful discussions.

REFERENCES

1. **Dwek, R. A.,** *Nuclear Magnetic Resonance (N.M.R.) in Biochemistry. Applications to Enzyme Systems,* Oxford University Press, Oxford, 1973.
2. **James, Th. L.,** *Nuclear Magnetic Resonance in Biochemistry. Principles and Applications,* Academic Press, New York, 1975.
3. **Berliner, L. J. and Reuben, J., Eds.,** *Biological Magnetic Resonance,* Vol. 3, Plenum Press, New York, 1981.
4. **Campbell, I. D. and Dobson, C. M.,** The application of high resolution nuclear magnetic resonance to biological systems, *Meth. Biochem. Anal.,* 25, 1, 1979.
5. **Levine, B. A., Moore, G. R., Ratcliffe, R. G., and Williams, R. J. P.,** Nuclear magnetic resonance studies of the solution structure of proteins, *Int. Rev. Biochem.,* 24, 77, 1979.
6. **Fee, J. A.,** Copper proteins. Systems containing the "blue" copper center, *Struct. Bonding (Berlin),* 23, 1, 1975.
7. **Cass, A. E. G. and Hill, H. A. O.,** Copper proteins and copper enzymes, in *Biological Roles of Copper,* Ciba Foundation Symp. No. 79, Excerpta Medica, Amsterdam, 1980, 71.
8. **Colman, P. M., Freeman, H. C., Guss, J. M., Murata, M., Norris, V. A., Ramshaw, J. A. M., and Venkatappa, M. P.,** X-ray crystal structure analysis of plastocyanin at 2.7 Å resolution, *Nature (London),* 272, 319, 1978.
9. **Adman, E. T., Stenkamp, R. E., Sieker, L. C., and Jensen, L. H.,** A crystallographic model for azurin at 3 Å resolution, *J. Mol. Biol.,* 123, 35, 1978.
10. **Blumberg, W. E. and Peisach, J.,** The optical and magnetic properties of copper in *Chenopodium album* plastocyanin, *Biochim. Biophys. Acta,* 126, 269, 1966.
11. **Boden, N., Holmes, M. C., and Knowles, P. F.,** Binding of water to "types I and II" Cu^{2+} in proteins, *Biochem. Biophys. Res. Commun.,* 57, 845, 1974.
12. **Koenig, S. H. and Brown, R. D.,** Anomalous relaxation of water protons in solutions of copper-containing proteins, *Ann. N.Y. Acad. Sci.,* 222, 752, 1973.
13. **Beattie, J. K., Fensom, D. J., Freeman, H. C., Woodcock, E., Hill, H. A. O., and Stokes, A. M.,** An NMR investigation of electron transfer in the copper-protein, plastocyanin, *Biochim. Biophys. Acta,* 405, 109, 1975.
14. **Ulrich, E. L. and Markley, J. L.,** Blue-copper proteins: nuclear magnetic resonance investigations, *Coord. Chem. Rev.,* 27, 109, 1978.
15. **Markley, J. L., Ulrich, E. L., Berg, S. P., and Krogmann, D. W.,** Nuclear magnetic resonance studies of the copper binding sites of blue copper proteins: oxidized, reduced and apoplastocyanin, *Biochemistry,* 14, 4428, 1975.
16. **Hill, H. A. O., Leer, J. C., Smith, B. E., Storm, C. B., and Ambler, R. P.,** A possible approach to the investigation of the structures of copper proteins: ^{1}H n.m.r. spectra of azurin, *Biochem. Biophys. Res. Commun.,* 70, 331, 1976.
17. **Ugurbil, K. and Bersohn, R.,** Nuclear magnetic resonance study of exchangeable and nonexchangeable protons in azurin from *Pseudomonas aeruginosa, Biochemistry,* 16, 3016, 1977.
18. **Hill, H. A. O. and Smith, B. E.,** Characteristics of azurin from *Pseudomonas aeruginosa* via 270-MHz ^{1}H nuclear magnetic resonance spectroscopy, *J. Inorg. Biochem.,* 11, 79, 1979.
19. **Silvestrini, M. Ch., Brunori, M., Wilson, M. T., and Darley-Usmar, V. M.,** The electron transfer system of *Pseudomonas aeruginosa:* a study of the pH-dependent transitions between redox forms of azurin and cytochrome c_{551}, *J. Inorg. Biochem.,* 14, 327, 1981.
20. **Adman, E. T., Canters, G. W., Hill, H. A. O., and Kitchen, N. A.,** The effect of pH and temperature on the structure of the active site of azurin from *Pseudomonas aeruginosa, FEBS Lett.,* 143, 287, 1982.

21. **Freeman, H. C., Norris, V. A., Ramshaw, J. A. M., and Wright, P. E.,** High resolution proton magnetic resonance studies of plastocyanin, *FEBS Lett.,* 86, 131, 1978.
22. **Hill, H. A. O. and Lee, W.-K.,** Investigation of the structure of stellacyanin by ^{1}H nuclear-magnetic-resonance spectroscopy, *Biochem. Soc. Trans.,* 7, 733, 1979.
23. **Hill, H. A. O. and Lee, W. K.,** Investigation of the structure of the blue copper protein from *Rhus vernicifera* stellacyanin by ^{1}H nuclear magnetic resonance spectroscopy, *J. Inorg. Biochem.,* 11, 101, 1979.
24. **Hill, H. A. O., Smith, B. E., Storm, C. B., and Ambler, R. P.,** The proton magnetic resonance spectra of a cobalt(II) azurin, *Biochem. Biophys. Res. Commun.,* 70, 783, 1976.
25. **McMillin, D. R. and Tennet, D. L.,** Physical studies of azurin and some metal replaced derivatives, in *ESR and NMR of Paramagnetic Species in Biological and Related Systems,* Bertini, I. and Drago, R. S., Eds., D. Reidel, Dordrecht, Holland, 1980, 369.
26. **Markley, J. L., Ulrich, E. L., and Krogmann, D. W.,** Spinach plastocyanin: comparison of reduced and oxidized forms by natural abundance carbon-13 nuclear magnetic resonance spectroscopy, *Biochem. Biophys. Res. Commun.,* 78, 106, 1977.
27. **Ugurbil, K., Norton, R. S., Allerhand, A., and Bersohn, R.,** Studies of individual carbon sites of azurin from *Pseudomonas aeruginosa* by natural-abundance carbon-13 nuclear magnetic resonance spectroscopy, *Biochemistry,* 16, 886, 1977.
28. **Wherland, S. and Gray, H. B.,** Electron transfer mechanisms employed by metalloproteins, in *Biological Aspects of Inorganic Chemistry,* Addison, A. W., Cullen, W. R., Dolphin, D. H., and James, B. R., Eds., Wiley-Interscience, New York, 1977, 289.
29. **Lappin, A. G., Segal, M. G., Weatherburn, D. C., and Sykes, A. G.,** Identification of binding sites in reactions of blue copper proteins with inorganic complexes and implications of such findings, *J. Chem. Soc., Chem. Commun.,* 38, 1979.
30. **Cookson, D. J., Hayes, M. T., and Wright, P. E.,** NMR study of the interaction of plastocyanin with chromium(III) analogues of inorganic electron transfer reagents, *Biochim. Biophys. Acta,* 591, 162, 1980.
31. **Cookson, D. J., Hayes, M. T., and Wright, P. E.,** Electron transfer reagent binding sites on plastocyanin, *Nature (London),* 283, 682, 1980.
32. **Handford, P. M., Hill, H. A. O., Lee, R. W.-K., Henderson, R. A., and Sykes, A. G.,** Investigation of the binding of inorganic complexes to blue copper proteins by ^{1}H nmr spectroscopy. I. The interaction between the $[Cr(phen)_3]^{3+}$ and $[Cr(CN)_6]^{3-}$ ions and the copper(I) form of parsley plastocyanin, *J. Inorg. Biochem.,* 13, 83, 1980.
33. **Marwedel, B. J., Kurland, R. J., Kosman, D. J., and Ettinger, M. J.,** Fluoride ion as an NMR relaxation probe of paramagnetic metalloenzymes. The binding of fluoride to galactose oxidase, *Biochem. Biophys. Res. Commun.,* 63, 773, 1975.
34. **Marwedel, B. J., Kosman, D. J., Bereman, R. D., and Kurland, R. J.,** Magnetic resonance studies of cyanide and fluoride binding to galactose oxidase copper(II): evidence for two exogenous ligand sites, *J. Am. Chem. Soc.,* 103, 2842, 1981.
35. **Kurland, R. J. and Marwedel, B. J.,** Fluoride ion as a nuclear magnetic resonance probe of galactose oxidase. An analysis of the fluorine-19 nuclear magnetic relaxation rates, *J. Phys. Chem.,* 83, 1422, 1979.
36. **Marwedel, B. J. and Kurland, R. J.,** Fluoride ion as an NMR relaxation probe of galactose oxidase-substrate binding, *Biochim. Biophys. Acta,* 657, 495, 1981.
37. **Winkler, M. E., Bereman, R. D., and Kurland, R. J.,** Kinetic and magnetic resonance studies of substrate binding to galactose oxidase copper(II), *J. Inorg. Biochem.,* 14, 223, 1981.
38. **Kluetz, M. D. and Schmidt, P. G.,** Proton relaxation study of the hog kidney diamine oxidase active center, *Biochemistry,* 16, 5191, 1977.
39. **Barker, R., Boden, N., Cayley, G., Charlton, S. C., Henson, R., Holmes, M. C., Kelly, I. D., and Knowles, P. F.,** Properties of cupric ions in benzylamine oxidase from pig plasma as studied by magnetic-resonance and kinetic methods, *Biochem. J.,* 177, 289, 1979.
40. **Gaber, B. P., Brown, R. D., Koenig, S. H., and Fee, J. A.,** Nuclear magnetic relaxation dispersion in protein solutions. V. Bovine erythrocyte superoxide dismutase, *Biochim. Biophys. Acta,* 271, 1, 1972.
41. **Fee, J. A. and Gaber, B. P.,** Anion binding to bovine erythrocyte superoxide dismutase. Evidence for multiple binding sites with qualitatively different properties, *J. Biol. Chem.,* 247, 60, 1972.
42. **Rigo, A., Stevanato, R., Viglino, P., and Rotilio, G.,** Competitive inhibition of Cu,Zn superoxide dismutase by monovalent anions, *Biochem. Biophys. Res. Commun.,* 79, 776, 1977.
43. **Terenzi, M., Rigo, A., Franconi, C., Mondovi, B., Calabrese, L., and Rotilio, G.,** pH dependence of the nuclear magnetic relaxation rate of solvent water protons in solutions of bovine superoxide dismutase, *Biochim. Biophys. Acta,* 351, 230, 1974.
44. **Boden, N., Holmes, M. C., and Knowles, P. F.,** Properties of the cupric sites in bovine superoxide dismutase studied by nuclear-magnetic-relaxation measurements, *Biochem. J.,* 177, 303, 1979.
45. **Bertini, I., Luchinat, C., and Messori, L.,** A water ^{17}O NMR study of the pH dependent properties of superoxide dismutase, *Biochem. Biophys. Res. Commun.,* 101, 577, 1981.

46. **Viglino, P., Rigo, A., Stevanato, R., Ranieri, G. A., Rotilio, G., and Calabrese, L.,** The binding of fluoride ion to bovine cuprozinc superoxide dismutase as studied by ^{19}F magnetic relaxation, *J. Magn. Reson.,* 34, 265, 1979.

47. **Rigo, A., Viglino, P., Argese, E., Terenzi, M., and Rotilio, G.,** Nuclear magnetic relaxation of ^{19}F as a novel assay method of superoxide dismutase, *J. Biol. Chem.,* 254, 1759, 1979.

48. **Viglino, P., Rigo, A., Argese, E., Calabrese, L., Cocco, D., and Rotilio, G.,** ^{19}F relaxation as a probe of the oxidation of Cu,Zn superoxide dismutase. Studies of the enzyme in steady-state turnover, *Biochem. Biophys. Res. Commun.,* 100, 125, 1981.

49. **Forsén, S. and Lindman, B.,** Ion binding in biological systems as studied by NMR spectroscopy, *Meth. Biochem. Anal.,* 27, 289, 1981.

50. **Fee, J. A. and Ward, R. L.,** Evidence for a coordination position available to solute molecules on one of the metals at the active center of reduced bovine superoxide dismutase, *Biochem. Biophys. Res. Commun.,* 71, 427, 1976.

51. **Stokes, A. M., Hill, H. A. O., Bannister, W. H., and Bannister, J. V.,** Nuclear magnetic resonance spectra of human and bovine superoxide dismutases, *FEBS Lett.,* 32, 119, 1973.

52. **Stokes, A. M., Hill, H. A. O., Bannister, W. H., and Bannister, J. V.,** The active site of bovine superoxide dismutase, *Biochem. Soc. Trans.,* 2, 489, 1974.

53. **Richardson, J. S., Thomas, K. A., Rubin, B. H., and Richardson, D. C.,** Crystal structure of bovine Cu,Zn superoxide dismutase at 3 Å resolution: chain tracing and metal ligands, *Proc. Natl. Acad. Sci. U.S.A.,* 72, 1349, 1975.

54. **Cass, A. E. G., Hill, H. A. O., Smith, B. E., Bannister, J. V., and Bannister, W. H.,** Investigation of the structure of bovine erythrocyte superoxide dismutase by ^{1}H nuclear magnetic resonance spectroscopy, *Biochemistry,* 16, 3061, 1977.

55. **Cass, A. E. G., Hill, H. A. O., Smith, B. E., Bannister, J. V., and Bannister, W. H.,** Carbon-2 proton exchange at histidine-41 in bovine erythrocyte superoxide dismutase, *Biochem. J.,* 165, 587, 1977.

56. **Cass, A. E. G. and Hill, H. A. O.,** Anion binding to copper(I) superoxide dismutase: a high resolution ^{1}H nuclear magnetic resonance spectroscopic study, in *Chemical and Biochemical Aspects of Superoxide and Superoxide Dismutase,* Bannister, J. V. and Hill, H. A. O., Eds., Elsevier/North Holland, New York, 1980, 290.

57. **Jones, J. R. and Taylor, S. E.,** Isotopic hydrogen exchange in purines — mechanisms and applications, *Chem. Soc. Rev.,* 10, 329, 1981.

58. **Rowan, N. S., Storm, C. B., and Rowan, R., III,** Properties of metal-ion coordinated imidazoles: nmr and C-2 H exchange in Co(III) complexes, *J. Inorg. Biochem.,* 14, 59, 1981.

59. **Baldwin, G. S., Galdes, A., Hill, H. A. O., Smith, B. E., Waley, S. G., and Abraham, E. P.,** Histidine residues as zinc ligands in β-lactamase II, *Biochem. J.,* 175, 441, 1978.

60. **Cass, A. E. G., Hill, H. A. O., Bannister, J. V., Bannister, W. H., Hasemann, V., and Johansen, J. T.,** The exchange of histidine C-2 protons in superoxide dismutases. A novel method for assigning histidine-metal ligands in proteins, *Biochem. J.,* 183, 127, 1979.

61. **Valentine, J. S., Pantoliano, M. W., McDonnell, P. J., Burger, A. R., and Lippard, S. J.,** pH-dependent migration of copper(II) to the vacant zinc-binding site of zinc-free bovine erythrocyte superoxide dismutase, *Proc. Natl. Acad. Sci. U.S.A.,* 76, 4245, 1979.

62. **Dunbar, J. C., Johansen, J. T., Cass, A. E. G., and Hill, H. A. O.,** Assignment of the metal-histidine ligands from the tritium exchange rate of the histidine C-2 protons in the Cu(II),Zn(II)-superoxide dismutase from *Saccharomyces cerevisiae, Carlsberg Res. Commun.,* 45, 349, 1980.

63. **Malinowski, D. P. and Fridovich, I.,** Chemical modification of arginine at the active site of bovine erythrocyte superoxide dismutase, *Biochemistry,* 18, 5909, 1979.

64. **Stoesz, J. D., Malinowski, D. P., and Redfield, A. G.,** Nuclear magnetic resonance study of solvent exchange and the nuclear Overhauser effect of the histidine protons of bovine superoxide dismutase, *Biochemistry,* 18, 4669, 1979.

65. **Lippard, S. J., Burger, A. R., Ugurbil, K., Pantoliano, M. W., and Valentine, J. S.,** Nuclear magnetic resonance and chemical modification studies of bovine erythrocyte superoxide dismutase: evidence for zinc-promoted organization of the active site structure, *Biochemistry,* 16, 1136, 1977.

66. **Burger, A. R., Lippard, S. J., Pantoliano, M. W., and Valentine, J. S.,** Nuclear magnetic resonance study of the exchangeable histidine protons in bovine and wheat germ superoxide dismutases, *Biochemistry,* 19, 4139, 1980.

67. **Stoesz, J. D., Redfield, A. G., and Malinowski, D. P.,** Cross relaxation and spin diffusion effects on the proton nmr of biopolymers in H$_2$O. Solvent saturation and chemical exchange in superoxide dismutase, *FEBS Lett.,* 91, 320, 1978.

68. **Cass, A. E. G., Hill, H. A. O., Bannister, J. V., and Bannister, W. H.,** Zinc(II) binding to apo-(bovine erythrocyte superoxide dismutase), *Biochem. J.,* 177, 477, 1979.

69. **Beem, K. M., Rich, W. E., and Rajagopalan, K. V.,** Total reconstitution of copper-zinc superoxide dismutase, *J. Biol. Chem.,* 249, 7298, 1974.

70. **Bailey, D. B., Ellis, P. D., and Fee, J. A.,** Cadmium-113 nuclear magnetic resonance studies of cadmium-substituted derivatives of bovine superoxide dismutase, *Biochemistry,* 19, 591, 1980.

71. **Armitage, I. M., Schoot Uiterkamp, A. J. M., Chlebowski, J. F., and Coleman, J. E.,** ^{113}Cd nmr as a probe of the active site of metalloenzymes, *J. Magn. Reson.,* 29, 375, 1978.

72. **Cass, A. E. G., Galdes, A., Hill, H. A. O., McClelland, C. E., and Storm, C. B.,** Heavy metal binding to biological molecules: identification of ligands by observation of ^{199}Hg-^{1}H coupling, *FEBS Lett.,* 94, 311, 1978.

73. **Cass, A. E. G., Hill, H. A. O., Hasemann, V., and Johansen, J. T.,** ^{1}H nuclear magnetic resonance spectroscopy of yeast copper-zinc superoxide dismutase. Structural homology with the bovine enzyme, *Carlsberg Res. Commun.,* 43, 439, 1978.

74. **Hill, H. A. O., Lee, W.-K., Bannister, J. V., and Bannister, W. H.,** Investigation of human erythrocyte superoxide dismutase by ^{1}H nuclear-magnetic-resonance spectroscopy, *Biochem. J.,* 185, 245, 1980.

75. **Bannister, J. V., Bannister, W. H., Cass, A. E. G., Hill, H. A. O., and Johansen, J. T.,** A comparison of the active site structures of bovine, human, swordfish and yeast copper-zinc superoxide dismutases, in *Chemical and Biochemical Aspects of Superoxide and Superoxide Dismutase,* Bannister, J. V. and Hill, H. A. O., Eds., Elsevier/North Holland, New York, 1980, 284.

76. **Fielden, E. M., Roberts, P. B., Bray, R. C., Lowe, D. J., Mautner, G. N., Rotilio, G., and Calabrese, L.,** The mechanism of action of superoxide dismutase from pulse radiolysis and electron paramagnetic resonance. Evidence that only half the active sites function in catalysis, *Biochem. J.,* 139, 49, 1974.

77. **Rigo, A., Viglino, P., and Rotilio, G.,** Kinetic study of O_2^- dismutation by bovine superoxide dismutase. Evidence for saturation of the catalytic sites by O_2^- *Biochem. Biophys. Res. Commun.,* 63, 1013, 1975.

78. **Blumberg, W. E., Eisinger, J., Aisen, P., Morell, A. G., and Scheinberg, I. H.,** Physical and chemical studies on ceruloplasmin. I. The relation between blue color and the valence states of copper, *J. Biol. Chem.,* 238, 1675, 1963.

79. **Goldberg, M., Vuk-Pavlović, S., and Pecht, I.,** Proton and oxygen-17 magnetic resonance relaxation in *Rhus* laccase solutions: proton exchange with type 2 copper(II) ligands, *Biochemistry,* 19, 5181, 1980.

Chapter 4

STRUCTURAL STUDIES OF COPPER PROTEINS USING X-RAY ABSORPTION SPECTROSCOPY

Man Sung Co and Keith O. Hodgson

TABLE OF CONTENTS

I. INTRODUCTION

X-ray absorption spectroscopy (XAS) is a technique that reveals local atomic structure of materials of all types, from heterogeneous catalysts to biological macromolecules.[1-5] As early as the 1930s, Kronig[6] had observed that the absorption cross-section in the X-ray region had complex oscillations as a function of energy which extended several hundred electron volts above the absorption threshold. Many attempts were subsequently made to present an adequate theory to explain such structure. In 1970, Sayers et al.[7] presented a derivation using a short-range single-electron single-scattering theory that successfully interpreted experimental data collected on a copper foil with a conventional X-ray source. However, the intensity of the X-ray source (Bremsstrahlung radiation produced by a conventional X-ray tube) was not sufficient to accumulate data within a reasonable time for systems with low metal concentrations. The application of the technique to studies on metalloproteins and metalloenzymes was not possible until 1974, when synchrotron radiation became available at the Stanford Positron Electron Accelerating Ring (SPEAR)[8] and provided a stable, highly collimated, high flux X-ray source. Since then, several more synchrotron storage rings have been built.[9] The increase in flux from a storage ring of approximately 10^4 to 10^6 compared to a rotating anode X-ray tube allows systems of low metal content in metalloproteins and metalloenzymes to be studied using the technique.

Information is contained in two regions of an X-ray absorption spectrum: the absorption edge and the extended X-ray absorption fine structure (EXAFS). When applied to metalloproteins or metalloenzymes, the absorption edge region usually reveals information about the oxidation state and the site symmetry of the absorbing metal atom, and in certain cases, even the type of the liganding atoms. This information is derived from measurements of the energies and intensities of the electronic transitions as the incident photon energy is scanned from below the edge to ≈ 50 eV above the edge. The EXAFS spectrum (extending several hundred eV above the the edge) provides information concerning the types of ligand atoms and the corresponding interatomic distances. This information is usually obtained by resolving the EXAFS spectrum into individual sine waves using Fourier transformation or numerical curve-fitting techniques. Analysis of the frequency of the sine wave provides information concerning interatomic distances, while the amplitude reveals the number and the type of scatterers.

The XAS technique has become a useful tool for elucidating metal-site structures of proteins and enzymes containing metals. Most of the proteins and enzymes studied contain iron, copper, or molybdenum at their active centers. XAS has contributed to a better understanding of the active sites of the iron in rubredoxin, hemoglobin, chloroperoxidase, cytochrome *P*-450, cytochrome c_{551}, and cytochrome *c* oxidase, the copper in hemocyanin, blue copper proteins, laccase, and cytochrome *c* oxidase, as well as the molybdenum in nitrogenase and its Fe-Mo cofactor, sulfite oxidase, nitrate reductase, and xanthine oxidase. In this chapter, we shall review the XAS work that has been done on the copper-containing proteins and enzymes.

II. THEORY AND PRACTICE

A. Extended X-Ray Absorption Fine Structure

The EXAFS spectrum is usually obtained as absorption coefficient vs. X-ray photon energy. The absorption coefficient is directly related to the probability that an electron is excited by an X-ray photon, i.e., it depends on the initial state and the final state of the electron. The initial state is that of the core level corresponding to the absorption edge (a 1s electron in the case of K-absorption edge) and the final state is that of the ejected photoelectron which can be described by an outgoing spherical wave originating at the absorbing atom. If there are other neighboring atoms around the absorbing atom, the photoelectron will be backscattered by the neighboring atoms producing incoming waves. The final state is then the sum of the outgoing and incoming waves. It is the interference of the outgoing and the incoming waves that gives rise to the modulations of the absorption coefficient on the high energy side of the threshold that are termed EXAFS.

Theoretical derivations of EXAFS have been given in detail elsewhere.[7,10-13] Here we describe qualitatively the physical phenomenon of EXAFS and the physical significance of the terms in the derived EXAFS expression:

$$\chi(k) = \frac{\mu - \mu_s}{\mu_0} = \frac{1}{k} \sum_s \frac{N_s |f_s(\pi,k)|}{R_{as}^2} e^{-\sigma_{as}^2 k^2} e^{-2\mu R_{as}} \sin \left[2kR_{as} + \alpha_{as}(k) \right] \quad (1)$$

where k is the wave vector defined by $[2m(E - E_o)/\hbar^2]^{1/2}$, E_o the threshold energy for ionization of the electron, and m the mass of the photoelectron.

The quantity known as EXAFS, denoted by χ, is simply the relative modulation of the absorption coefficient μ of a particular atom compared to the smooth background coefficient μ_s, normalized by the absorption coefficient μ_0, that would be observed for the free atom. The expression shows the summation over all atoms in the vicinity of the absorber. Each term corresponds to a scatterer at a certain absorber-scatterer distance and is represented by a sine wave.

The frequency of each sine wave depends on the distance (R_{as}) between the absorbing atom and the scattering atom. Actually, the photoelectron wave experiences a phase shift δ_1 for leaving the absorber, α_s for reflection from the scatterer, and δ_1 again for reentering the potential of the absorber. The phase also changes by $2kR_{as}$ as the photoelectron travels from the absorber to the scatterer and back. Thus the EXAFS wave oscillates as $\sin[2kR_{as} + 2\delta_1(k) + \alpha_s(k)]$. The absorber and scatterer phase shifts are often combined to form a total phase shift α_{as} as given in Equation (1). The total phase shift can be theoretically calculated or empirically measured in model compounds and thus R_{as} can be determined to high accuracy (usually ± 0.02 Å).[2]

The amplitude of each sine wave depends upon N_s, the number of equivalent scatterers, and $[f_s(\pi,k)]$, the backscattering amplitude of the scatterers. The shape of $[f_s(\pi,k)]$ is characteristic of the atomic number of the scatterer and is used to identify the type of scattering atoms. The amplitude is inversely related to the square of the distance R_{as} between the absorber and the scatterer and is modified by a Debye-Waller-like factor $\exp(-\sigma_{as}^2 k^2)$, where σ_{as}^2 represents the relative mean square displacement of the scatterer with respect to the absorber. Another exponential term, $\exp(-2\mu R_{as})$ is also included to account for the decrease of the amplitude due to inelastic scattering of the photoelectron. In general, the amplitude reveals the type of scattering atoms and predicts the number of scattering atoms to within about 25%.[2]

Structural determinations by EXAFS depend on the feasibility of resolving the data into individual waves corresponding to the different types of scatterers of the absorbing atom. A typical data reduction procedure is described diagrammatically in Figure 1, exemplified by a transmission spectrum collected on $Cu(Im)_4(ClO_4)_2$ (Im = imidazole). Unlike many other kinds of spectra, the EXAFS spectrum is not unambiguously determined. Figure 1a shows an experimentally acquired spectrum, μ_{total} including pre-edge, edge, and extended fine structure. The monotonically decreasing background is a result of residual absorption by the other elements present (and in some cases by the L edges of the same element), as well as the energy dependence of the detector system (spectrometer baseline). The background absorption μ_{back} is usually removed by fitting a polynomial[14] to the pre-edge portion of the spectrum μ_{total} and subtracted from the entire spectrum to give μ, as shown in Figure 1a. Subtraction of μ_s from μ, commonly termed "background subtraction", is done by

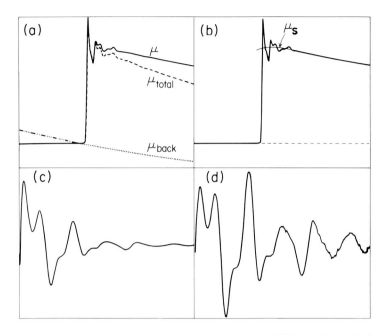

FIGURE 1.   EXAFS data processing, illustrated by $Cu(Im)_4(ClO_4)_2$. (a) Removal of residual absorption by other elements to isolate the copper component of the absorption. (b) Fitting a cubic spline μ_s to the oscillations and normalization of the oscillations with respect to the size of edge. (c) EXAFS as observed after background subtraction and normalization. (d) EXAFS weighted by k^2 to enhance visibility of oscillations at high k. Abscissa is in eV in (a) and (b), and in k in (c) and (d).

means of a cubic spline routine.[15] In this procedure, the absorption spectrum above the edge is divided into several regions, each of which is fit with a third-order polynomial, normally with a k^2 or k^3 weighting scheme. The individual polynomials are constrained to meet with equal slope at the spline points and combine to produce an overall curve for μ_s. To scale the amplitude of the EXAFS oscillations, it is necessary to normalize them to the absorption edge of that element. Usually this is done by normalizing the modulations with respect to μ_s, as shown in Figure 1b. Subtracting μ_s from μ and dividing by μ_0 yields the EXAFS, shown in Figure 1c. However, the EXAFS is normally weighted by k^2 or k^3 to enhance the visibility of the oscillations at high k, as shown in Figure 1d.

Once the normalized fine structure has been extracted from the absorption spectrum, it is Fourier transformed to reveal the major frequency components. Use of the Fourier transform to analyse EXAFS data was first introduced by Stern et al.[10a] They have shown that the major peaks in the transform correspond to the important absorber-scatterer distances, but shift to lower R by typically a few tenths of an angstrom. A Fourier transform is given in Figure 2A for $Cu(Im)_4(ClO_4)_2$. The major peak corresponds to the liganded nitrogen atoms from imidazoles and the minor peaks are from the two shells of atoms in the imidazole ligands. The major peak can be Fourier filtered and backtransformed into the k space (shown in Figure 2B) for curve-fitting analysis.

Structural information is usually obtained by Fourier filtering the transformed peak(s) back into k space and performing a curve-fitting analysis.[15,16] The essence of EXAFS curve-fitting analysis is to propose a parameterized function that will model the observed EXAFS, and then to adjust the structure-dependent parameters in this theoretical expression until the fit with the experimentally observed EXAFS is optimized by nonlinear least squares methods. The final value of the optimized parameters should then yield structural information about the molecule under study. The parameterized function can either be theoretically calculated

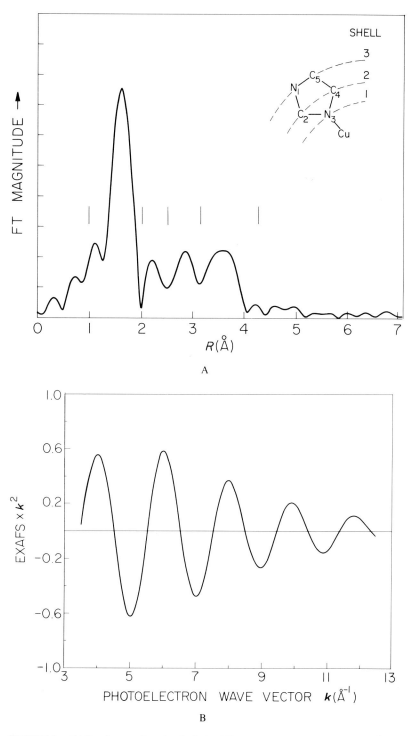

FIGURE 2. (A) Fourier transform for $Cu(Im)_4 (ClO_4)_2$ over the k range of 3.5 to 12.5 Å^{-1}, with k^3 weighting. The major peak is attributed to the scattering from the coordinated nitrogen atom (N_3) of the imidazole (shell 1). The first minor peak is the scattering from the oxygen atom of the anion. The second and the third minor peaks are the backscattering peaks due to shell 2 (C_2, C_4) and shell 3(N_1, C_5) of the imidazoles. The bars indicate the windows used for filtering out each peak. (Reprinted with permission from Co., M. S., Scott, R. A., and Hodgson, K. O., *J. Am. Chem. Soc.,* 103, 986, 1981. Copyright 1981, American Chemical Society.) (B) EXAFS observed after Fourier filtering the major peak.

or empirically obtained from structurally known model compounds. In our laboratory, we typically employ a parameterized function:

$$\chi(k) = \frac{c_0 e^{-c_1 k^2}}{k^{c_2}} \sin(a_0 + a_1 k + a_2/k) \qquad (2)$$

where the amplitude part is modeled by $c_0 e^{-c_1 k^2}/k^{c_2}$ and the phase part by $a_0 + a_1 k + a_2/k$ (in some cases, k^2 will be used instead of k^{-1}). Identification of the type of scatterer requires a reasonable fit with this expression with parameters derived from suitable models. The distance information is readily obtained from a_1 (since the absorber-scatterer distance is k-dependent) and the number of scatterers determined from c_0 (after adjustment for the inverse square distance dependence).

Recently, a more accurate representation of rigid coordinated ligands, known as the "group-fitting" technique,[17] was developed and proved useful. The group-fitting approach accounts for all scatterers from a rigid ligand, allowing more accurate modeling of the phase and amplitude behavior of outer shell atoms from the ligand. An account of the group-fitting technique involving imidazole ligands as applied to metalloproteins and metalloenzymes will be discussed in the following section.

B. Absorption Edges

The absorption edge is considerably more difficult to interpret than the EXAFS region. In Figure 3, the general features of the edge region are seen to be (1) the existence of one or more distinct absorption lines (broadened by the core-hole lifetime) in the -20 to 0 eV; (2) an abrupt rise in cross-section at the edge; and (3) a series of fairly narrow resonant peaks superimposed on the continuum cross-section in the range 0 to ≈ 50 eV. The absorption lines in the pre-edge region are interpreted as "bound $\rightarrow$ bound" transitions,[18,19] and for an isolated atom are traditionally labeled in terms of angular momentum quantum numbers, e.g., for a Cu(I) ion (d^{10} configuration) the lowest available transitions would be $1s \rightarrow 4s$ and $1s \rightarrow 4p$. For the case of Cu(II) ion (d^9 configuration) there would also be a $1s \rightarrow 3d$ transition. For isolated atoms, the strength of these transitions would be determined by the dipole selection rule, so that of the above three transitions, only the $1s \rightarrow 4p$ transition would be allowed on this basis.

For a metal in an environment of ligands, however, a correct analysis must be based on the symmetry group (point group) and bonding of the molecule in question. Thus for a molecule with inversion symmetry (e.g., a square-planar Cu-nitrogen complex) parity violating transitions such as $1s \rightarrow 4s$ and $1s \rightarrow 3d$ would still be forbidden to zero order in atomic displacements, while for a molecule without inversion symmetry (e.g., a tetrahedral Cu-nitrogen complex) the 4p and 4s orbitals can become hybridized to a molecular T_2 state so that both $1s \rightarrow 4s$ and $1s \rightarrow 4p$ atomic transitions have nonzero projection for dipole-allowed molecular transitions. Some of these features are illustrated in Figure 3 where it may be seen that the small "3d-like" absorption about 8 eV below the edge is absent for Cu(I) and present for Cu(II). Some authors have speculated that it is vibronic in origin.[18] A recent polarized X-ray absorption edge study of a single crystal of the creatinium salt of a square planar tetrachlorocuprate(II) complex has shown that the transition can be primarily accounted for with a quadrupole-allowed mechanism.[20]

The oxidation states of a metal ion have been shown to affect the edge position.[21] By simple electrostatics, it will require more energy to remove a core electron as the positive charge on the ion is increased. Thus the absolute position of the continuum edge is observed to increase by 3 eV or so for each successive one-electron oxidation of a given ion. The problem of quantitatively interpreting edge position is that the actual position of the edge is hard to determine unambiguously as it frequently overlaps bound $\rightarrow$ bound and resonance

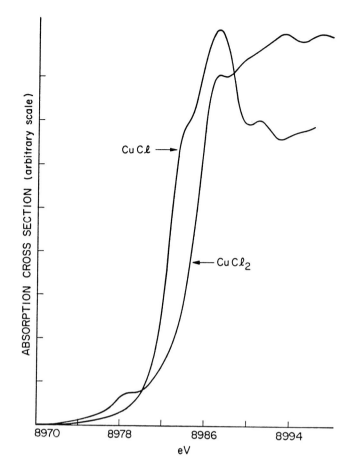

FIGURE 3. Cu K-absorption edges for CuCl and CuCl₂.

features of the spectrum. Nevertheless, a study of a series of molybdenum complexes showed that the edge position correlates reasonably well with other measures of ionicity.[21] In the following section, the edge positions are frequently used as indicators for oxidation states of copper in proteins and enzymes.

III. APPLICATIONS TO COPPER PROTEINS

A. Hemocyanin

Hemocyanin[22] (Hc) is a copper-containing protein which functions as a dioxygen carrier in molluscs and arthropods. The smallest functional units, with a relative molecular mass of approximately 50,000 for molluscs and 75,000 for arthropods, contain two copper atoms and bind one molecule of dioxygen reversibly. The two copper ions in OxyHc are EPR nondetectable, present as a type-3 binuclear site. Many studies have been aimed at elucidating the structure of the active site. Recent resonance Raman studies[23] have established that the dioxygen is bound in the form of peroxide in OxyHc and suggested that imidazole ligands are coordinated to the copper site,[24,25] consistent with earlier spectroscopic, acid-base titration, and photooxidation studies. However, these studies fail to establish an acceptable number of coordinating imidazoles. Recent EPR studies[26] on several derivatives of Hc have provided a better view of the active-site structure. However, a structural model for the active site is still not completely clear. As a great deal of kinetic data have been collected on Hc over the years,[27] a good structural model for the active site could definitely provide insight into

Table 1
X-RAY ABSORPTION EDGE DATA FOR
Cu(I) COMPOUNDS

Compounds	Transition energies (eV) "1s → 4s"	Ref.
Cu₂O	8,983.5	15
CuCN	8,982.9	15, 19
CuSCN	8,983.6	15, 19
CuCl	8,984.6	15, 19
CuI	8,984.0	19
[Cu(HB(pz)₃)]₂[a]	8,984.8	19
CuCO(HB(pz)₃)	8,985.2	19
Cu(Me₂PIMI)CO[b]	8,982.2	19
Average	8,983.8 ± 1.0	

[a]   HB(pz)₃ = hydrotris(1-pyrazolyl) borate.
[b]   Me₂PIMI = 1,3-bis[2-(4-methylpyridyl)imino] isoindoline.

the mechanism of oxygen binding and perhaps explain the observed cooperativity of oxygen binding.

X-ray absorption spectroscopic studies (absorption edge and EXAFS) have been performed on several states of Hc[15,28-31] and have provided valuable information on the structure of the dimeric active site. For determination of the copper oxidation state in Hc, absorption edge spectra were compared to a series of Cu(I) and Cu(II) models. The K-absorption edge spectral data of Cu(I) and Cu(II) models are presented in Tables 1 and 2, respectively. The most significant feature distinguishing the two oxidation states is the presence of the 1s → 3d transition at 8,979 eV, never seen for the Cu(I) state. However, the absence of this transition does not necessarily rule out the presence of Cu(II) since more covalent Cu(II) compounds sometimes give very weak 1s → 3d transitions, preventing them from being distinguished from the noise. Another useful feature in recognizing the presence of Cu(I) is a distinct transition at approximately 8,984 eV, with a strength about half that of the principal peak. The corresponding transition, labeled "1s → 4s" in early literature, usually appears as a "bump" on the edge at approximately 8,987 eV for Cu(II).

K-absorption edge spectra[31] for four states (Oxy, Deoxy, Dimer, and half-Apo) of Hc from *Megathura crenulata* are presented in Figure 4. The OxyHc and Dimer Hc spectra both show a weak pre-edge transition at 8,979 eV and a "bump" at 8,987 eV, consistent with the assignments of 1s → 3d and "1s → 4s" for Cu(II). The unique Cu(I) transition is not observed. This experiment hence suggests that only Cu(II) is present in OxyHc and Dimer Hc. Examination of the copper absorption edge spectra of DeoxyHc and half-ApoHc reveals a distinct peak, about half the strength of the principal peak, at approximately 8,983 eV. These observations, together with the absence of the unique 1s → 3d transition, are consistent with only Cu(I) being present in DeoxyHc and half-ApoHc. Absorption edge spectra for OxyHc and DeoxyHc from *Busycon canaliculatum*[28] and *Helix pomatia*[31] show similar behavior.

Copper EXAFS spectra on Hc's were collected using synchrotron radiation at the Stanford Synchrotron Radiation Laboratory. Since each EXAFS spectrum records the sum of EXAFS as seen by each absorbing atom and then normalized per absorbing atom, all results quoted herein are average figures per copper atom. Table 3 shows the curve-fitting results of the first Fourier filtered peak of EXAFS on OxyHc and DeoxyHc. The most significant result is the reduction of coordination number as the protein is deoxygenated. Hc's from both *M.*

Table 2
X-RAY ABSORPTION EDGE DATA FOR Cu(II)
COMPOUNDS

Compounds	Transition energies (eV)		Ref.
	"1s → 3d"	"1s → 4s"	
CuS	8,979.3	8,987.5	15
CuO	8,979.4	8,987.5	15
CuCl$_2$·2H$_2$O	8,978.8	8,988.2	19
Cu(MNT)$_2^{2-}$ [a]	8,980.5	8,985.5	15
Cu(diethyldithiocarbamate)$_2$	8,979.5	8,985.2	15
Cu(GGH)[b]	8,980.1	8,986.8	15
Cu(Im)$_4$I$_2$	8,979.7	8,987.0	15
Cu(C$_2$H$_5$COO)$_2$	8,978.5	8,987.0	15
Cu(CH$_3$COO)$_2$·H$_2$O	8,979.0	8,987.2	15
Cu(NH$_3$)$_4$SO$_4$·H$_2$O	8,979.0	8,987.3	15
Cu(PAA)$_2$en[c]	8,979.9	8,986.5	19
K$_2$Cu(bi)$_2$[d]	8,979.8	8,986.2	19
CuCl(HB(pz)$_3$)	8,979.1	8,988.1	19
Cu(tren)Im(PF$_6$)$_2$	8,979.2	8,987.0	19
Cu(dien)(NO$_3$)$_2$	8,979.4	8,987.6	19
Average	8,979.4 ± 0.5	8,987.0 ± 0.9	

[a] MNT = maleonitrile dithiolate.
[b] GGH = glycylglycyl-O-methyl-L-histidine.
[c] (PAA)$_2$en = the Schiff base complex of ethylenediamine with 1-phenyl-1,3,5-hexanetrione.
[d] bi = biuret.

crenulata and *B. canaliculatum* indicate that the coordination number changes from about 4 to 2 as the protein is deoxygenated.

EXAFS Fourier transform peaks beyond the first coordination shell were also analysed. Fourier transforms of copper EXAFS were performed and are presented in Figure 5 for Oxy-,Deoxy-, Dimer, and half-ApoHc from *M. crenulata*. An unusually large peak at 3.3 Å (uncorrected for phase shift, usually 0.3 to 0.6 Å is added to obtain the actual distance) is observed in the transform of OxyHc (both *M. crenulata*[29] and *B. canaliculatum*[28]). This peak has been interpreted to represent a copper backscatterer at about 3.6 Å. Curve-fitting results give a Cu-Cu distance of 3.67 Å for *B. canaliculatum*[28] and 3.55 Å for *M. crenulata*.[29] Such a short metal-metal distance provides a pathway (probably through a bridging atom) for the antiferromagnetic coupling that leads to the diamagnetism in OxyHc. The distance discrepancy could result from an intrinsic difference between species or an inconsistency between data processing and analysis.

Despite the close agreement of results between OxyHc from *M. crenulata* and *B. canaliculatum*, the DeoxyHc from the two species shows conflicting results. Data on DeoxyHc from *B. canaliculatum*[28] were interpreted to observe a copper backscatterer at 3.4 Å, while data on DeoxyHc from *M. crenulata*[30] showed no evidence of a copper backscatterer within 4 Å. At first, the lack of observed copper backscatterer in *M. crenulata* was suspected to be the result of a larger Debye-Waller factor (caused by the uncorrelated movements between the two unbridged copper atoms) since the original data were collected at room temperature. Low temperature (−70°C) EXAFS data were recorded[30] for *M. crenulata* DeoxyHc (low temperature should reduce the Debye-Waller factor and enhance the possibility of detecting

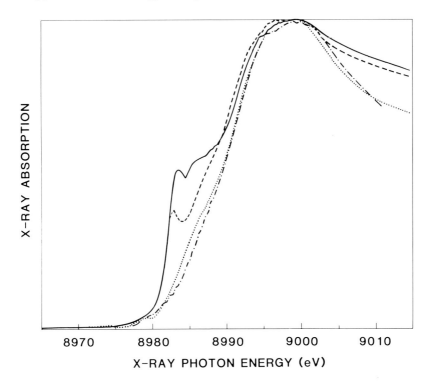

FIGURE 4.   Cu K-absorption edge spectra for *Megathura crenulata* hemocyanin: Oxy
(-·-·-), Deoxy (—), Dimer (. . .), and half-Apo (---). From comparison with Cu(I) and
Cu(II) models, it has been concluded that the copper site is oxidized Cu(II) in Oxy- and
Dimer Hc, and reduced Cu(I) in Deoxy- and half-ApoHc.

Table 3
CURVE-FITTING RESULTS ON FIRST
FOURIER-FILTERED PEAKS OF
HEMOCYANINS[a]

Protein	Cu-N (or-O)[b]		
	Distance (Å)	Number	Ref.
Megathura crenulata			
OxyHc	1.98	4.0	29
DeoxyHc	1.95	2.0	30
Busycon canaliculatum			
OxyHc	1.96	3.2—3.4	28
DeoxyHc	1.95	1.4—2.0	28
Helix pomatia			
OxyHc (α-component)	1.98	4.0	29
OxyHc (β$_c$-component)	1.98	4.0	29

[a]   Since oxygen and nitrogen atoms are not readily distinguish-
able in a curve-fitting analysis, only a shell of nitrogen is
fitted to the filtered EXAFS.

[b]   Errors are estimated to be about ±0.02 Å in distances and
about 25% in coordination numbers.

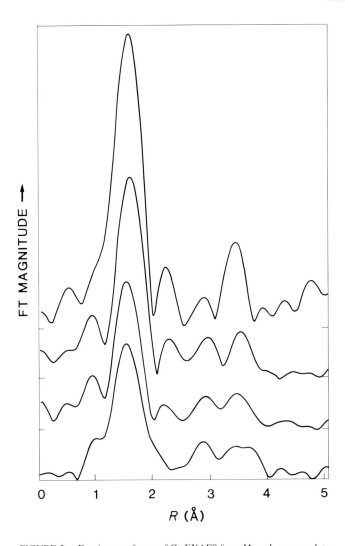

FIGURE 5.   Fourier transforms of Cu EXAFS from *Megathura crenulata*
hemocyanin. The transforms are taken over a k range of 4 to 12 Å, $^{-1}$ with
k^3 weighting. From top to bottom: Oxy-, Dimer, Deoxy-, half-ApoHc.
The relative size of the amplitude of the major transform peaks corresponds
to change in coordination numbers, with amplitude in OxyHc approximately twice that of DeoxyHc. A relatively large peak at 3.3 Å is observed
in OxyHc, but absent in other derivatives.

a Cu backscatterer if it were present), but still no evidence of a copper backscatterer was
observed. Therefore, it is postulated that in DeoxyHc from *M. crenulata* the two copper
atoms move away from each other as the bridging ligand is broken or are at minimum held
in a much less structurally rigid environment. It should be observed, however, that the lower
temperature used for *B. canaliculatum* Hc (77 K) still leaves some uncertainty as to whether
a discrepancy exists. Data should be collected under the same conditions for both proteins
in order to make a valid comparison.

A comparative investigation of the copper site of Hc from different derivatives of the
proteins has been performed on *M. crenulata*.[31] Four different derivatives of the protein
were examined, using the group-fitting technique.[17] The group-fitting approach allows more
accurate modeling of the phase and amplitude behavior of second- and third-shell atoms of
the imidazole group, allowing other atoms of interest to be more easily identified. When

Table 4

**GROUP-FITTING RESULTS FOR *MEGATHURA CRENULATA*
HEMOCYANINS[31]**

	Cu-Im		Cu-O		Cu-Cu	
Protein	Distance (Å)	Number	Distance (Å)	Number	Distance (Å)	Number
OxyHc	2.01	2.0	1.92	2.0	3.55	1.1
DeoxyHc	1.95	1.6	1.95	0.5	—	—
Dimer Hc[a]	2.01	1.6	2.05	1.4	—	—
Half-ApoHc	1.94	2.1	2.05	0.8	—	—

Note: Errors are estimated to be about ± 0.02 Å in distances and about 25% in coordination numbers.

[a]   EPR-detectable MetHc.

the imidazole group-fitting technique is applied to Hc, all the atoms of the imidazole ring can be described as a single group. The distance from the copper atom to the group and the number of such groups constitute the minimal two variables necessary to describe the imidazoles in the copper environment. Results on Hc were obtained fitting the EXAFS data with imidazole groups and other appropriate atoms to provide a reasonable fit. Table 4 presents the group-fitting results on the four derivatives of Hc. In each case, an average of two imidazoles (within about $\pm 25\%$ experimental uncertainty) were found coordinated to each copper atom. This is the first precise determination of the number of coordinated imidazoles.

As mentioned above, the use of the imidazole group-fitting technique enhances the possibility of identifying other atoms of interest. Table 4 also lists the types and the numbers of atoms other than imidazoles identified in the curve-fitting procedure. Two interesting points are observed. First, evidence for a copper backscatterer within 4 Å was observed only in OxyHc. The lack of evidence for a copper backscatterer in half-ApoHc and in Dimer Hc is not unexpected since only one copper atom in each active site is present in the former[32] and the two copper atoms in the latter have been previously calculated to be approximately 6 Å apart,[33] beyond the detection limit of the EXAFS technique. The lack of evidence of a copper backscatterer in DeoxyHc is interesting since it implies a substantial rearrangement in the active-site structure during the oxygenation cycle. In fact, such conformational changes are consistent with numerous observations on oxygen binding, such as the dissociation of subunits,[34,35] changes in the 250-nm band in circular dichroism,[34,36] and a slow relaxation time in kinetic studies.[27] Second, the change in coordination number in the Oxy-Deoxy cycle is significant. A reduction of coordination number from 4 to 2 from OxyHc to DeoxyHc is also consistent with the suggestion of substantial structural rearrangement. Such rearrangement probably involves removal of the bound dioxygen and breaking of the bridging ligand bonds. The presence of two ligands in Cu(I) is not uncommon among Cu(I) compounds.[37] The unsaturation of the site in DeoxyHc may indeed serve the purpose of binding dioxygen. In half-ApoHc, an additional low Z atom was found coordinated to the Cu(I) site.[31] The additional ligand is probably cyanide ion since one cyanide ion has been shown to be coordinated to a copper atom in Hc after dialysis of Hc at low cyanide ion concentrations.[38]

With the recent work by techniques such as resonance Raman, EPR, and XAS, the structure of the dimeric copper active site has become better understood. It is hopeful that this knowledge of the active-site structure can advance our understanding of the mechanism of dioxygen binding, and finally of the observed cooperativity.

B. Blue Copper Proteins

Proteins that contain the "blue" (or type-1) copper site[39] have long offered a challenge to chemists to explain the unique spectroscopic and redox properties. Some of the basic aspects of the structure of such blue sites have already become clear, as revealed by interpretation of several types of spectroscopic data. The ligand-field spectrum implies a distorted tetrahedral coordination of the copper.[40] A thiolate sulfur from cysteine was proposed as one of the ligands.[40] Two imidazole nitrogens from histidines were also implicated as ligands, from NMR studies.[41,42] The fourth ligand of copper is less certain. The X-ray protein structures of plastocyanin (Pc)[43] and azurin(Az)[44] have been resolved to 2.7 and 3.0 Å resolution, respectively. The copper in each protein is found to be coordinated by cysteine, two histidine imidazoles, and methionine. However, such crystallographic results do not provide evidence on subtle changes in bond distances and coordination geometry upon electron transfer. The X-ray absorption technique, which provides the advantage of elucidating the *local* structure of an absorbing atom to a high degree of accuracy, was applied to three blue copper proteins by Tullius and Hodgson[45-47] to complement the information not available from other techniques.

X-ray absorption edge spectra of Pc, Az, and stellacyanin (St) were first examined. As discussed for Hc, it is possible to distinguish Cu(I) from Cu(II) by inspection of absorption edge spectra. In their oxidized state, all three blue copper proteins show a weak pre-edge peak at 8,979 eV and a "bump" on the edge at 8,988 eV, typical of the "1s → 3d" and "1s → 4s" transitions for Cu(II). Absorption edge spectra on reduced Pc and St were also recorded. In these spectra, the unique pre-edge peak vanished, and the "bump" on the edge shifted to 8,984 eV, typical of the "1s → 4s" transition for Cu(I). The absorption edge experiments hence reflect the obvious fact that the electron-transfer properties of the blue copper proteins are mediated at the copper site. The addition of an electron to the oxidized proteins reduces the copper atom.

To obtain a more explicit picture of the copper active site during the electron-transfer process, copper EXAFS spectra of the oxidized proteins and the reduced state of two of the proteins were recorded and compared.[45-47] The EXAFS spectra of Pc are given in Figure 6. Distinct differences in amplitude and phase are observed. The difference in phase indicates a change of the copper-ligand bond lengths upon reduction. For quantitative comparisons, the first shell in each Fourier transform was filtered and curve fit using a least squares procedure. Results are given in Table 5 for oxidized Pc, Az, and St, as well as reduced Pc and St. An interesting observation is that a sulfur ligand is coordinated at the copper site at a very short distance (2.11 Å) which is lengthened to 2.22 Å upon reduction.[45-47] There are also two to three nitrogen atoms coordinated to copper at about 1.98 Å which is slightly lengthened (to 2.05 Å) upon reduction. Calculation of a fractional number of sulfur atoms in the reduced proteins is probably caused by a larger Debye-Waller factor due to a weaker Cu-S bond as a result of change in copper oxidation state.

The transition state through which the reactants pass during the electron-transfer process must of necessity possess metal-ligand bond lengths intermediate between those found in the members of the redox couple. Since the greatest change in ligand distance for blue copper is 0.1 Å, only a modest expansion of the bonds (around 0.05 Å) is required to reach this transition state. A reasonable description of the details of electron transfer in blue copper proteins might then involve little angular rearrangement of the ligands, with only small changes in metal-ligand distances being necessary to accommodate the change in copper oxidation state.[47] The long-hypothesized "tetrahedral" geometry of the oxidized blue copper site, as confirmed by X-ray crystallography for Pc[43] and Az,[44] is consistent with these postulations.

Imidazole group-fitting techniques were again applied to the blue copper proteins.[48] The group-fitting technique provides a better fit to the EXAFS. The results show that, within

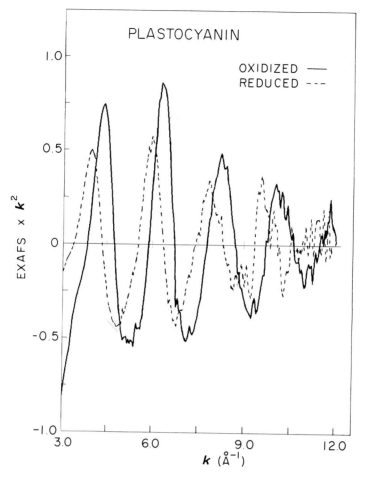

FIGURE 6.   Change in the Cu EXAFS spectrum upon reduction of the copper atom of plastocyanin. A clear shift in phase and amplitude of the EXAFS is seen upon reduction, indicating a change in average copper-ligand bond distance.

Table 5
TWO-SHELL FITS TO FIRST FOURIER-FILTERED PEAKS OF EXAFS ON BLUE COPPER PROTEINS[46,47]

	Cu-N		Cu-S	
Protein	**Distance (Å)**	**Number**	**Distance (Å)**	**Number**
Plastocyanin				
Oxidized	1.97	2.3	2.11	1.1
Reduced	2.05	2.2	2.22	0.8
Stellacyanin				
Oxidized	2.04	2.9	2.11	1.1
Reduced	2.07	2.7	2.25	0.7
Azurin				
Oxidized	1.97	3.3	2.11	0.9

Note: Errors are estimated to be about ±0.02 Å in distances and about 25% in coordination numbers.

the experimental uncertainty of ± 0.02 Å for the distance prediction and approximately 25% for coordination number estimation, each copper of the oxidized protein is coordinated to one sulfur at 2.11 Å and to two imidazoles at 1.98 Å. Upon reduction, the Cu-S distance and the Cu-N distance from imidazoles are lengthened to 2.21 Å and 2.05 Å, respectively. The fourth ligand of the copper atom, the thioether sulfur from methionine as revealed from X-ray crystallographic work on Pc and Az, was not detectable from the EXAFS.[49] The lack of evidence for this long-distance sulfur scatterer could be explained by a large Debye-Waller factor as a result of a much weaker copper thioether bond.

C. Cytochrome *c* Oxidase

Cytochrome *c* oxidase[50] is the membrane-bound enzyme involved in the terminal step of mitochondrial respiration. The enzyme plays an extremely important role in electron transport (catalysing the reduction of dioxygen to water) and in bioenergetics (as a proton pump). It is a relatively large enzyme (M_r approximately 150,000) containing two heme irons and two copper ions. The two iron sites are denoted as cytochrome *a* and cytochrome a_3, and the copper sites denoted as Cu_A and Cu_B. Cu_B and the heme iron of cytochrome a_3, which are both EPR silent, are believed to be antiferromagnetically coupled. X-ray absorption studies[19,51-55] have been reported on both copper and iron, but only the copper work will be reviewed here.

Early X-ray absorption work on the coppers of cytochrome *c* oxidase was limited to studies of the absorption edges.[19,51,52] The earliest experiments were performed by Hu et al.,[19] who concluded from the shape and position of the edge, that one copper of the resting enzyme was actually in the reduced form. Upon reduction of the protein with dithionite, both coppers were found to be Cu(I). Hu et al.[51] also investigated the effects of cyanide treatment on the iron and copper absorption edges. They found that CN^- has no effect on the copper edge spectrum of the oxidized protein and that both copper atoms of the CN^- complex in the presence of excess dithionite are reduced. Later work by Powers et al.[52] disputed the previous Cu(I) assignment in the resting state enzyme and attributed some findings of the work by Hu et al. to adventitious copper and to photoreduction (which was subsequently refuted by Brudvig et al.[56]). They compared the copper edge spectra in the oxidized and reduced states of cytochrome *c* oxidase with those of St and came to the conclusion that the EPR-silent Cu_B is a ''blue'' type copper similar to St. However, the designation of Cu_B as ''blue'' was subsequently challenged by Brudvig and Chan,[57] and later by Beinert et al.[58] on the basis of chemical and other spectroscopic experiments. Hence, we must consider the results of these experiments as still leaving ambiguous the nature of the Cu_B site.

EXAFS work on cytochrome *c* oxidase has been continued by Chance and co-workers[54] and Scott et al.[53,55] Recent publications by the two groups have revealed more detail about the active-site structure of this large enzyme complex. Scott et al.[53] reported the first direct evidence for sulfur ligation to copper. The EXAFS curve-fitting results on the resting state enzyme show that there are 1 to 1.5 sulfurs per copper at an average Cu-S distance of 2.27 Å and 2 nitrogens (or oxygens) per copper at an average Cu-N (-O) distance of 1.97 Å. Later work by Scott[55] was extended to include copper EXAFS on the fully reduced and mixed valence-formate derivatives. The comparison of filtered EXAFS curves (Figure 7) indicates that the EXAFS derived from the two major transform peaks looks very similar for the mixed valence-formate and the fully reduced enzyme. Both of these data sets look substantially different from the resting state EXAFS. The fits in Table 6 suppport these observations. The average Cu-S distance changes from 2.26 Å in the resting state to 2.32 Å in the fully reduced and mixed valence-formate states. The average Cu-N (or -O) distance does not change significantly upon reduction. The differences observed in coordination numbers in Table 6 are accounted for as a result of Debye-Waller factor changes upon reduction of Cu(II) to Cu(I).

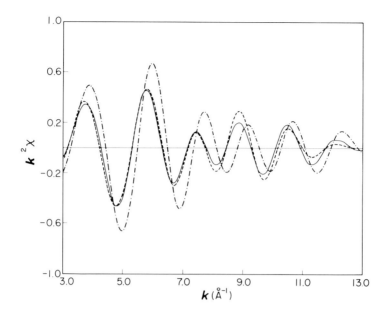

FIGURE 7. Comparison of filtered EXAFS of cytochrome *c* oxidase derivatives: resting state (-··-··), fully reduced (---), and mixed valence-formate (—). The average phase shift of the resting state EXAFS is significantly different from the phase shift of the fully reduced and mixed valence-formate EXAFS. This is accounted for by a significant change in the average Cu-S distance (see Table 6).

Table 6
TWO-SHELL FITS TO FILTERED EXAFS ON
CYTOCHROME *c* OXIDASE[55]

Protein	Cu-N (or -O)		Cu-S	
	Distance (Å)	No.	Distance (Å)	No.
Resting state	1.95	1.8	2.26	1.7
Fully reduced	1.95	1.2	2.32	1.4
Mixed valence-formate	1.96	1.3	2.32	1.4

Note: Errors are estimated to be ±0.02 Å in distances and about 25% in coordination numbers.

Taking the common belief that Cu_B remains oxidized and Cu_A reduced in the mixed valence-formate state,[58] Scott[55] attributed the lengthening of the average Cu-S distance (by 0.06 Å) to the reduction of Cu_A. Upon reduction of Cu_B, no further change is observed in the average Cu-S distance. Hence, he assigned most (if not all) of the redox-sensitive sulfur ligands to Cu_A, and postulated the following assignment of copper ligands: Cu_A has two (or three) sulfur ligands with an average Cu-S distance of 2.26 Å which changes upon reduction to 2.32 Å. Cu_A also has two (or one) nitrogen (or oxygen) ligands. Cu_B has mostly nitrogen (or oxygen) ligands with a coordination sphere which does not change significantly upon reduction.

Chance and co-workers[54] recently reported their EXAFS work on both copper and iron, of which only copper work will be reviewed here. Their work included four different redox states of the enzyme (fully oxidized, CO reduced, mixed valence-formate, and mixed valence-CO). Using St as a model for Cu_B in both oxidized and reduced states, they examined the first-shell EXAFS in a manner analogous to that used for the edge data. They concluded that, for the oxidized Cu_B, the first shell is coordinated to two nitrogens and one sulfur at an average distance of 1.97 and 2.18 Å, respectively. A second-shell contribution appeared to contain most likely a sulfur at 2.82 Å, but nitrogen or carbon at 2.95 Å is also feasible. In the reduced state, one nitrogen at 2.07 Å and two sulfurs at an average distance of 2.25 Å were obtained. The second nitrogen was not observed. For Cu_A, the oxidized copper is coordinated to one nitrogen (or oxygen) at 1.97 Å and three sulfurs at an average distance of 2.27 Å. Upon reduction, the corresponding distances change to 2.00 and 2.35 Å without a change in coordination numbers. However, they also noted that fitting with a combination of two nitrogens and two sulfurs at the same average distances was also feasible within the error. Analysing the higher-shell contribution from both copper and iron EXAFS, they also reported evidence of a Cu-Fe interaction at 3.75 Å. With the copper and iron EXAFS analyses, they presented a model which is described in Reference 54. This model affords the basis for an oxygen reduction mechanism involving oxy- and peroxy-intermediates. However, the acceptability of the model has to await resolution of the controversial assignment of Cu_B as a ''blue-like'' copper.

D. Laccase and Superoxide Dismutase

Laccase[59] is a copper-containing oxidase which, like cytochrome *c* oxidase, can reduce dioxygen to water. It contains four copper atoms per molecule. In the oxidized enzyme, two of these coppers (type 1 and type 2) are detectable by EPR. The two EPR-nondetectable copper atoms (type 3) are believed to constitute an antiferromagnetically coupled binuclear unit. All copper ions in laccase become EPR detectable upon anaerobic denaturation and appear to be in oxidation state equivalent to four copper(II) ions.

LuBien et al.[60] reported the copper absorption edge spectra for three states (native, type 2-depleted [T2D], and T2D treated with excess H_2O_2) of *Rhus vernicifera*. The T2D derivative is invaluable in understanding the role of the type-2 copper in the mechanism and the active-site structure of the enzyme. Figure 8 presents copper absorption edge spectra for the three laccase derivatives. Comparison of the spectra for native laccase with a Cu(II)-imidazole complex shows that the copper atoms are all oxidized. Native laccase clearly exhibits the pre-edge transition at 8,979 eV and no strong transition in the 8,983 eV region. In the case of T2D laccase, the edge is strikingly different from native laccase and closely resembles a Cu(I)-imidazole complex. The presence of the strong transition at 8,983 eV is definite evidence for the presence of a significant amount of Cu(I). Upon treatment of T2D laccase with excess H_2O_2, this strong transition disappears, giving rise to an absorption edge which is virtually superimposable upon the edge of native laccase. These experiments clearly show that the type-3 copper is obtained in a reduced state after the preparation of T2D laccase, and is stable to oxidation by oxygen. Treatment with excess H_2O_2 results in the copper atoms being oxidized.

Superoxide dismutases[61] in eukaryotes contain copper and zinc in equal amounts. X-ray crystallographic analysis[62] has shown that bovine superoxide dismutase (SOD) is a dimeric polypeptide with two active sites, each site containing a copper atom surrounded by four imidazole ligands, one of which is also ligated by a zinc atom. It is not known whether both metals change their valence during catalysis. To address this problem, Blumberg et al.[63] examined absorption edge spectra for Cu and Zn in oxidized and reduced SOD.

Comparing with oxidized and reduced copper imidazole complexes, the copper atom in SOD is found in the Cu(II) form in the oxidized enzyme and in the Cu(I) form in the reduced enzyme, while the zinc edge remained virtually unchanged upon reduction. Therefore, they concluded that the added electron affects the copper site almost exclusively in the catalytic process.

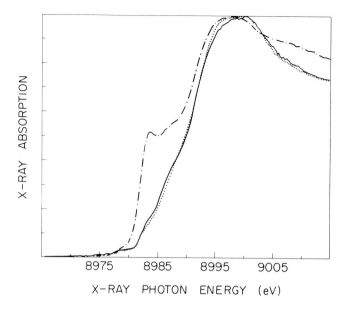

FIGURE 8. X-ray absorption edges for native and T2D laccase derivatives: native (—), T2D (-·-·), and T2D + excess H_2O_2 (. . .). The presence of the strong transition at 8,983 eV for T2D derivative is definite evidence for the presence of a significant amount of Cu(I). (Reprinted with permission from LuBien, C. D., Winkler, M. E., Thamann, T. J., Scott, R. A., Co, M. S., Hodgson, K. O., and Solomon, E. I., *J. Am. Chem. Soc.*, 103, 7014, 1981. Copyright 1981, American Chemical Society.)

IV. CONCLUSIONS

The information obtained from XAS studies — the oxidation state and coordination geometry of the metal atom, the type, number, and distance of ligands coordinated to the metal atom — is a large part of what bioinorganic chemists seek in understanding the structure/function relationship of proteins and enzymes containing metal atoms. Furthermore, being sensitive to short-range order in atomic arrangements rather than long-range crystalline order, XAS studies can focus on the local environment of a specific absorbing atom. This means that only a small number of variables need be defined to obtain the local structure, whereas the accuracy of diffraction results depends on refining all of the large number of atomic positions in the crystal. The relatively easy sample preparation is another advantage. The material can be examined in any form or state, ranging from crystalline and amorphous solids, liquids, and solutions, to gaseous states. The selection rules are such that as an absorption edge and associated fine structure always exist, paramagnetic or isotopically enriched samples are unnecessary. Finally, X-ray absorption spectra can be collected in minutes or hours, and once the data are obtained, the results may be available within days or weeks.

There is a severe deficiency of EXAFS, however, in that it does not provide the three-dimensional structure. Only a one-dimensional radial distribution function about the metal atom is available. It is hoped that better understanding of the edge and near-edge spectra could complement information about the coordination geometry of the metal and help provide angular structural information about the metal site. Another drawback about the technique is that when more than one absorbing atom is present, XAS gives only an average spectrum of all the absorbing atoms. However, coupled with other chemical or spectroscopic information, it is possible to use XAS to investigate a particular absorbing atoms for valuable information. The recent work on the copper K-absorption edge of laccase[60] is a good example of this. It is anticipated that the technique can provide useful information on the more complicated multicopper-containing proteins in the near future.

ACKNOWLEDGMENTS

We thank Dr. Tom Tullius and Dr. Robert A. Scott for Figures 6 and 7 and for discussion of results on blue copper proteins and on cytochrome *c* oxidase before publication. This work was supported by the National Science Foundation through Grant PCM-82-08115. Synchrotron radiation time for the XAS work was provided by the Stanford Synchrotron Radiation Laboratory, supported by the National Science Foundation Grant DMR 77-27489 in cooperation with the Stanford Linear Accelerator Center and the U.S. Department of Energy, as well as the National Institutes of Health RR-01209.

REFERENCES

1. **Doniach, S., Eisenberger, P., and Hodgson, K. O.,** X-ray absorption spectroscopy of biological molecules, in *Synchrotron Radiation Research,* Winick, H. and Doniach, S., Eds., Plenum Press, New York, 1980, chap. 13.
2. **Cramer, S. P. and Hodgson, K. O.,** X-ray absorption spectroscopy: a new structural method and its applications to bioinorganic chemistry, *Prog. Inorg. Chem.,* 25, 1, 1979.
3. **Teo, B.-K.,** Chemical applications of extended X-ray absorption fine structure (EXAFS) spectroscopy, *Acc. Chem. Res.,* 13, 412, 1980.
4. **Eisenberger, P. and Kincaid, B. M.,** EXAFS: new horizons in structure determinations, *Science,* 200, 1441, 1978.
5. **Teo, B.-K.,** Extended X-ray absorption fine structure (EXAFS) spectroscopy. Techniques and applications, in *EXAFS Spectroscopy,* Teo, B.-K. and Joy, D. C., Eds., Plenum Press, New York, 1981, 13.
6. **Kronig, R. de L.,** Zur Theorie der Feinstruktur in den Röntgenabsorptionsspektren, *Z. Phys.,* 70, 317, 1931; Zur Theorie der Feinstruktur in den Röntgenabsorptionsspektren. II, *Z. Phys.,* 75, 191, 1932.
7. **Sayers, D. E., Lytle, F. W., and Stern, E. A.,** Point scattering theory of X-ray K-absorption fine structure, *Adv. X-ray Anal.,* 13, 248, 1970.
8. **Winick, H. and Bienenstock, A.,** Synchrotron radiation research, *Annu. Rev. Nucl. Part. Sci.,* 28, 33, 1978.
9. **Winick, H.,** Synchrotron radiation sources, research facilities, and instrumentation, in *Synchrotron Radiation Research,* Winick, H. and Doniach, S., Eds., Plenum Press, New York, 1980, chap. 3.
10. **Stern, E. A.,** Theory of the extended X-ray absorption fine structure, *Phys. Rev. B.,* 10, 3027, 1974.
10a. **Stern, E. A., Sayers, D. E., and Lytle, F. W.,** Extended X-ray absorption fine structure technique. III. Determination of physical parameters, *Phys. Rev. B,* 11, 4836, 1975.
11. **Ashley, C. A. and Doniach, S.,** Theory of extended X-ray absorption fine structure (EXAFS) in crystalline solids, *Phys. Rev. B.,* 11, 1279, 1975.
12. **Lee, P. A. and Pendry, J. B.,** Theory of the extended X-ray absorption fine structure, *Phys. Rev. B,* 11, 2795, 1975.
13. **Brown, G. S. and Doniach, S.,** The principles of X-ray absorption spectroscopy, in *Synchrotron Radiation Research,* Winick, H. and Doniach, S., Eds., Plenum Press, New York, 1980, chap. 10.
14. **Lytle, F. W., Sayers, D. E., and Stern, E. A.,** Extended X-ray absorption fine structure technique. II. Experimental practice and selected results, *Phys. Rev. B.,* 11, 4825, 1975.
15. **Eccles, T. K.,** *X-Ray Absorption Studies of Hemocyanin,* Ph.D. thesis, Stanford University, Stanford, Calif., 1977.
16. **Cramer, S. P.,** Structure determination by X-ray absorption spectroscopy, in *Synchrotron Radiation Applied to Biophysical and Biochemical Research,* Castellani, A. and Quercia, I. F., Eds., Plenum Press, New York, 1979, 291.
17. **Co, M. S., Scott, R. A., and Hodgson, K. O.,** Metalloprotein EXAFS. A group fitting procedure for imidazole ligands, *J. Am. Chem. Soc.,* 103, 986, 1981.

18. **Shulman, R. G., Yafet, Y., Eisenberger, P., and Blumberg, W. E.,** Observation and interpretation of X-ray absorption edges in iron compounds and proteins, *Proc. Natl. Acad. Sci. U.S.A.,* 73, 1384, 1976.

19. **Hu, V. W., Chan, S. I., and Brown, G. S.,** X-ray absorption edge studies on oxidized and reduced cytochrome *c* oxidase, *Proc. Natl. Acad. Sci. U.S.A.,* 74, 3821, 1977.

20. **Hahn, J. E., Scott, R. A., Hodgson, K. O., Doniach, S., Desjardins, S. R., and Solomon, E. I.,** Observation of an electric quadrupole transition in the X-ray absorption spectrum of a Cu(II) complex, *Chem. Phys. Lett.,* 88, 595, 1982.

21. **Cramer, S. P., Eccles, T. K., Kutzler, F. W., Hodgson, K. O., and Mortenson, L. E.,** Molybdenum X-ray absorption edge spectra. The chemical state of molybdenum in nitrogenase, *J. Am. Chem. Soc.,* 98, 1287, 1976.

22. **Van Holde, K. E. and Miller, K. I.,** Haemocyanins, *Q. Rev. Biophys.,* 15, 1, 1982.

23. **Thamann, T. J., Loehr, J. S., and Loehr, T. M.,** Resonance Raman study of oxyhemocyanin with unsymmetrically labeled oxygen, *J. Am. Chem. Soc.,* 99, 4187, 1977.

24. **Freedman, T. B., Loehr, J. S., and Loehr, T. M.,** A resonance Raman study of the copper protein, hemocyanin. New evidence for the structure of the oxygen-binding site, *J. Am. Chem. Soc.,* 98, 2809, 1976.

25. **Larrabee, J. A. and Spiro, T. G.,** Structural studies of the hemocyanin active site. II. Resonance Raman spectroscopy, *J. Am. Chem. Soc.,* 102, 4217, 1980.

26. **Himmelwright, R. S., Eickman, N. C., LuBien, C. D., and Solomon, E. I.,** Chemical and spectroscopic comparison of the binuclear copper active site of mollusc and arthropod hemocyanins, *J. Am. Chem. Soc.,* 102, 5378, 1980.

27. **Brunori, M.,** Kinetics of oxygen binding by *Octopus* hemocyanin, *J. Mol. Biol.,* 46, 213, 1969.

27a. **Van Driel, R., Brunori, M., and Antonini, E.,** Kinetics of the co-operative and non-co-operative reaction of *Helix pomatia* haemocyanin with oxygen, *J. Mol. Biol.,* 89, 103, 1974.

28. **Brown, J. M., Powers, L., Kincaid, B., Larrabee, J. A., and Spiro, T. G.,** Structural studies of the hemocyanin active site. I. Extended X-ray absorption fine structure (EXAFS) analysis, *J. Am. Chem. Soc.,* 102, 4210, 1980.

29. **Co, M. S., Hodgson, K. O., Eccles, T. K., and Lontie, R.,** Copper site of molluscan oxyhemocyanins. Structural evidence from X-ray absorption spectroscopy, *J. Am. Chem. Soc.,* 103, 984, 1981.

30. **Co, M. S. and Hodgson, K. O.,** Copper site of deoxyhemocyanin. Structural evidence from X-ray absorption spectroscopy, *J. Am. Chem. Soc.,* 103, 3200, 1981.

31. **Co, M. S. and Hodgson, K. O.,** The binuclear copper site of hemocyanin. Structural evidence from X-ray absorption spectroscopy, in *Frontiers in Protein Chemistry,* Yasunobu, K. T., Ed., Elsevier/North Holland, New York, in press.

32. **Himmelwright, R. S., Eickman, N. C., and Solomon, E. I.,** Preparation and characterization of met apo hemocyanin: a single copper(II) active site, *Biochem. Biophys. Res. Commun.,* 81, 243, 1978.

33. **Schoot Uiterkamp, A. J. M., Van der Deen, H., Berendsen, H. C. J., and Boas, J. F.,** Computer simulation of the EPR spectra of mononuclear and dipolar coupled copper(II) ions in nitric oxide- and nitrite-treated hemocyanins and tyrosinase, *Biochim. Biophys. Acta,* 372, 407, 1974.

34. **DePhillips, H. A., Nickerson, K. W., Johnson, M., and Van Holde, K. E.,** Physical studies of he-mocyanins. IV. Oxygen-linked disassociation of *Loligo pealei* hemocyanin, *Biochemistry,* 8, 3665, 1969.

35. **Van Driel, R. and Van Bruggen, E. F. J.,** Oxygen-linked association-dissociation of *Helix pomatia* hemocyanin, *Biochemistry,* 13, 4079, 1974.

36. **Wood, E. J. and Dalgleish, D. G.,** *Murex trunculus* haemocyanin. II. The oxygenation reaction and circular dichroism, *Eur. J. Biochem.,* 35, 421, 1973.

37. **Zuberbühler, A. D.,** Interaction of Cu(I) complexes with dioxygen, *Metal Ions in Biological Systems,* Vol. 5, Sigel, H., Ed., Marcel Dekker, New York, 1976, 325.

38. **De Ley, M. and Lontie, R.,** The reversible reaction of cyanide with *Helix pomatia* haemocyanin, *Biochim. Biophys. Acta,* 278, 404, 1972.

39. **Malkin, R. and Malmström, B. G.,** The state and function of copper in biological systems, *Adv. Enzymol.,* 33, 177, 1970; **Fee, J. A.,** Copper proteins. Systems containing the blue copper center, *Struct. Bonding (Berlin),* 23, 1, 1975; **Gray, H. B. and Solomon, E. I.,** Electronic structures of blue copper centers in proteins, in *Copper Proteins, Metal Ions in Biology,* Vol. 3, Spiro, Th. G., Ed., John Wiley & Sons, New York, 1981, chap. 1.

40. **Solomon, E. I., Hare, J. W., and Gray, H. B.,** Spectroscopic studies and a structural model for blue copper centers in proteins, *Proc. Natl. Acad. Sci. U.S.A.,* 73, 1389, 1976.

41. **Markley, J. L., Ulrich, E. L., Berg, S. P., and Krogmann, D. W.,** Nuclear magnetic resonance studies of the copper binding sites of blue copper proteins: oxidized, reduced, and apoplastocyanin, *Biochemistry,* 14, 4428, 1975.

42. **Ugurbil, K., Norton, R. S., Allerhand, A., and Bersohn, R.,** Studies of individual carbon sites of azurin from *Pseudomonas aeruginosa* by natural-abundance carbon-13 nuclear magnetic resonance spectroscopy, *Biochemistry,* 16, 886, 1977.

43. **Colman, P. M., Freeman, H. C., Guss, J. M., Murata, M., Norris, V. A., Ramshaw, J. A. M., and Venkatappa, M. P.,** X-ray crystal structure analysis of plastocyanin at 2.7 Å resolution, *Nature (London),* 272, 319, 1978.

44. **Adman, E. T., Stenkamp, R. E., Sieker, L. C., and Jensen, L. H.,** A crystallographic model for azurin at 3 Å resolution, *J. Mol. Biol.,* 123, 35, 1978.

45. **Tullius, T. D., Frank, P., and Hodgson, K. O.,** Characterization of the blue copper site in oxidized azurin by extended X-ray absorption fine structure: determination of a short Cu-S distance, *Proc. Natl. Acad. Sci. U.S.A.,* 75, 4069, 1978.

46. **Tullius, T. D.,** Structures of Metal Complexes in Biological Systems: EXAFS Studies of Blue Copper Proteins, Xanthine Oxidase and Vanadocytes, Ph.D. thesis, Stanford University, Stanford, 1979.

47. **Tullius, T. D. and Hodgson, K. O.,** EXAFS studies of the blue copper site. Structural changes upon electron transfer in plastocyanin and stellacyanin, *J. Am. Chem. Soc.,* in press.

48. **Co, M. S. and Hodgson, K. O.,** unpublished results.

49. **Scott, R. A., Hahn, J. E., Doniach, S., Freeman, H. C., and Hodgson, K. O.,** Polarized X-ray absorption spectra of oriented plastocyanin single crystals. Investigation of methionine-copper coordination, *J. Am. Chem. Soc.,* 104, 5364, 1982.

50. For a recent review, see: **Malmström, B. G.,** Cytochrome *c* oxidase. Structure and catalytic activity, *Biochim. Biophys. Acta,* 549, 281, 1979.

51. **Hu, V. W., Chan, S. I., and Brown, G. S.,** X-ray absorption edge studies on cyanide-bound cytochrome *c* oxidase, *FEBS Lett.,* 84, 287, 1977.

52. **Powers, L., Blumberg, W. E., Chance, B., Barlow, C. H., Leigh, J. S., Jr., Smith, J., Yonetani, T., Vik, S., and Peisach, J.,** The nature of the copper atoms of cytochrome *c* oxidase as studied by optical and X-ray absorption edge spectroscopy, *Biochim. Biophys. Acta,* 546, 520, 1979.

53. **Scott, R. A., Cramer, S. P., Shaw, R. W., Beinert, H., and Gray, H. B.,** Extended X-ray absorption fine structure of copper in cytochrome *c* oxidase: direct evidence for copper-sulfur ligation, *Proc. Natl. Acad. Sci. U.S.A.,* 78, 664, 1981.

54. **Powers, L., Chance, B., Ching, Y., and Angiolillo, P.,** Structural features and the reaction mechanism of cytochrome oxidase. Iron and copper X-ray absorption fine structure, *Biophys. J.,* 34, 465, 1981.

55. **Scott, R. A.,** Extended X-ray absorption fine structure of the copper sites in cytochrome *c* oxidase, in *The Biological Chemistry of Iron,* Dunford, H. B., Dolphin, D. H., Raymond, K. N., and Sieker, L. C., Eds., D. Reidel, Boston, 1982, 475.

56. **Brudvig, G. W., Bocian, D. F., Gamble, R. C., and Chan, S. I.,** Evidence for the absence of photo-reduction of the metal centers of cytochrome *c* oxidase by X-irradiation, *Biochim. Biophys. Acta,* 624, 78, 1980.

57. **Brudvig, G. W. and Chan, S. I.,** Cu_{a_3} of cytochrome *c* oxidase is not a type 1 (blue) copper, *FEBS Lett.,* 106, 139, 1978.

58. **Beinert, H., Shaw, R. W., Hansen, R. E., and Hartzell, C. R.,** Studies on the origin of the near-infrared (800-900 nm) absorption of cytochrome *c* oxidase, *Biochim. Biophys. Acta,* 591, 458, 1980.

59. **Reinhammar, B.,** The copper-containing oxidases, in *Advances in Inorganic Biochemistry,* Vol. 1, Eichhorn, G. L. and Marzilli, L. G., Eds., Elsevier/North Holland, New York, 1979, 91.

60. **LuBien, C. D., Winkler, M. E., Thamann, T. J., Scott, R. A., Co, M. S., Hodgson, K. O., and Solomon, E. I.,** Chemical and spectroscopic properties of the binuclear copper active site in *Rhus* laccase: direct confirmation of a reduced binuclear type 3 copper site in type 2 depleted laccase and intramolecular coupling of the type 3 to the type 1 and type 2 copper sites, *J. Am. Chem. Soc.,* 103, 7014, 1981.

61. **McCord, J. M. and Fridovich, I.,** Superoxide dismutase. An enzymic function for erythrocuprein (hemocuprein), *J. Biol. Chem.,* 244, 6049, 1969.

62. **Richardson, J. S., Thomas, K. A., Rubin, B. H., and Richardson, D. C.,** Crystal structure of bovine Cu,Zn superoxide dismutase at 3 Å resolution: chain tracing and metal ligands, *Proc. Natl. Acad. Sci. U.S.A.,* 72, 1349, 1975.

63. **Blumberg, W. E., Peisach, J., Eisenberger, P., and Fee, J. A.,** Superoxide dismutase, a study of the electronic properties of the copper and zinc by x-ray absorption spectroscopy, *Biochemistry,* 17, 1842, 1978.

Chapter 5

STRUCTURAL INFORMATION ON COPPER PROTEINS FROM RESONANCE RAMAN SPECTROSCOPY

Thomas M. Loehr and Joann Sanders-Loehr

TABLE OF CONTENTS

I. INTRODUCTION

The application of resonance Raman (RR) spectroscopy to the investigation of biological chromophores, typically the focus of chemical reactivity and interest, is becoming increasingly widespread. This technique is highly structure sensitive because the frequencies of specific bond vibrations of such chromophores are observable, free from interference by the bulk of the sample. In addition, the biological activity of a specimen is, in general, not impaired by this nondestructive light-scattering technique and only a minute quantity of sample, but at a relatively high concentration, is required for an experiment.

Structural studies of metalloproteins using RR spectrometry are now over 10 years old since the first application to rubredoxin was reported by Long and Loehr.[1] Since that time, major efforts in RR spectroscopy have gone into the study of the molecular and electronic structures of hemes and hemoproteins, nonheme iron and copper proteins, conjugated chromophores such as carotenoids, visual pigments, and flavins, and inorganic complexes.[2,3]

Among copper-containing proteins, the blue copper centers characterized by an intense absorption at $\approx$625 nm (type 1) and the binuclear copper centers where reaction with molecular oxygen is believed to occur (type 3) are the chromophores which have been most extensively investigated by RR spectroscopy. These copper proteins are the principal subject of this review.

II. SPECTROSCOPIC TECHNIQUE

A. Theory

When ultraviolet or visible photons are incident upon a sample, a tiny fraction ($\leqslant 10^{-6}$) of the energy is inelastically scattered and suffers a change in wavelength. The small difference in the energies of the incident and scattered photons corresponds to a change in the vibrational energy state of the sample. A Raman spectrum is the plot of the intensity of scattered light vs. its wavelength which, for convenience, is expressed as the *difference* between the excitation and scattered frequencies since this value is directly proportional to the vibrational energy. This spectroscopic effect is named after Raman who, with Krishnan in 1928, was first to experimentally verify this phenomenon which had earlier been theoretically predicted.[4]

A simplified energy-level diagram (Figure 1a) serves to illustrate ordinary Raman scattering for the case when the excitation wavelength lies outside of any optical absorption bands of the sample. The molecule undergoes a vibrational transition, generally by one quantum level, by which the scattered photon, $h\nu_R$, is red-shifted with respect to the incident photon, $h\nu_o$. Energy conservation is expressed by the change in vibrational state, $\Delta E = h\nu_o - h\nu_R$.

Similar vibrational transitions within the electronic ground state can occur via vibronic coupling with electronically excited state(s). Thus, in Figure 1b, Raman scattering is illustrated for a molecule for which $h\nu_o$ is close in energy to or within an absorption band. The *intensity* of the scattered photon, $h\nu_{RR}$, is now subject to *resonance enhancement*, as is apparent from an adiabatic treatment of the scattering mechanism which gives for the Raman intensity, I_R,

$$I_R = \frac{8\pi\nu_R^4}{9c^4} I_o \sum_{ij} \left| (\alpha_{ij})_{mn} \right|^2 \tag{1}$$

where

$$(\alpha_{ij})_{mn} = \frac{2\pi}{h} \sum_e \left[\frac{(M_j)_{me}(M_i)_{en}}{\nu_e - \nu_o + i\Gamma_e} + \frac{(M_i)_{me}(M_j)_{en}}{\nu_e + \nu_o + i\Gamma_e} \right] \tag{2}$$

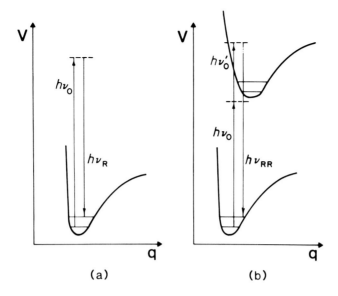

FIGURE 1. Diagrams of potential energy vs. normal coordinate for a molecule undergoing vibrational excitation by (a) the Raman effect or (b) resonance Raman effects.

for a transition between vibrational states m and n.[5] I_o is the intensity of the incident photon of frequency ν_o, ν_R is the frequency of the Raman-scattered photon, and α_{ij} is the ij-th element of the scattering tensor (the classical equivalent of the molecular polarizability). The general form of α_{ij} is given as the sum over all vibronic states of the molecule. The Ms are the electric dipole transition moments connecting the ground state with the electronically excited states, e, characterized by absorption energies, $h\nu_e$. The resonant denominator is evident in the first term of Equation (2) as ν_o approaches ν_e. The denominator is prevented from becoming zero by a damping term, $i\Gamma_e$, where Γ_e is a measure of the width of the absorption band. A detailed and rigorous account of the theory of the RR effect is beyond the scope of this review and the interested reader is referred to several excellent treatises.[3,5-9] The relatively novel technique known as coherent anti-Stokes Raman scattering (CARS)[10] is not discussed here because this more complicated and costly two-photon excitation method is beyond the reach of most laboratories and no CARS experiments on copper proteins have yet been reported.

The experimentalist studies the changes in the intensities of the Raman spectral peaks (relative to a nonabsorbing internal standard) as a function of excitation wavelength by obtaining a series of Raman spectra on the same sample but each time varying the incident light. It is desirable to have such a resonance Raman enhancement profile (RREP) across the full width of the optical absorption band of the chromophore under study. Under optimum conditions, the scattering power of a sample at resonance may be many orders of magnitude greater than for a nonresonant situation. This can result in the observation of a select number of resonance-enhanced modes due to vibrations of a chromophore free from solvent or matrix Raman scattering. One of the most dramatic demonstrations of this selective enhancement observed in the authors' laboratory was the total absence of Raman peaks due to solvent hexane in a $\approx 10~\mu M$ β-carotene solution which, when excited by 514.5-nm laser radiation, produced very strong RR bands arising from the conjugated chain.

The vibrational modes most strongly resonance enhanced, i.e., those which most effectively couple their ground and excited electronic state vibrational wave functions (Franck-Condon overlaps), are those whose normal modes closely approximate the nuclear displacements of the excited state relative to the ground state.[11] To be RR active, therefore, the

Table 1
PRINCIPAL WAVELENGTHS AND FREQUENCIES
OF COMMON CONTINUOUS-WAVE GAS LASERS[a]

Kr ion		Ar ion		Other	
λ_o(nm)	ν_o(cm^{-1})	λ_o(nm)	ν_o(cm^{-1})	λ_o(nm)	ν_o(cm^{-1})
676.4	14,784	514.5	19,436	632.8	15,803[b]
647.1	15,454	501.7	19,932	441.6	22,645[c]
568.2	17,599	496.5	20,141		
530.9	18,836	488.0	20,492		
520.8	19,201	476.5	20,986		
476.2	21,000	457.9	21,839		
413.1	24,207	454.5	22,002		
406.7	24,588	363.8	27,488		
356.4	28,058	351.1	28,482		
350.7	28,514				

[a]   Emission at certain wavelengths may require special laser optics.
[b]   He-Ne Laser.
[c]   He-Cd Laser.

driving electronic transition must possess some degree of charge-transfer (CT) character between the atoms undergoing vibrational motions in order that vibronic coupling may take place. Various intra- and intermolecular CT mechanisms have been reviewed with respect to their effectiveness in RR scattering.[7]

B. Data Collection

The Raman effect is intrinsically such a weak scattering phenomenon that high sample concentrations have been a general requirement. Water and glass are extremely poor Raman scattering substances and are, therefore, ideal as solvent and sample container, respectively. This is a principal reason that structural studies of molecules in aqueous solution have historically been carried out by Raman rather than infrared vibrational spectroscopy. The availability of He-Ne laser radiation in the early to mid 1960s generated an enormous awakening of the utility of the Raman technique by greatly increasing light intensities in very small spot sizes (decreasing the need for large sample volumes) and facilitating beam and sample placement relative to an optical monochromator. Photomultipliers, with their excellent gain characteristics, became standard detectors whose amplified output was typically converted to an analog signal and recorded on a strip chart recorder.[12] Digital processing of the photomultiplier signal and the utilization of computerized data analysis, although costly and remote, presented no new technological barriers. The development of microprocessors and microcomputers into economically attractive devices is having a major impact on the sophistication of the computer-controlled laser Raman spectrophotometers of today.[13,14] The availability of convenient and reliable laser radiation at a large number of wavelengths (Ar and Kr ion laser sources; Table 1) in the 1970s permitted the study of resonance effects and the subsequent fertile new research into the structures of optical chromophores by RR spectroscopy.[2] Although the resonant intensities were greatly increased for certain vibrational modes, some new complications accompanied the RR experiment: sample heating and possible decomposition by absorption of the excitation energy, as well as self-absorption of Raman-scattered light. A variety of special experimental refinements and useful information for planning RR experiments are detailed below. Additional information may be found in a recent review by Shriver.[15]

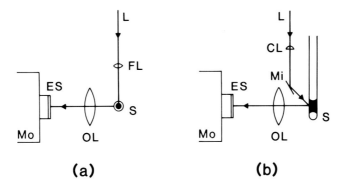

FIGURE 2.   Representative geometries for (a) 90° scattering from a transverse sample cell and (b) 180° backscattering from a vertically mounted capillary cell. The incident laser beam (L) is focused onto the sample (S) by a lens (FL) which produces a spot focus or a cylindrical lens (CL) which produces a line focus. Scattered light is collected by an objective lens (OL) and focused onto the entrance slit (ES) of a monochromator (Mo). In (b) the mirror (Mi) is positioned out of the scattered beam; this arrangement is referred to as a quasi-backscattering or ≈180° geometry.

1. Scattering Geometries and Sample Cells

The two most common arrangements, named according to the angle between the incident and scattering rays, are the 90 and 180° geometries illustrated in Figure 2 for a capillary sample cell. The same arrangements are easily adapted to other cell types. The most widely used sample cell is the common melting point capillary tube (1.1 to 1.6 mm i.d.), preferred for its small sample volume (5 to 10 $\mu\ell$), insignificant cost, and disposable nature. We have found that commercial melting point capillaries often contain fluorescent impurities and we recommend that such capillaries be washed with detergent before use. Intensely colored samples can absorb so much of the incident and scattered radiation, even with small capillaries used as in Figure 2a, that the backscattering technique, requiring little penetration of the incident light and a minimal pathlength for the scattered light (Figure 2b), is oftentimes preferred. When the incident wavelength is far away from any absorption bands and when colorless solutions are used, more efficient sample cells have been devised which use internal reflections of the laser beam to increase the effective pathlength and, hence, the number of scatterers. Gains of 10 to 50 over a single pass can be realized. However, such specialty cells have little utility for the RR spectroscopist studying absorbing samples and they require considerably larger sample volumes.

2. Temperature Control

Biological samples are generally held below ambient temperature for storage and experimental measurements. Particularly with RR spectroscopy, thermal control becomes even more critical since the incident radiation is more efficiently absorbed by the sample. Three basic temperature controls are routinely employed to maintain sample integrity: (1) spinning samples, (2) cold gas or cold-finger heat exchangers with stationary samples, and (3) flowing samples.

The first method was pioneered by Kiefer and Bernstein[16] and subsequently many laboratories have devised their own variations. Common to all is that the sample container is rapidly rotated (3,000 to 5,000 rev/min) in order to distribute the laser power input over a larger area. The method may be employed with solids or solutions and in conjunction with additional heating or cooling by convective means. A sophisticated rotating sample stage has been perfected in Germany and is now commercially available.[17]

The second method includes flowing cooled dry air or N_2 over a sample container (identical

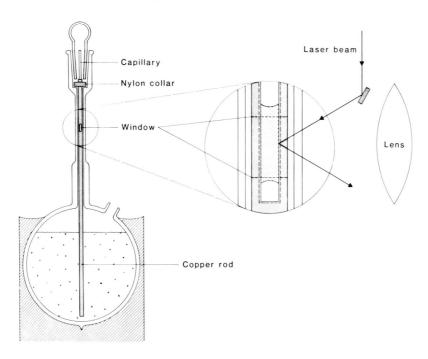

FIGURE 3.   General purpose backscattering Dewar and "cold-finger" sample holder for low-temperature control of capillary tubes. The detail at right shows the illumination of the sample (cross-hatched area within capillary) with the copper rod turned by 90°. The overall height of the Dewar is 36 cm and the volume of the flask is ≈1 ℓ.

to temperature controllers used in EPR or NMR spectrometers) or contacting the sample cell with a cold finger immersed in the appropriate refrigerant. Both of these techniques have the disadvantage that the sample cell must be surrounded by an evacuated Dewar (to prevent moisture condensation on the cell) and, therefore, presents two additional glass surfaces to both the incident and scattered beams. The former is adaptable to both 90 and ≈180° scattering geometries. In our experience, however, the flowing gas controllers not only use considerable quantities of liquid N_2, but are also unreliable at sample temperatures close to the freezing point of water. We, therefore, designed a general purpose backscattering sample Dewar (Figure 3) which has worked satisfactorily in our research for sample control between ice and liquid N_2 temperatures. With this equipment, we are able to perform RR studies of reactive intermediates in frozen solutions.[18]

Third, flowing sample devices have been described in which fresh sample is continually renewed under the probing laser beam.[19] Such designs are extremely useful when photodecomposition rather than thermal decomposition of a sample is suspected.[20] However, a significant drawback is the requirement of a large sample size.

3. Laser Excitation

The RR laboratory ideally should have access to all wavelengths between, say, 200 and 800 nm. While this desire may appear to be utopian, we may recall that 20 years ago the only coherent light source was the He-Ne laser with a single high gain line at 632.8 nm. Since that time, the availability of practical excitation lines has increased dramatically from continuous gas lasers alone (Table 1). With the continuing development of tunable dye lasers, metal vapor lasers, solid-state lasers (e.g., Nd:YAG), and frequency doubling crystals, the availability of continuous wave and pulsed excitation sources at a generous number of discrete wavelengths will be assured.

For most biological samples, laser power is generally not a limiting factor of light sources today. The power of the lines listed in Table 1 is in the 10- to 1,000-mW range for suitable commercial lasers. A sensitive sample, even with temperature control, will not survive long at power levels above 10 to 30 mW whereas a robust specimen, such as cytochrome *c*, can withstand exposure of 10 times these levels for extended periods. It is better to err in the low direction with laser power than to destroy a precious sample which may have taken months of labor to prepare. Signal averaging (see below) permits improvement of the Raman data by accumulation of multiple scans at safe power levels. (Array detectors or vidicons, which view large segments of the desired spectral region without requiring the point-by-point scanning of gratings, achieve a similar signal improvement with respect to data collection time.[21])

Most laser light is accompanied by plasma emission of the excited gas atoms. Such emission lines can be used as practical wavelength markers if permitted to enter the monochromator, however, their intensities vary with sample scattering and make unreliable external standards. They may be removed or minimized by specific line interference filters or broadband prism monochromators. Both filter systems reduce the effective laser line power by at least 50%. In our experience, the former spike filters are initially more effective, efficiently blocking all wavelengths beyond 10 to 20 Å of the source; however, these filters are expensive, require a large inventory since one filter can serve for only one laser line, and worst of all, they deteriorate with use so that only a fraction of the desired line is passed. We have had filters drop to 1% transmittance! On the other hand, prism monochromators are less efficient at removing plasma lines and some ''leakage'' must be suspected. None of the published tables of the wavelengths of gas emissions account for all possible plasma lines. Thus, the Raman spectroscopist must be continually wary not to mistake a plasma emission line for a Raman line.

4. The Sample

Sample volumes clearly depend upon the choice of the cell. Small capillaries require only a few microliters of volume; an amount in excess of the laser-illuminated volume is not necessary. Rapidly spinning devices which centrifuge the sample solution against the wall of the cell also have low volume requirements for backscattering geometries. However, the often more critical ingredient is the sample concentration. The literature reports RR experiments where concentrations range from ≈ 1 to over 100 mg/mℓ. Our rules of thumb for RR samples are (1) the sample must be perceptibly ''colored'' to the eye and (2) it should have an absorbance in the range 0.5 to 5 at the laser wavelength closest to λ_{max} of the sample. Each individual compound or metalloprotein will have its own optimum concentration for successful experiments. It is entirely possible that a sample is too concentrated, in which case dilution may be called for.

One cannot stress carefully enough the need for an internal standard in performing quantitative RR studies of the intensities. Generally suitable standards for aqueous solutions are the unreactive oxoanions such as ClO_4^-, SO_4^{2-}, or NO_3^- whose symmetric stretching modes at 933, 981, and 1,050 cm^{-1} respectively, are sharp and distinct when used at concentrations of 0.1 to 0.3 *M*. Each has higher and lower frequency components in its Raman spectrum which should be known along with any spectral contributions due to solvent and buffer.

In some instances it may not be possible to achieve a sufficiently high concentration of an enzyme in solution without encountering unacceptable solvent conditions. A case in point in our experience was with ribonucleotide reductase from *Escherichia coli*. The so-called B2 subunit, a binuclear iron protein, must be stabilized at high concentrations by glycerol. However, that co-solvent generated its own interfering Raman spectrum. We were still able to carry out our experiments by working with $(NH_4)_2SO_4$ precipitates of the enzyme to obtain high concentrations without glycerol.[22]

5. Digital Data Collection

Although some biological chromophores may give adequately satisfactory RR spectra in a single slow scan, the advantages of digital data collection with the attendant benefits of signal averaging and rapid data processing far outweigh the former classical routine.[13-15] We have previously described the modifications to our own instrumentation and the facility of data analysis via interactive graphics.[14] Commercial Raman instruments now offer many of these features and older systems can be upgraded in similar fashion.

Some of the gains to be realized by computer-controlled data handling are illustrated in Figure 4. The sample of the iron-containing enzyme, protocatechuate 3,4-dioxygenase, was highly fluorescent (as are many other protein preparations) and the weak Raman spectrum from the enzyme is superimposed on an intense fluorescent background (Figure 4A). On a chart recorder, this signal would have been off-scale and beyond the reach of ordinary zero suppression. Furthermore, in the absence of multiple scanning the data would have had to be taken 12 times more slowly in order to obtain the same spectral quality. Since the intensity of the fluorescent background often decreases with time, the slow scanning might also have resulted in a spectrum which was too steeply sloped to be handled on one scale. In the computerized system, background subtraction allows expansion of the intensity scale to give a 30-fold vertical gain in Figure 4B. A 9-point cubic smooth of the data gives a 3-fold improvement in signal/noise in Figure 4C. Additional features of the digitized system are the ability to read out peak heights and areas, to expand or contract the frequency scale, to subtract solvent or matrix spectra from sample spectra, and to add spectra obtained from separate samples. The last item is particularly useful for samples which have a limited lifetime in the laser beam.

C. Data Analysis

The third major task, following successful sample preparation and spectral measurement, is the interpretation of the Raman spectral data. The analysis of the location, shapes, and intensities of the observed peaks contained in a Raman spectrum, and in the case of RR spectroscopy, in a series of excitation wavelength-dependent spectra, will yield both molecular and electronic structural details of the system under investigation. The interpretation of RR data on metalloproteins relies heavily on (1) the availability of vibrational spectroscopic information of suitable model compounds for their metal-ligand (M-L) and intraligand (L) vibrational frequencies; (2) confirmatory vibrational assignments from frequency shifts upon metal-ion substitution as well as metal and ligand isotopic substitutions; (3) studies of the RR intensity variations of structurally identified vibrations with excitation wavelength; and (4) to some degree, normal coordinate analysis (NCA) as a theoretical and calculational approach to the assignment of spectral features. Each of these aids will be discussed in this section with particular emphasis on their application to the structural investigation of metalloproteins, particularly copper proteins.

1. Metal-Ligand and Ligand Frequencies

The low frequency region of the Raman (or infrared) spectrum below $\approx 1,000$ cm^{-1} contains the M-L vibrational frequencies in addition to deformation modes of the M-L and L skeletons. Vibrational frequencies are principally dependent upon the strength of the bond force constant and the masses of the atoms participating in the vibration. These quantities are related by the formula:

$$2\pi\nu = (k/\mu)^{1/2} \tag{3}$$

where ν is the frequency, k the bond force constant, and μ the reduced mass ($M_1 M_2/[M_1 + M_2]$). Note the inverse dependence upon the mass, predicting a lower frequency for a given

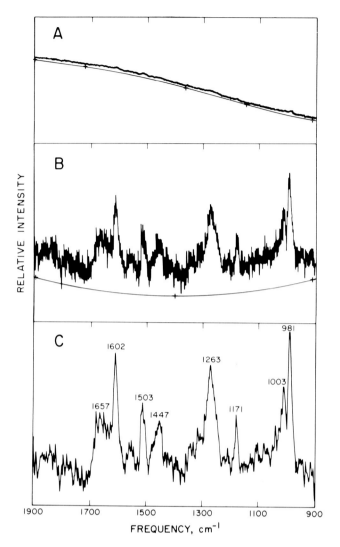

FIGURE 4. Raman spectrum of protocatechuate 3,4-dioxygenase (0°C, Tris-HCl buffer
solution) obtained in a backscattering geometry with 514.5 nm (200 mW) excitation. Slits
(fixed) 8.0 cm⁻¹; scan rate 1.0 cm⁻¹/sec; digitizing increment 1.0 cm⁻¹; number of scans
12. (A) The weak Raman-scattered signal sits on top of a strong background fluorescence.
The smooth curve is a quartic polynomial generated *in situ* for background subtraction.
Relative plot gain factor 0.001. (B) Spectrum after background subtraction. Plot gain factor
0.03. (C) Spectrum after further background correction and 9-point cubic smooth. Plot gain
factor 0.05. This spectrum represents a threefold improvement in signal/noise over that in
(B) and a ≈10-fold improvement over a single scan obtained under the given conditions.
(From Loehr, Th. M., Keyes, W. E., and Pincus, P. A., *Anal. Biochem.*, 96, 456, 1979.
With permission.)

vibrational mode involving a heavier substituent, as in an isotopic exchange. M-L deformation modes have lower force constants than M-L stretching modes and, therefore, occur
at frequencies which are too low (<200 cm⁻¹) to be generally used in the study of
metalloproteins.

Copper-ligand vibrational frequencies that have been observed in RR spectra of copper
proteins and assigned by reference to model complexes of divalent metal ions are listed in
Table 2. It is interesting to note that Cu-N(imidazole), Cu-S(cysteine), and Cu-S(methionine)
are the only M-L systems for which vibrational spectroscopic evidence exists from the study

Table 2
TYPICAL METAL-LIGAND VIBRATIONAL FREQUENCIES FOR DIVALENT METAL IONS IN PROTEINS AND METAL COMPLEXES

Ligand	Type of binding	Range of ν(M-L)[a]		Ref.
Imidazolate	M—N (imidazole ring) N—H	245—260	Imidazole	23, 24
		330—400	Pyrazole	25
		225—285	Oxyhemocyanin	23, 24
		405—425	Azurin	26
Amine	M—NH$_2$R	300—330		27
	M—NH$_2$—CH—COO⁻ (R)	440—460		27
	M—N≡C— (O, R)	290—320		28
Thiolate	M—S—R	250—300		25, 27
		≈375		29, 30
		370	Azurin	26
Thioether	M—S (R, R)	245—280		31
		260	Azurin	26
Carboxylate	M—O—C—CH—NH$_3^+$ (O, R)	275—350		32
	M—O ⋯ >C—R (M—O)	300—380		33
Phenolate	M—O—(phenyl)	545—560		34

[a] ν = stretching mode; frequencies in cm^{-1}.

of copper-containing proteins. Although specific M-L vibrations due to amine, carboxylate, or phenolate have not yet been identified in any metalloproteins, they are all reasonable candidates for copper ligation. Amines and particularly amides are excellent Cu(II) ligands in model complexes,[28] and peptide nitrogen has been observed to coordinate to copper in tetraglycine, vasopressin, and serum albumin-Cu(II) complexes.[35] Carboxylates are well-known for their ability to form binuclear Cu(II) complexes,[36] and protein carboxylate has been identified by X-ray crystallography at, e.g., the binuclear Fe(III) site in hemerythrin[37] and at the mononuclear Zn(II) site in carboxypeptidase.[38] Phenolate coordination to copper has been detected in Cu(II)-substituted transferrin[39] and lactoferrin[40] from the appearance of resonance-enhanced tyrosine ring vibrations (Table 3). The transferrin spectra show addi-

Table 3
TYPICAL RESONANCE-ENHANCED LIGAND VIBRATIONS IN
METALLOPROTEINS AND METAL COMPLEXES

Ligand	Source	Frequency (cm^{-1})[a]	Assignment[b]	Ref.
Imidazolate	[Co(histidinate)$_2$]$^{2-}$	1,330 m	δ(Ring)	41
		1,275 s	δ(C$_2$-H)	
		1,035 w	δ(Ring)	
Phenolate	Cu(II)-lactoferrin	1,605 s	δ(Ring)	39, 40
		1,500 m	δ(Ring)	
		1,275 s	δ(Ring) + ν(C-O)	
		1,170 m	δ(Ring) + δ(C-H)	
Peroxide	Oxyhemocyanin	745 s	ν(O-O)	23

[a]   s = strong; m = medium; w = weak.
[b]   δ = deformation mode; ν = stretching mode.

tional features in the 300- to 600-cm^{-1} region which could be due to Cu-N(imidazole) and Cu-O(phenolate) vibrations, but this remains to be verified.

Metal-coordinated ligands of amino-acid side chains of proteins, such as imidazolates and phenolates, are expected to show resonance-enhanced ligand vibrations at higher frequencies, ≈1,000 to 1,700 cm^{-1}, due principally to deformation modes of the conjugated rings (Table 3). The Fe(III)-transferrins and -lactoferrins as well as several Fe(III)-dioxygenases have all been classified as iron-tyrosinate proteins[42] on the basis of their highly characteristic phenolate ring vibrations which undergo strong resonance enhancement (Table 3 and Figure 4). When Cu(II) is substituted for Fe(III) in transferrin and lactoferrin, these four marker bands remain intact; the ring modes are enhanced by vibrational coupling to the ligand → metal charge transfer (LMCT) and are not strongly dependent on the nature of the transition metal ion. Although imidazole ring modes show some enhancement in model complexes (Table 3), no RR intensification of these modes has ever been observed in histidine-coordinated metalloproteins. The reasons for this "spectroscopic silence" are not yet understood. Substrates such as O$_2$ in the respiratory protein, hemocyanin (Hc), also are subject to RR enhancement. In this case it is the O-O stretching frequency which is intensity enhanced by coupling to the O$_2^{2-}$ → Cu(II) CT (Table 3).

2. Metal, Ligand, and Isotope Substitutions

Generally, the most satisfactory identification of vibrational spectroscopic features is through isotopic exchange in which it is assumed that bond-force constants are unaltered but mass effects give rise to a predictable frequency shift by application of Equation (3), e.g., OxyHc has been studied in both the ^{16}O$_2$- and ^{18}O$_2$-forms and the only frequency shift is that of the peak at 744 cm^{-1} (Table 3) to 704 cm^{-1}, respectively.[23] The ratio of these frequencies, 1.057, is in excellent agreement with the theoretical value of 1.061 for ν(^{16}O$_2$)/ν(^{18}O$_2$), and unambiguously identifies these vibrations as ν(O-O) of the protein-bound dioxygen. Deuteration of histidine residues has been achieved by exposing OxyHc to D$_2$O.[24] The resultant frequency shifts of 0.6 and 1.5 cm^{-1} in the 226 and 267 cm^{-1} Cu-L vibrations support the assignment of these modes to ν[Cu-N(imidazole)]. However, shifts of this magnitude are close to the limit of resolution of Raman spectra of proteinaceous materials.

In other instances, it may be possible to change the identity of a ligand atom and so alter the reduced mass of the vibrating system. This has been successfully carried out with the iron-sulfur protein, adrenodoxin, in which the acid-labile sulfides were substituted by selenide ions and large frequency shifts for the Fe-S to Fe-Se conversion were observed.[43] Much more difficult, however, were attempts to replace the proposed azurin (Az) ligand, methi-

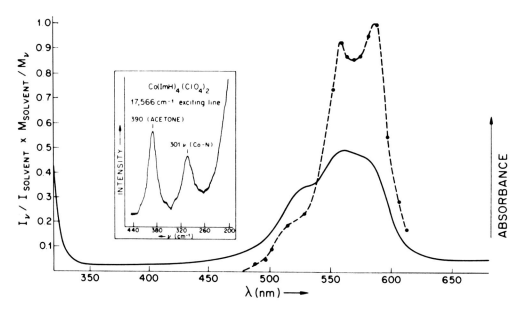

FIGURE 5.   Absorption spectrum (solid line) of an acetone solution of Co(ImH)$_4$(ClO$_4$)$_2$ and the RREP (dashed line) of its asymmetric Co-N(imidazole) vibration at 301 cm^{-1}. Inset: RR spectrum of 9 mM solution in acetone obtained with dye-laser excitation operating at 569.3 nm; the acetone line at 390 cm^{-1} serves as an internal intensity standard for the enhancement profile. (Reprinted with permission from Salama, S. and Spiro, Th. G., *J. Am. Chem. Soc.*, 100, 1105, 1978. Copyright 1978, American Chemical Society.)

onine, by selenomethionine, since this required growing a methionine auxotroph of *Pseudomonas aeruginosa* on selenomethionine. Because of the poor growth on the methionine analog, it was difficult to be certain that the bacteria actually incorporated selenomethionine at the amino-acid location assigned to the methionine ligand.[44]

Substitution of the metal ion in a metalloprotein is often difficult and may result in alterations of the structure and/or function of the metal center. Although isotope substitutions are preferable, the relatively small change in mass makes frequency shifts more difficult to detect. In OxyHc, seven Raman peaks between 100 and 400 cm^{-1} are believed to be metal-related, but only the three most intense peaks can be shown to shift definitively (by 2 to 3 cm^{-1}) upon replacement of ^{63}Cu by ^{65}Cu.[24] The observed shifts in ν(Cu-L) imply a low symmetry for the metal site which permits copper movement during the vibration. Lack of copper-isotope dependence in the low-frequency peaks ($<$200 cm^{-1}) leads to their assignment as bending modes. The replacement of Cu(II) by Ni(II) in Az leads to 25- to 50-cm^{-1} decreases in M-L vibrational frequencies.[45,46] Since there is little mass difference between the two metal ions, the magnitude of the shifts must be due to changes in coordination geometry and bond strengths.

3. Intensities of Resonance Raman Peaks and Resonance Raman Enhancement Profiles (RREP)

What criteria may be used to establish that observed Raman peaks are indeed resonance-enhanced features of the spectrum? Without doubt, the safest demonstration is from a wavelength-dependent study of the Raman intensities in the form of a RREP, e.g., Salama and Spiro[41] studied a number of Co(II)-imidazole complexes as models for Co(II)-substituted Zn-proteins. Figure 5 shows the optical absorption spectrum of tetrahedral Co(II)(imidazole)$_4$(ClO$_4$)$_2$ and the RREP of the 301-cm^{-1} asymmetric Co-N(imidazole) vibration. For this sample an acetone solvent peak at 390 cm^{-1} served as an internal (non-resonant) intensity standard. A representative RR spectrum of the sample excited with a dye-laser near λ_{max} is shown in the inset to Figure 5. Fifteen separate Raman experiments were performed to plot the excitation profile.

The remarkable feature of such RREPs is that they not only reveal the electronic state(s) responsible for RR enhancement (in Figure 5, all three components of the broad absorption band), but that they oftentimes exhibit greater fine structure than the parent absorption band. The strong enhancement of cobalt-ligand modes produced via coupling with the ligand-field absorptions of tetrahedral $Co(ImH)_4^{2+}$ is ascribed to substantial mixing of the d-d transitions with higher energy d-p and/or LMCT transitions which substantially increases the ligand field oscillator strength and, thus, the visible absorption of the molecule. In this case, the imidazole $\rightarrow$ Co(II) CT contribution to the visible electronic spectrum is probably small, and imidazole ring modes are only slightly enhanced in this region.

These arguments can be supported by a theoretical treatment of the resonance behavior. Although a variety of scattering mechanisms and accompanying factors affect the intensity of resonant Raman peaks,[3,5-9,47] this problm may be treated in a practical fashion if one limits intensity enhancement to one of two mechanisms, as suggested by Albrecht and Hutley.[47] In the first, only a single electronic state is important and the frequency dependence of the resonant vibrational mode ($\Delta\nu$) is given by the *A*-term:

$$F_A^2 = \left[\frac{\nu^2 (\nu_e^2 + \nu_o^2)}{(\nu_e^2 - \nu_o^2)^2} \right]^2 \tag{4}$$

where ν_o is the excitation frequency, ν_e is the frequency of the electronic state responsible for resonance enhancement, and ν is the frequency of the scattered photon ($\nu = \nu_o - \Delta\nu$). Equation (4) permits one to calculate relative enhancement factors which may be compared with the experimentally observed intensities to determine ν_e and, thus, to identify the electronic state in question. *A*-term scattering is greatest for strongly allowed electronic transitions and for the enhancement of totally symmetric vibrational modes. Note that the F_A expression itself has a resonant denominator, such that the Raman intensity should increase dramatically as ν_o approaches ν_e. *A*-term dependence has been well demonstrated for the copper-ligand vibrations in resonance with the $\approx$345-nm peroxide $\rightarrow$ Cu(II) CT band in OxyHc (see Figures 10 and 13).[23,24]

In the second case, *B*-term scattering, two electronic states are involved in resonance enhancement. The *B*-term frequency factors are given by:

$$F_B^2 = \left[\frac{2\nu^2 (\nu_e\nu_e' + \nu_o^2)}{(\nu_e^2 - \nu_o^2)(\nu_e'^2 - \nu_o^2)} \right]^2 \tag{5}$$

where ν_e' is the frequency of the second electronic transition involved. Since both states ν_e and ν_e' are coupled via the appropriate vibrational mode(s), nontotally symmetric (asymmetric) vibrations can also be enhanced by this mechanism. In the case of $Co(ImH)_2Cl_2$, the enhancement of the 1,254 cm^{-1} imidazole ring mode is best matched by a combination of F_A terms, while the enhancement of the 305-cm^{-1} Co-Cl symmetric stretch shows a dependence on both F_A and F_B terms.[41]

4. Normal Coordinate Analysis

Most vibrational spectroscopic assignments are based either on group frequencies or comparisons with spectra of related molecules for which assignments have been proposed. The concept of group frequencies works satisfactorily because of the relative insensitivity of frequency to molecular environment. This approach is often referred to as "fingerprinting" of a molecule for study of its vibrational spectrum, e.g., the dramatic change in bond-stretching force constants with bond order for the series O_2, O_2^-, and O_2^{2-} gives rise to their characteristic group frequencies of 1,555, $\approx$1,100, and $\approx$800 cm^{-1}, respectively. The use of previously published assignments adapted to one's own molecule is certainly widely used, however, it should be recognized that while some of these assignments would be sound, others may border on pure guesswork. What may have been published as a "tentative" identification of a frequency by one author may be quoted by another as an established fact. A number of characteristic M-L frequencies are given in Table 2, where it may be noted that nearly all entries fall within a narrow range of 250 to 450 cm^{-1}. It would surely be difficult to make a frequency assignment for some unidentified 300-cm^{-1} peak solely by comparison with such a data base.

Another method available for identification and assignment of spectral peaks is through calculations within a theoretical framework. In NCA a molecule of *n*-atoms is treated as a classical mechanical system of masses interconnected by springs possessing characteristic force constants. The problem is to set up expressions for the kinetic and potential energies of the molecule in terms of internal coordinates (bond displacements and bond angle changes) which describe the $3n$ - 6 vibrational normal modes. Typically, a number of trial force constants not exceeding the number of observed frequencies is chosen, and the energy expressions are used to derive secular equations from which frequencies of normal modes are calculated. Force constants may then be refined by an iterative process until the differences between observed and calculated frequencies are at a minimum. The classic treatment of the theory of molecular vibrations is without doubt the authoritative book of the same title by Wilson et al.[48] A new comprehensive work on this subject has recently become available and is recommended for the interested reader wishing to pursue the theory in detail.[49]

In general, exact solutions by normal mode analysis are not possible because the mechanical problem is underdetermined, i.e., the number of reliable frequencies is generally far less than the number of force constant parameters (the *force field*) required for a full description of the atomic interactions. Simplifying assumptions must therefore be introduced to make the calculations tractable, but at the same time, the resultant best-fit between observed and calculated frequencies may contain some arbitrariness due to the adjustable nature of the force field. Unfortunately, force constants are not fixed quantities from like-bond to like-bond in different molecules. These quantities are strongly dependent upon bond lengths and the strengths of nonbonded interactions, and relatively slight alterations in the force field produce changes in the calculated frequencies. Moreover, this arbitrariness is revealed by the fact that an observed frequency is likely to be matched by a calculated value simply by accident. Familiarity with the calculational procedure, experience with vibrational spectroscopy and bond-force constants, intuition in structural chemistry, and not the least, some optimism, may lead to some useful insights and proposed assignments for frequencies.

In addition to the calculated frequencies, the computation reveals the composition of each of the normal modes in terms of internal coordinates. Thus, each mode may be described as having one or more contributions of certain nuclear motions to a specified degree, e.g., one normal mode may be essentially pure bond stretching or angle deformation, whereas another mode may be a mixture of stretching or bending vibrations involving different bonds or angles, respectively.

In small, symmetrical molecules, this entire procedure is highly refined and remarkably accurate, especially if additional observed frequencies are available from isotopic substitutions which do not affect the force field. For very large molecules of low symmetry, NCA is still a sophisticated form of guesswork. However, in the area of RR spectrometry of biological chromophores, this approach, although still in its infancy, will find practical application. NCA ought to be applicable since the optical chromophore may itself be a relatively small molecular entity. In heme proteins, where only vibrations of the porphyrin ring system are observed, frequency calculations have already been reported.[50,51] Recently, we have undertaken NCA of some oxamide and biuret complexes of divalent and trivalent copper.[28] With this initial experience we cautiously approached the application of normal mode analysis to the structural elucidation of the type-1 site of Az and the assignment of its RR spectrum. We present some of our current result on Az in the following section.

III. BLUE (TYPE-1) COPPER SITES

The blue or type-1 copper site in "blue" copper proteins is characterized by an intense absorption at ≈625 nm with a molar absorption coefficient per Cu(II) of 3,000 to 5,000 M^{-1} cm^{-1} and an unusually small value of the EPR hyperfine coupling constant, $A_{\|}$.[52,53] Both of these properties are special to the divalent copper in this class of protein and are quite atypical of inorganic Cu(II) complexes. The type-1 site is the sole metal cofactor in Az, plastocyanin (Pc), stellacyanin (St), and umecyanin; it appears in conjunction with type-2 (nonblue) plus type-3 (binuclear) copper sites in the multicopper oxidases ascorbate oxidase, ceruloplasmin, and laccase. In both classes of proteins, the blue copper center is believed to function as an electron-transfer agent between other protein-bound metal centers. It shows no tendency to bind substrates or other exogenous ligands and, thus, appears to be coordinatively saturated.

Az, easily isolatable from bacterial sources, and Pc, from plant leaves, have received considerable attention with regard to the molecular structures of their polypeptide chains and, particularly, their metal centers in an attempt to provide the molecular basis for the unusual physical properties. A long and intense effort by Gray and co-workers[54,55] on their electronic spectroscopy along with NMR,[56] RR (see below), and EXAFS[57] experiments and the reports of their crystal structures[58,59] have now provided a solid foundation. The two protein structures are both 8-stranded β-barrels with the sole copper atom located toward one end of the cylindrically shaped molecule. The copper coordination is close to tetrahedral in both Az and Pc and from their known amino-acid sequences four ligands have been identified. Two histidine Ns and a cysteine S have normal, or perhaps slightly shortened, bond lengths; a fourth ligand appears to be a S-coordinated methionine. The intense blue color is accounted for by the (Cys)S → Cu(II) CT transition in these proteins, whereas the unusual EPR properties are explained by the distortion of Cu(II) away from planarity.[60] An analysis of the sequence homologies between Az's and Pc's in their carboxyl terminal regions had led to a similar proposal of histidine, cysteine, and methionine ligands of the type-1 copper.[61] It should be noted, however, that methionine is not a totally conserved residue[62] (it is completely absent from St, e.g.[63]) and it, therefore, is expected that the principal structural features of the blue copper site will be the near tetrahedral coordination of two histidines, one cysteine, and a fourth variable ligand.

A. Azurin

1. Protein Studies

The blue copper proteins have been well studied by RR spectroscopy because in their oxidized form they possess an intense absorption in the visible region for which laser excitation wavelengths are available (Table 1). Because of the absence of a prosthetic group, RR spectra of these proteins were expected to provide direct information on the nature of the Cu(II) ligands. However, despite generally good quality spectra with a rich information content, the interpretation of these data was hampered by a lack of suitable model complexes to explain the unusually high frequencies of the major vibrational bands of the blue centers.

Figure 6 shows the RR spectrum of Az isolated from *Pseudomonas aeruginosa* obtained in the authors' laboratory with 647.1-nm excitation. It greatly exceeds the spectral detail of two previously published Az RR spectra.[46,64] The most intense vibrational modes occur at 424, 404, and 369 cm^{-1}. Since the ≈625-nm absorption band of Az is now generally

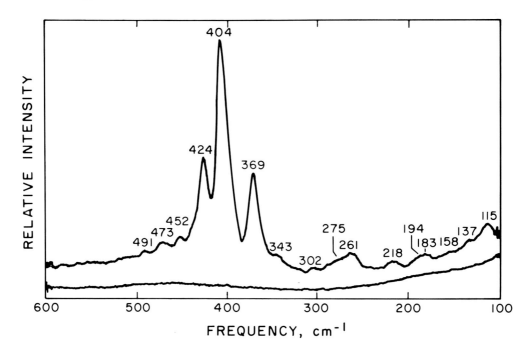

FIGURE 6.   Upper trace: resonance Raman spectrum of azurin ($\approx$1 m*M* in 50 m*M* ammonium acetate, pH 6.0) obtained with 647.1-nm excitation (120 mW at the Dewar containing the sample) in a backscattering geometry (see Figure 3). The sample was maintained at $\approx$2°C for 67 repetitive scans each at a scan rate of 2.0 cm^{-1}/sec, slitwidth of 4.5 cm^{-1}, and a digitizing increment of 0.5 cm^{-1}. Lower trace: Raman spectrum of dithionite-reduced azurin under identical conditions; 56 scans. Both spectra have been subjected to a 25-point smooth.

believed to be a (Cys)S $\rightarrow$ Cu(II) transition,[54] the RR spectrum would be expected to be dominated by the Cu-S vibration. However, Cu-S bond vibrations had generally been believed to have frequencies at or below 300 cm^{-1}. The other ligand candidate, histidine, was also expected to have its Cu-N vibrational frequencies below 300 cm^{-1}, by analogy with Cu(II)-imidazole complexes and the copper proteins, Hc and tyrosinase (Table 2 and Section V). Moreover, in these latter proteins, the Cu(II)-N(His) vibrations were maximally enhanced by excitation in the near-UV, whereas the blue copper center gave its maximal RR intensities by excitation within the (Cys)S $\rightarrow$ Cu(II) CT. Another complication related to the presence of methionine as a possible ligand. Woodruff and co-workers[31] presented a solid argument based upon Cu(II)-polythiaether complexes that a Cu-S(Met) vibrational frequency could well account for the $\approx$260-cm^{-1} band observed in all type-1 site RR spectra, but the absence of methionine from St[63] and the very long Cu-S(Met) bond lengths of up to 2.9 Å make this assignment less than rigorous.

    With these difficulties in mind, we set out to interpret the RR spectrum of Az (Figure 6) by a NCA of the Cu[N(His)]$_2$ [S(Cys)][S(Met) (?)] skeleton.[26] The results of such an analysis for a tetrahedral coordination geometry are given in Table 4. Several interesting observations may be made from these results. First, the force field yields calculated frequencies which provide a very satisfactory fit to the observed frequencies; the fit to the principal bands is excellent. Second, the values of the refined force constants for this structural model using the Cu-ligand bond lengths derived from the EXAFS study of Az[57] are entirely satisfactory and are included in Table 4. "Satisfactory" means that the values of bond-stretching force constants and angle-deformation force constants are reasonable in magnitude for M-L bonds and typical of such values in the literature. Third, the ratio of the Cu-S(Cys) to Cu-S(Met) refined force constants is nearly 2:1, reflecting the weaker bond due to the larger distance (fixed at 2.75 Å in the calculations) of the proposed methionine ligand.

Table 4
NORMAL COORDINATE ANALYSIS OF THE CuN₂SS′ TETRAHEDRAL
MODEL. OBSERVED AND CALCULATED FREQUENCIES, REFINED
FORCE CONSTANTS, AND BAND ASSIGNMENTS

Observed (cm⁻¹)	Calculated (cm⁻¹)	Band assignments
491 w		
473 w	}	Combination bands (see text)
452 w		
424 s	424	97% ν(Cu-N)
404 vs	404	81% ν(Cu-N) + 17% ν(Cu-S)
369 s	370	62% ν(Cu-S) + 15% ν(Cu-N) + 11% ν(Cu-S′)
343 w		
302 w		
275 (sh), w		
261 m	260	58% ν(Cu-S′) + 19% ν(Cu-S) + 11% δ(N-Cu-S)
218 w	210	80% δ(N-Cu-S) + 19% δ(N-Cu-S′)
194 (sh), w	202	41% δ(N-Cu-S) + 30% ν(Cu-S′) + 27% δ(N-Cu-S′)
183 w		
158 vw	160	80% δ(N-Cu-S′) + 18% δ(N-Cu-S)
137 vw	130	37% δ(N-Cu-S′) + 35% δ(N-Cu-N) + 22% δ(N-Cu-S)
115 m	102	69% δ(S-Cu-S′) + 27% δ(N-Cu-N)

Stretching force constants (mdyn Å = 10^2 N m⁻¹): K (Cu-N) = 1.1; K(Cu-S) = 1.6; K(Cu-S′) = 0.84.
Bending force constants: H(N-Cu-N) = 0.12; H(N-Cu-S) = 0.68; H(N-Cu-S′) = 0.59; H(S-Cu-S′) = 0.29.

Note: v = very; s = strong; m = medium; w = weak; sh = shoulder. Contributions to band assignments less than 9% are not listed. S = (Cys)S at 2.10 Å; S′ = (Met)S at 2.75 Å; N = His at 1.97 Å.

Another important point derived from the NCA is that most of the vibrations are strongly coupled. Two of the strong modes at 404 and 370 cm⁻¹ are assigned as mixtures of Cu-S(Cys) and Cu-N(His) vibrations. Such mixing is necessary to understand why the excitation within the S → Cu(II) CT band should simultaneously enhance Cu-N motions, even if there is no appreciable Cu-N CT character within the ≈625-nm visible absorption band. Ferris et al.[46] drew a similar conclusion about the nature of the mixed Cu-S/Cu-N stretching modes from observation of overtone and combination bands in the RR spectrum of Az in the 750- to 820-cm⁻¹ region. Finally, although only preliminary in nature, we have observed quite strong low frequency modes in Az at ≈60 and 80 cm⁻¹ whose origin is still uncertain; these could form combination bands with the principal RR bands at 424, 404, and 369 cm⁻¹ and, thus, account for the reproducible features at 491, 473, and 452 cm⁻¹. In this manner, nearly all of the observed Az RR peaks may be accounted for and would derive largely from motions of the CuN₂SS′ skeleton.

The NCA study[26] considered several different structural models (tetrahedral, trigonal, square planar, trigonal bipyramidal) as well as several different ligand-atom combinations. The best fit, however, was obtained for the CuN₂SS′ tetrahedral model. It should be noted that the NCA provided the first detailed assignment of the RR spectrum of Az and these results should be transferable to other type-1 sites (see Section III.B). However, the identification of the ≈260-cm⁻¹ peak as a Cu-S(Met) vibrational mode remains somewhat in question. In St (which lacks methionine) the ≈262-cm⁻¹ peak might fortuitously correspond to another Cu-L mode, such as a disulfide from cystine.[46] The alternative of no fourth ligand contributing significantly to the RR spectrum and the ≈260-cm⁻¹ peak representing a bending mode must also be considered.

McMillin and co-workers[45] have prepared a Ni(II)-substituted Az and obtained its RR spectrum[46] with 457.9-nm excitation, close to the 440-nm maximum of the modified protein. Three major vibrational bands were observed at 345, 356, and 395 cm^{-1} and correspond closely in relative intensities to the features at 369, 404, and 424 cm^{-1} of the native form of the protein, respectively. The fact that the frequencies are lower was taken to suggest that the Ni(II) coordination environment is more nearly a perfect tetrahedron, since Ni-L frequencies are known to be very sensitive to coordination geometry and the observed shifts are far too great to be accounted for on the basis of the mass difference between Ni and Cu.[46]

2. Model Compounds for the Type-1 Site

Until recently, few satisfactory type-1 site model compounds existed. It was recognized early on that tetrahedral distortion away from the square planar chemistry of Cu(II) could give rise to the EPR and optical properties necessary to mimic the copper center of the proteins.[65] Experimental verification of the satisfactory EPR parameters was demonstrated in several complexes,[66-69] but these lacked the optical and/or the RR properties of the blue copper proteins, e.g., γ-irradiation of Cu(I)-acetonitrile complexes[68] or Cu(I)-thioacetamide complexes[69] produces tetrahedral Cu(II)N$_4$ or Cu(II)S$_4$ centers, respectively, at liquid nitrogen temperatures having requisite values of $g_\parallel$ and $A_\parallel$, yet their optical absorptions did not effectively model the proteins.

Other workers concentrated their efforts upon finding suitable complexes exhibiting strong absorptions near 600 nm, and a large variety of ligand donor sets and geometries were found to accomplish this requirement.[70-72] A tetrahedral CuS$_4$ complex exhibits an intense 575-nm absorption ($\epsilon \approx 3,600 \ M^{-1} \ cm^{-1}$),[70] but so does a macrocyclic, planar polythiaether Cu(II) complex.[71] With a series of complexes of varying geometries Amundsen et al.[72] stressed that at least one thiolate sulfur ligand is important in generating absorption due to CT; furthermore, the λ_{max} of the band moves to lower energy, and therefore serves to model the blue proteins, as the geometry approaches tetrahedral from planar. In addition, Schugar and co-workers[73] have characterized pseudotetrahedral Cu(II) tetrapyrazolyl complexes in which π(pyrazole) → Cu(II) CT transitions in the 400- to 500-nm region ($\epsilon = 1,900 \ M^{-1} \ cm^{-1}$) could account for some of the additional features observed in the optical spectra of the blue proteins, assuming a pyrazolyl behavior similar to that of the imidazole moiety of histidine.

A third model-compound approach has been the study of the vibrational spectra of thiolate-containing Cu(II) complexes in an attempt to relate the frequencies of the simpler models to those observed in the blue proteins. The Cu(II) complex of D-penicillamine is particularly noteworthy in that it gives resonance-enhanced Raman spectra with peaks at ≈375 and 425 cm^{-1} close to those observed in the type-1 site proteins.[29,30] Although there is some disagreement on the assignments of these bands, the peak at 375 cm^{-1} has been assigned as a Cu(II)-S vibration. The reason for the occurrence of a Cu-S vibration at such a high frequency (Table 2) is uncertain because its Cu(II)-S distances of ≈2.27 Å are not unusually short.[74] By contrast, the infrared and Raman spectra of bis(imidotetraphenyldithiodiphosphino-S,S')Cu(II) obtained for both ^{63}Cu and ^{65}Cu centers have been used to assign the Cu-S vibrations at ≈285 cm^{-1};[75] the Raman band is very strongly resonance enhanced by the S(π) → Cu CT band at 575 nm.[70]

Another interesting series of models of the type-1 site has been described based on Cu(I) and Cu(II) complexes with a tetrahedrally constrained CuN$_3$S formulation.[25,76] These complexes consist of the hydrotris(3,5-dimethyl-1-pyrazolyl)borate ligand forming a trigonal base of nitrogens with a sulfur ligand, p-nitrobenzenethiolate, completing the coordination sphere (Figure 7). This complex exhibits an optical absorption spectrum that is virtually superimposable upon the spectrum of Az in the 450- to 700-nm region. Moreover, the complexes give Raman spectra with four principal stretching vibrations assigned as the three

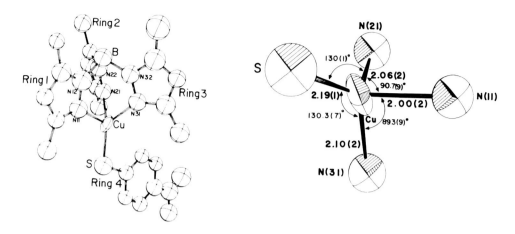

FIGURE 7. The molecular structure of the copper complex [Cu(HB(3,5-Me₂pz)₃) (SC₆H₄NO₂)]⁻ (left) and the detailed coordination geometry about the copper atom showing the CuN₃S pseudotetrahedral structural arrangement (right). (Reprinted with permission from Thompson, J. S., Marks, T. J., and Ibers, J. A., *J. Am. Chem. Soc.*, 101, 4180, 1979. Copyright 1979, American Chemical Society.)

Table 5
RAMAN FREQUENCIES AND ASSIGNMENTS OF THE SKELETAL MODES OF SEVERAL TYPE-1 SITE MODEL COMPLEXES

A	B	C	D	Proposed assignment
270	275	276		ν(Cu-S)
332	335	339	311	ν(Cu-N)
351	369	360	340	ν(Cu-N)
375	399	385	369	ν(Cu-N)
			400	ν(Cu-O)

Note: Frequencies in cm⁻¹; A: CuN₃S complex, K[Cuᴵ(HB(3,5-Me₂pz)₃)(SC₆H₄NO₂)]·2C₃H₆O; B: CuN₃S complex, K[Cuᴵ(HB(3,5-Me₂pz)₃)(SCH₂CH(NH₂)(COOC₂H₅)]; C: CuN₃S complex, Cuᴵᴵ(HB(3,5-Me₂pz)₃)(SC₆H₄ NO₂); D:CuN₃O complex, Cuᴵᴵ(HB(3,5-Me₂pz)₃) (OC₆H₄NO₂).

Reprinted with permission from Thompson, J. S., Marks, T. J., and Ibers, J. A., *J. Am. Chem. Soc.*, 101, 4180, 1979. Copyright 1979, American Chemical Society.

Cu-N modes and the Cu-S or Cu-O mode (Table 5). The presence of a new band at 400 cm⁻¹ and the absence of the 275-cm⁻¹ band in the *p*-nitrophenolate complex argues in favor of a 275-cm⁻¹ assignment to Cu-S in the *p*-nitrobenzenethiolate copper complex. The discrepancy in the 275-cm⁻¹ frequency of ν(Cu-S) in this model compound, compared to the Az value of 370 cm⁻¹, can be ascribed to the longer Cu-S distance of 2.19 Å relative to 2.10 Å in Az.

The EPR spectra of the pyrazolylborate-Cu(II) complexes with S-ligands have *g*-values that match those of the blue copper proteins, although their copper hyperfine $A_\parallel$-values are normal rather than very small, as for the proteins. At this point, one could hope for the successful synthesis of similar *imidazolate* complexes, although the electronic spectral studies on the pyrazolyl complexes revealed that the pyrazolyl group gives similar CT transitions to that of imidazolate.[73] NCA would be valuable to establish the amount of mixing in the CuN$_3$S vibrational modes of these novel type-1 site model complexes.

B. Plastocyanin, Stellacyanin, Ascorbate Oxidase, Ceruloplasmin, and Laccase

The RR spectra of several other "blue" copper site proteins obtained by excitation within their intense Cys(S) → Cu(II) CT band at ≈625 nm include Pc,[64,77] St,[77,78] ascorbate oxidase,[77] ceruloplasmin,[64,77-79] and *Rhus vernicifera* laccase.[77,78] The RR spectra obtained by Siiman et al.[77] on these proteins are shown in Figure 8. Except for some variability in relative peak intensities, these spectra are remarkably similar. They are all dominated by a series of two to three intense peaks in the rather narrow 350- to 430-cm^{-1} region, and possess a number of weaker features at both higher and lower energies. All of the proteins exhibit a medium intensity peak close to 260 cm^{-1} which has been ascribed to a Cu(II)-S(Met) vibration,[31,46] but this interpretation is still open to question (see Section III.A).

The multicopper oxidases, ceruloplasmin and ascorbate oxidase, might be expected to exhibit more complex RR spectra than the single copper enzymes or laccase because they have multiple type-1 copper sites.[53] However, the spectra in Figure 8 imply that the type-1 copper sites must be quite similar to one another in a given protein. The spectrum of St (one site), e.g., matches feature for feature that of ceruloplasmin (two sites) and ascorbate oxidase (three sites). Of considerable interest are the dramatic intensity variations in these spectra, which are a measure of electronic structural differences among their copper sites.

Pc and Az offer a good example for comparison of their RR spectra, since their structures are being investigated by X-ray crystallography.[58,59] Their principal RR peak positions are quite close (Figures 6 and 8): 426, 407, 379, 262 cm^{-1} in Pc vs. 424, 404, 369, and 261 cm^{-1} in Az. However, the intensity of the dominant 404-cm^{-1} Az peak [assigned as 81% ν(Cu-N) + 17% ν(Cu-S) in Table 4] is markedly decreased in Pc. Both proteins are reported to have the same ligand set, CuN$_2$SS'. Presumably, these intensity variations arise from different Franck-Condon factors among the blue copper centers brought about by subtle changes in the copper coordination geometry. Such changes would affect the degree of vibrational mixing in both ground and excited electronic states. Despite the intensity variations, the close correspondence in the number and frequency of the vibrational modes in all of the type-1 copper sites suggests that the NCA of Az (Table 4) is applicable to the other type-1 sites. Thus, the predominant Cu-S vibration at 369 cm^{-1} in Az is also strongly enhanced in all the other blue copper sites (frequency varies from 379 to 400 cm^{-1}), as expected for vibronic coupling to the Cys(S) → Cu(II) CT band.

IV. NONBLUE (TYPE-2) COPPER SITES

The nonblue designation refers to Cu(II) centers in proteins with EPR and electronic absorption properties characteristic of copper in inorganic complexes. As such, this category encompasses a wide variety of copper proteins ranging from the multicopper oxidases (laccase, ceruloplasmin, ascorbate oxidase) to the monocopper oxidases (amine oxidase, galactose oxidase, dopamine β-monooxygenase), and the Cu/Zn-containing superoxide dismutase (SOD).[52] The optical properties of the nonblue copper in the oxidases are generally masked by the presence of other chromophores but are probably similar to those of Cu/Zn-SOD which has its maximal visible absorption at 675 nm (ϵ = 250 M^{-1} cm^{-1}).[80] The low molar absorption coefficient of the nonblue site makes it a marginal candidate for RR

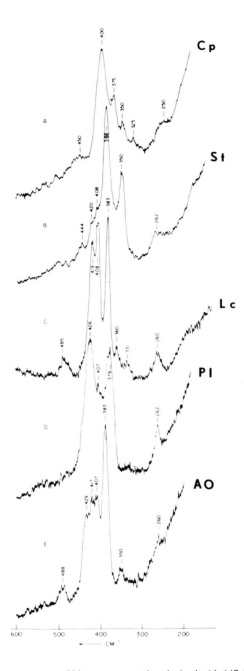

FIGURE 8.   Resonance Raman spectra of blue copper proteins obtained with 647.1-nm excitation by irradiation of their (Cys)S → Cu(II) charge transfer bands at ≈620 nm. (A) Ceruloplasmin (Cp) human plasma, 11.6 mg/ml, 50 mM acetate, pH 5.5. (B) Stellacyanin (St), lacquer tree, 8.4 mg/ml, 50 mM phosphate, pH 5.5. (C) Laccase (Lc), lacquer tree, 42.5 mg/ml, 50 mM phosphate, pH 5.5. (D) Plastocyanin (Pl), spinach leaves, 1.5 mg/ml, 50 mM phosphate, pH 6.9. (E) Ascorbate oxidase (AO), zucchini squash, 42.4 mg/ml, 50 mM phosphate, pH 7.0. (Reprinted with permission from Siiman, O., Young, N. M., and Carey, P. R., *J. Am. Chem. Soc.*, 98, 744, 1976. Copyright 1976, American Chemical Society.)

spectroscopy, as copper concentrations in excess of 2 mM would be required. The likelihood of substantial metal-centered ligand-field character in these electronic transitions further weakens their potential for resonance enhancement and none has been reported to date.

The lack of intense absorption in the visible region would tend to rule out cysteine-S ligation (as is found in the blue copper centers) and tyrosinate-O ligation [as is found in Cu(II)-substituted transferrin[39] and lactoferrin[40] with $\epsilon_{435} \approx 2,000\ M^{-1}\ cm^{-1}$]. Comparison of the EPR parameters, $A_{\parallel}$ and $g_{\parallel}$, of nonblue copper proteins with those of model copper complexes shows a good fit with N and O coordination, and no evidence of S coordination.[60] Nitrogen ligation is further indicated by the appearance of nitrogen superhyperfine lines in the EPR spectrum. Cu/Zn-SOD treated with alkali or cyanide shows superhyperfine features attributable to 3-4 N-atoms,[81,82] and similar observations have been made from the type-2 EPR spectrum of cyanide-treated fungal laccase.[83]

Histidine is likely to be the most prevalent ligand in the nonblue copper centers in view of its already well-documented presence in a number of copper proteins including representatives of all three types of copper. The nonblue copper in Cu/Zn-SOD is coordinated to four histidines[84] and this could turn out to be a general structure for the type-2 copper site. In tetragonal copper-imidazole complexes, imidazole $\rightarrow$ Cu(II) CT bands in the visible region are weak and increase in intensity in the ultraviolet region ($\epsilon_{330} < 400\ M^{-1}\ cm^{-1}$).[73] Weak enhancement (2 to 3 times) of Raman intensities of imidazole ring modes is observed for tetragonal Cu-imidazole complexes using ultraviolet excitation.[24]

In addition to histidine N, peptide N should also be considered as a reasonable ligand for nonblue Cu(II). Peptide nitrogen coordination has been implicated in Cu(II) complexes of tetraglycine, vasopressin, and serum albumin.[35] The bisbiuret complex of Cu(II):

serves as a useful model for Cu(II) coordination to deprotonated amide nitrogens. It has a weak absorption in the visible region ($\epsilon_{505} = 45\ M^{-1}\ cm^{-1}$)[85] and a rich vibrational spectrum in the Raman and infrared.[28] Assignment of the vibrational frequencies by NCA reveals that they are generally complex mixtures of different vibrational modes. Thus, one would predict that resonance-enhanced Raman spectra from such a chromophore would yield considerable spectroscopic detail and this has, in fact, been observed in RR spectra of Cu(III)-amide complexes of tetraglycine[86] and oxalodihydrazide macrocycles[87] (see Section VI). However, the far lower intensity of electronic transitions in Cu(II)-amides relative to Cu(III)-amides greatly increases the difficulty of obtaining RR spectra from the Cu(II) species. Significant spectral information might be obtained if nonblue Cu(II) ions could be selectively oxidized to the Cu(III) state.

V. BINUCLEAR (TYPE-3) COPPER SITES

Binuclear copper centers in proteins are characterized by strong antiferromagnetic coupling of the copper ions in the Cu(II) state and an intense ultraviolet absorption band at 330 to 350 nm with $\epsilon_{Cu} = 1,000$ to $10,000\ M^{-1}\ cm^{-1}$.[52,53,88] The binuclear site is found as the sole copper constituent in Hc and tyrosinase and as one of the three types of copper present in the multicopper enzymes: laccase, ceruloplasmin, and ascorbate oxidase.

A. Hemocyanin

Of all the binuclear copper proteins, Hc has been the most intensively studied due to its ready availability as the respiratory protein of molluscs and arthropods and to the simpler nature of the chemical reactivity of its copper center compared to the multicopper enzymes. The binuclear copper center in Hc is responsible for the reversible binding of molecular oxygen. A number of chemical, physical, and spectroscopic studies have been undertaken to elucidate the structure of the copper complex and the mechanism of oxygen binding.[89,90] The RR experiments have been among the most successful.

1. Excitation within the 575-nm Absorption Band

The oxygenated form of Hc has proved to be particularly amenable to study by RR spectroscopy due to a visible absorption band at $\approx$575 nm ($\epsilon_{Cu} \approx 500\ M^{-1}cm^{-1}$) in addition to the $\approx$345-nm band ($\epsilon_{Cu} \approx 9,000\ M^{-1}cm^{-1}$) associated with the binuclear copper center.[89] Excitation into the $\approx$575-nm electronic absorption band results in the appearance of a resonance-enhanced Raman peak at $\approx$750 cm^{-1} (Figure 9, 530.9 nm). The $\approx$750-cm^{-1} band has been assigned to the O-O stretch of a bound peroxide species on the basis of the oxygen-isotope dependence of the vibrational frequency and its agreement with that of model compounds.[23,91] Oxygenation of Hc with $^{18}O_2$ in place of $^{16}O_2$ causes the Raman peak to shift $\approx$40 cm^{-1} to lower energy (Figure 9), as expected from the increased mass of the oxygen. Binuclear cobalt(III) peroxo complexes exhibit O-O stretching vibrations at 790 to 835 cm^{-1} in their RR spectra.[92,93] The lower frequency of the peroxide vibration in Hc is matched by the value of 738 cm^{-1} for anhydrous Na_2O_2,[94] suggesting a hydrophobic environment in the protein.[91] The identification of peroxide in OxyHc by RR spectroscopy provides definitive evidence that the binding of O_2 to DeoxyHc is accomplished by oxidation of the binuclear copper center from the Cu(I) to the Cu(II) state. These experiments rule out molecular oxygen or superoxide as the coordinated form of oxygen since the O-O stretching vibrations for these species occur at 1,555 (Table 6 below) and $\approx$1,100 cm^{-1},[95] respectively.

The RREP of the $\approx$750-cm^{-1} peroxide vibration in OxyHc coincides with the electronic absorption band at $\approx$575 nm (Figure 10B, D). Thus, the band at $\approx$575 nm can be assigned to $O_2^{2-} \rightarrow$ Cu(II) CT rather than a copper-localized d-d transition, in agreement with the high intensity of this absorption. The enhancement profile in Figure 10 shows an additional shoulder at $\approx$490 nm, indicating that there are actually two electronic transitions with which the O-O stretch is in resonance. The existence of two $O_2^{2-} \rightarrow$ Cu(II) transitions is well accounted for by a molecular orbital treatment for Cu-O$_2$-Cu with the atoms arranged as:

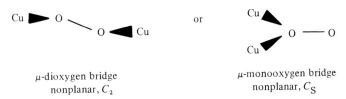

| μ-dioxygen bridge | μ-monooxygen bridge |
| nonplanar, C_2 | nonplanar, C_S |

According to the molecular orbital scheme in Figure 11, the $\approx$490- and $\approx$575-nm absorption bands are due to transitions from the filled, oxygen-localized 5a and 5b orbitals, respectively, to the one unfilled, Cu(II)-localized ϕ_A orbital. The pairing of Cu(II) electrons in ϕ_A accounts for the known diamagnetism of OxyHc.[96] The selective enhancement of the O-O vibration is explained since the $O_2^{2-} \rightarrow$ Cu(II) transitions remove electron density from antibonding orbitals on peroxide and, thereby, strengthen the O-O bond. (An alternative scheme discussed below assigns the two visible absorptions to transitions from the highest occupied oxygen orbital to separate half-filled d orbitals.[24])

The above analysis narrowed down the possibilities for copper-peroxide coordination in OxyHc to a μ-dioxygen bridge or a μ-monooxygen bridge. This issue has been further clarified by RR experiments with the mixed isotope, ^{16}O-^{18}O.[97,98] In a μ-dioxygen bridge the two oxygen atoms are equivalent and, thus, give rise to only a single O-O vibration; in

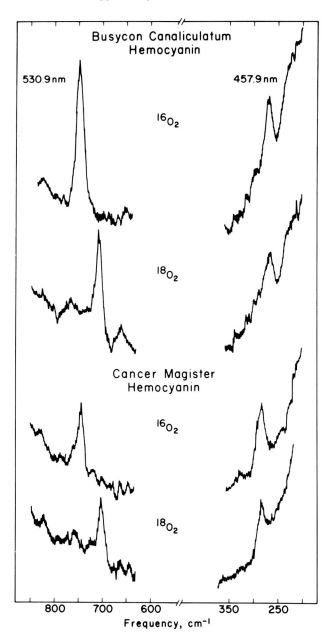

FIGURE 9.   Raman spectra of $^{16}O_2$- or $^{18}O_2$-hemocyanin, using 530.9- or 457.9-nm excitation; 20 cm^{-1}/min scanning rate, 5-sec time constant, 8-cm^{-1} spectral slit width; *Busycon canaliculatum* OxyHc, 165 mg/mℓ, pH 9.8; *Cancer magister* OxyHc, 60 mg/mℓ, pH 8.5. (Reprinted with permission from Freedman, T. B., Loehr, J. S., and Loehr, Th. M., *J. Am. Chem. Soc.*, 98, 2809, 1976. Copyright 1976, American Chemical Society.)

a µ-monooxygen bridge the two possible modes of attachment of ^{16}O-^{18}O lead to two O-O vibrations. As shown in Table 6, DeoxyHc exposed to an O_2 mixture containing ^{16}O-^{18}O produced the same relative peak intensities as the initial gas sample. A splitting of the ^{16}O-^{18}O vibration would have led to a broader and less intense peak. If anything, the relative height of the ^{16}O-^{18}O peak has increased. Although the peak widths are more difficult to quantitate, there does not appear to be significant broadening of the ^{16}O-^{18}O peak. These data favor the existence of the µ-dioxygen bridging arrangement in OxyHc.

Table 6
COMPARISON OF RAMAN PEAK DISTRIBUTION IN THE OXYGEN-ISOTOPE MIXTURE AND THE RESULTANT OXYHEMOCYANIN

	O_2 gas[a] (55.1 atom % ^{18}O)	Hemocyanin *(Busycon canaliculatum)*
Relative peak intensities		
^{16}O—^{16}O	0.71	0.66
^{16}O—^{18}O	1.63	1.63
^{18}O—^{18}O	1.00	1.00
Full band width at half-maximum intensity (cm^{-1})		
^{16}O—^{16}O	8.3	15.0
^{16}O—^{18}O	7.6	11.7
^{18}O—^{18}O	7.3	9.9

[a] $\nu(O—O) = 1,555, 1,520, 1,470\ cm^{-1}$ for $\nu(^{16}O—^{16}O)$, $\nu(^{16}O—^{18}O)$, $\nu(^{18}O—^{18}O)$, respectively.

Reprinted with permission from Thamann, Th. J., Loehr, J. S., and Loehr, Th. M., *J. Am. Chem. Soc.*, 99, 4187, 1977. Copyright 1977, American Chemical Society.

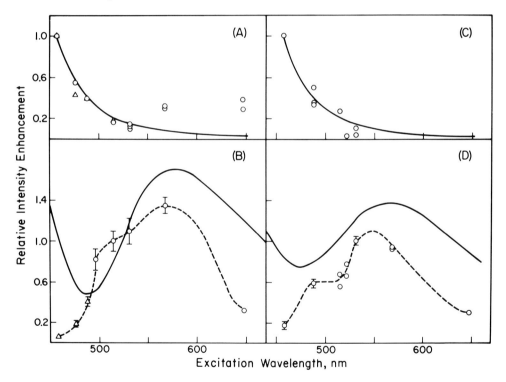

FIGURE 10.   Relative intensity enhancement profiles. (A) *Cancer magister* OxyHc, 282-cm^{-1} peak, normalized to relative intensity with 457.9-nm excitation. Solid line is the theoretical curve[47] for $\nu_e = 340$ nm, $\Delta\nu = 282$ cm^{-1}. (B) *C. magister* OxyHc, 744- and 704-cm^{-1} peaks, normalized to relative intensity with 514.5-nm excitation. Solid line is the absorption spectrum, arbitrary ordinate scale. (C) *Busycon canaliculatum* OxyHc, 267-cm^{-1} peak, normalized to relative intensity with 457.9-nm excitation. Solid line is the theoretical curve[47] for $\nu_e = 345$ nm, $\Delta\nu = 267$ cm^{-1}. (D) *B. canaliculatum* OxyHc, 749-cm^{-1} peak, normalized to relative intensity with 530.9-nm excitation. Solid line is the absorption spectrum, arbitrary ordinate scale. (Δ) $^{18}O_2$-hemocyanin; ($\bigcirc$) $^{16}O_2$-hemocyanin. Error bars indicate average deviations for three or more experiments. (Reprinted with permission from Freedman, T. B., Loehr, J. S., and Loehr, Th. M., *J. Am. Chem. Soc.*, 98, 2809, 1976. Copyright 1976, American Chemical Society.)

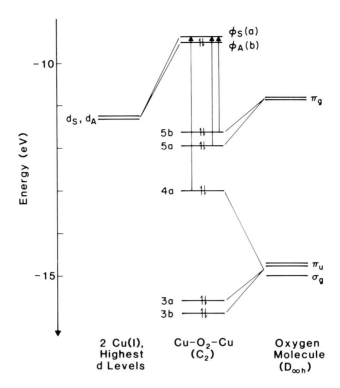

FIGURE 11.   Molecular orbital scheme for Cu-O$_2$-Cu, C_2 symmetry. Interaction of symmetric (d$_s$) and antisymmetric (d$_A$) combinations of the highest energy copper d levels with oxygen 3σ_g, π_u, and π_g molecular orbitals. Peroxide → Cu(II) charge transfer transitions 5b → ϕ_s, 5a → ϕ_s, and 4a → ϕ_s correspond to absorptions at 570, 490, and 345 nm, respectively. (Adapted with permission from Freedman, T. B., Loehr, J. S., and Loehr, Th. M., *J. Am. Chem. Soc.*, 98, 2809, 1976. Copyright 1976, American Chemical Society.)

2. Excitation into the 345-nm Absorption Band

When the Raman excitation is close to the ≈345-nm band associated with all type-3 binuclear copper sites, a different resonance enhancement pattern is observed. Now the predominant spectral feature is a peak at ≈280 cm^{-1} (Figure 9, 457.9 nm) which is unaffected by the mass of the bound O$_2$. The enhancement profile (Figure 10A, C) confirms that the ≈280-cm^{-1} peak is in resonance with an ultraviolet absorption band. Comparison with the spectra of model copper complexes (Table 2) and the suspected presence of imidazole as a copper ligand in Hc[99] led to the assignment of the ≈280-cm^{-1} feature as a Cu-N(imidazole) stretching vibration.[23]

Raman excitation within the ≈345-nm absorption band of OxyHc has produced considerably more spectral information.[24,100] As shown in Figure 12, OxyHc's exhibit a number of weaker Raman peaks in addition to the intense peak at ≈280 cm^{-1}. Arthropod and mollusc Hc's have similar peak frequencies (Table 7) and intensities (Figure 12), with the exceptions that the relative intensities of the 265- and 285-cm^{-1} peaks are reversed between the two phyla and that the 115- and 170-cm^{-1} peaks are less distinct in arthropod Hc's. Substitution with ^{65}Cu in place of ^{63}Cu results in ≈2-cm^{-1} shifts to lower energy for the peaks at 225, 265, and 285 cm^{-1}, indicating that these are Cu-related stretching vibrations. The lack of copper-isotope dependence of the peaks at 115 and 170 cm^{-1} leads to their assignment as bending modes, since these involve less motion of the metal atom.

Equilibration of OxyHc in D$_2$O leads to exchange of imidazole N1 and C2 protons and

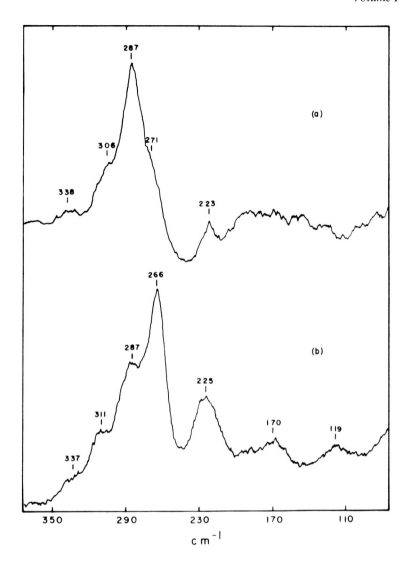

FIGURE 12. Raman spectra of oxyhemocyanins recorded with 351.1-nm excitation; 7-cm⁻¹ spectral slit. (a) *Limulus polyphemus* OxyHc, 1.5 m*M* in Cu, pH 7.0. (b) *Busycon canaliculatum* OxyHc, 1.0 m*M* in Cu, pH 9.8. (Reprinted with permission from Larrabee, J. A., Spiro, Th. G., Ferris, N. S., Woodruff, W. H., Maltese, W. A., and Kerr, M. S., *J. Am. Chem. Soc.*, 99, 1979, 1977. Copyright 1977, American Chemical Society.)

causes the peaks at 225 and 265 cm⁻¹ to shift ≈1 cm⁻¹ to lower energy. Although a shift of this size is approaching the precision limits of the technique, it does give an indication that these two vibrations involve copper-coordinated imidazoles. The peaks at 310 and 335 cm⁻¹ are too weak to determine their hydrogen-isotope dependence. It has been suggested that one or both of these peaks could be assigned to a vibration of the Cu-X-Cu group,[24] where X is a bridging ligand such as phenolate, carboxylate, or solvent μ-oxo.[72,90,101] If X were a μ-oxo group, then a symmetric vibration below 400 cm⁻¹ would be indicative of a Cu-O-Cu bridging angle in excess of 140°, with the asymmetric vibration close to 900 cm⁻¹ being more difficult to detect by Raman spectroscopy.[102] Frequencies in the 300- to 400-cm⁻¹ region would seem more plausible for ν(Cu-O) of carboxylate than of phenolate (Table 2), as would the absence of RR phenolate ring modes (Table 3).

Table 7
RAMAN PEAKS IN RESONANCE WITH THE ≈345-nm
ABSORPTION BAND OF OXYHEMOCYANIN[a] AND
OXYTYROSINASE[b]

Oxyhemocyanin			
Arthropod[c]	Mollusc[d]	Oxytyrosinase	Proposed assignment
	115		δ(Cu-L)
180 br	170	184 br	δ(Cu-L)
220	225	218	ν(Cu-N, imidazole)
265 sh	265	274	ν(Cu-N, imidazole)
285	285 sh		ν(Cu-N, imidazole)
305	310	296 sh	ν(Cu-X) or ν(Cu-L)
335	335	328	ν(Cu-X) or ν(Cu-L)
750	750	755	ν(O-O)
1,075 br	1,075 br		Singlet → triplet electronic transition

[a]   Frequencies (cm^{-1}) and assignments from Reference 24. br = broad; sh = shoulder; δ = bending mode; ν = stretching mode; L = unspecified ligand; X = bridging ligand.
[b]   Frequencies for *Neurospora crassa* oxytyrosinase.[113]
[c]   *Limulus polyphemus, Cancer magister, Cancer irroratus,* and *Cancer borealis.*
[d]   *Busycon canaliculatum* and *Megathura crenulata.*

All of the OxyHc vibrations listed in Table 7 appear to be in resonance with the ≈345-nm absorption band as indicated in Figure 13. Thus, it is of interest to determine the nature of the electronic transition at ≈345 nm and how it might be responsible for enhancement of a number of different Cu-L vibrations. In our earlier work we assigned it to an imidazole → copper CT,[23] and Larrabee et al.[100] assigned it to a simultaneous pair excitation of the antiferromagnetically coupled copper center. However, the recent observation of a MetHc which lacks the ≈345-nm absorption, but whose copper ligation to the protein and magnetic coupling remains intact,[101] makes it likely that the ≈345-nm absorption is directly related to oxygen binding and arises from a peroxide → Cu(II) CT. The question remains as to the nature of the orbitals involved in this transition. It should be emphasized at this point that the type of molecular orbital description one arrives at depends on the assumptions one makes concerning the relative energies of the different orbitals.[103]

The molecular orbital scheme for OxyHc in Figure 11 is based on the theoretical analysis of magnetically coupled copper dimers by Hay et al.[23,104] An alternative molecular orbital scheme has been proposed for OxyHc (Figure 14) based on a transition dipole vector coupling model.[101] The copper atomic orbitals $d_{Cu(1)}$ and $d_{Cu(2)}$ in Figure 14 appear to be similar to the Cu molecular orbitals $\phi_A(b)$ and $\phi_S(a)$ in Figure 11, whereas the oxygen orbitals π_S^* and π_A^* are similar to 5b and 5a. The major difference between the two schemes is that the Figure 14 description shows a greater splitting in the energies of the π^* orbitals and a lesser splitting in the energies of the d orbitals. Thus, in Figure 11 the molecular orbital energy gap is sufficient to pair the Cu-localized electrons in the lower energy $\phi_A(b)$ orbital, whereas in Figure 14 the electrons are spin-coupled but reside in separate d orbitals. Figure 11 predicts a single electronic transition in the near UV region (e.g., 345 nm), and Figure 14 predicts two near-UV transitions separated by about 30 nm. Since the enhancement profiles in Figure 13 show an excellent fit with the theoretical profile calculated for resonance with a *single* electronic state (i.e., an *A*-term dependence with the 345-nm absorption band) and there is no evidence of splitting in either the excitation profile or the absorption band, we prefer the scheme presented in Figure 11.

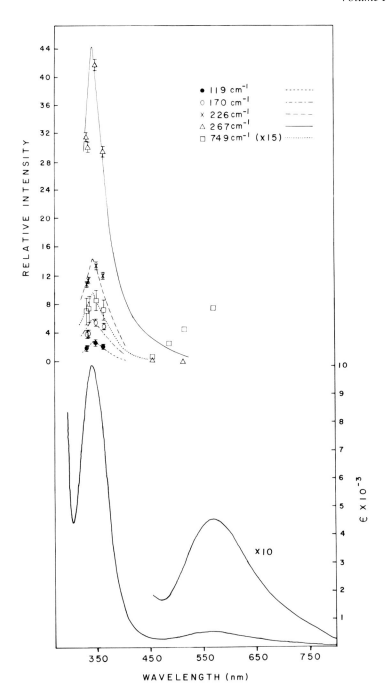

FIGURE 13.   Relative intensity enhancement profiles for *Busycon canaliculatum* OxyHc, 1 m*M* for UV, 4.5 m*M* for visible wavelengths, in 50 m*M* sodium carbonate buffer, pH 9.8. Intensities were measured relative to the v_1 (935 cm^{-1}) band of 0.2 *M* NaClO$_4$ present as an internal standard. Error bars indicate the range of values for three separate measurements. The curves drawn through the data points represent the *A*-term frequency dependence, using the actual peak frequency (29,000 cm^{-1}) of the 345-nm absorption band. (Reprinted with permission from Larrabee, J. A. and Spiro, Th. G., *J. Am. Chem. Soc.*, 102, 4217, 1980. Copyright 1980, American Chemical Society.)

In most other respects the consequences of the two orbital diagrams are very similar. Both predict a greater intensity in the ≈345-nm absorption band than in the visible absorption bands due to better orbital overlap; the 4a orbital in Figure 11 is viewed as an oxygen σ

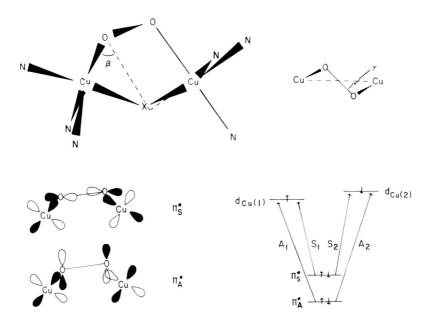

FIGURE 14.   Model of the oxygen binding site in hemocyanin and a molecular orbital scheme for Cu-O_2-Cu with C_2 symmetry. Interaction of the highest energy copper d levels, $d_{Cu(1)}$ and $d_{Cu(2)}$ with $3\pi_S^*$ and $3\pi_A$ molecular orbitals. Peroxide $\rightarrow$ Cu(II) charge-transfer transitions S_1 and S_2 correspond to the $\approx$570- and $\approx$490-nm absorptions, while transition A_1 and/or A_2 correspond(s) to the $\approx$345-nm absorption. (Reprinted with permission from Larrabee, J. A. and Spiro, Th. G., *J. Am. Chem. Soc.*, 102, 4217, 1980. Copyright 1980, American Chemical Society.)

molecular orbital with substantial admixture from the π_u molecular orbitals. In both cases the oxygen orbital responsible for the $\approx$345-nm transition, 4a or π^*_A, has considerable σ-bonding character so that the electrons in that orbital have become partially localized on the copper atoms. Thus, excitations from this orbital would show greater vibronic coupling with copper-ligand modes than with peroxide modes.[101] The broad 1,075-cm^{-1} peak (Table 7) attributed to an electronic Raman effect[24] can also be equally well explained by Figure 11 or 14 as a singlet $\rightarrow$ triplet transition of the d-orbital electrons. The energy difference of 1,075 cm^{-1} observed by RR spectroscopy is in agreement with the value of $2J \geqslant 1,100$ cm^{-1} obtained from magnetic susceptibility measurements of OxyHc.[96]

One possible structure of the Hc oxygen-binding site, which has emerged as a result of spectroscopic, magnetic, and EXAFS[105] data, is shown in Figure 14. In this model the copper atoms are 3.67 Å apart,[105] bridged by peroxide[97] and an endogenous ligand X.[101] There are also three histidine ligands to each copper.[24,105] However, the difficulties inherent in identifying Raman spectral peaks and in obtaining and interpreting EXAFS data suggest that such a model should still be considered tentative.

3. Species Comparisons

The most distinct Raman spectral difference between various OxyHc's is in the occurrence of the major spectral component at 285 cm^{-1} in arthropod Hc's and at 265 cm^{-1} in mollusc Hc's (Figure 12 and Table 7). As this effect appears to be restricted to two vibrational modes which have been tentatively assigned to histidine ligands, the most likely explanation is an alteration in imidazole-coordination geometry between the two phyla.[24,100] Only subtle spectral changes are observed in the Raman spectra of different subunits of *Limulus polyphemus* Hc.[100] These amount mainly to a loss of spectral detail in the 250- to 320-cm^{-1} region and as such are probably less informative than the shifts in absorption maxima from 560 to 590 nm in these same subunits.

The peroxide stretching vibration varies from 744 to 748 cm^{-1} in arthropod Hc's and from 745 to 752 cm^{-1} in mollusc Hc's.[23,24,106,107] It does not appear to be very sensitive to phylogenetic differences, indicating a fairly uniform hydrophobic environment for oxygen binding in all Hc's.[91] The 750-cm^{-1} O-O vibration is also insensitive to molecular changes accompanying tryptic digestion of *Helix pomatia* β_c-Hc into smaller functional fragments. These fragments do show a decreased intensity and broadening of the 270-cm^{-1} Cu-N(imidazole) vibration,[107] akin to that observed with subunits of *L. polyphemus*.[100]

B. Tyrosinase

Tyrosinase is a ubiquitous copper-containing enzyme which catalyses the oxidation of monophenols and *o*-diphenols to *o*-quinones.[108] All of the copper appears to be binuclear (type 3), with many analogies to Hc. Resting tyrosinase contains antiferromagnetically coupled Cu(II) with spectroscopic properties similar to MetHc.[109] Reduction to the Cu(I) form (similar to DeoxyHc) results in a species which can react with O$_2$ to yield oxytyrosinase (similar to OxyHc).[110] Oxytyrosinase will catalyse the oxidation of monophenols and diphenols.[111] Early studies showed that monophenol oxidation is accompanied by the incorporation of oxygen from O$_2$ into the organic substrate.[112] Thus, one of the reaction pathways for this enzyme appears to be

Oxytyrosinase + ⬡—OH + 2H$^+$ → ⬡—OH + Resting tyrosinase + H$_2$O

[2 Cu(II)–O$_2^{2-}$]   OH   [2 Cu(II)]

Resting tyrosinase + ⬡—OH → ⬡=O + Reduced tyrosinase + 2H$^+$

[2 Cu(II)]   OH   O   [2 Cu(I)]

Reduced tyrosinase + O$_2$ → oxytyrosinase

[2 Cu(I)]   [2 Cu(II)–O$_2^{2-}$]

Definitive evidence for the presence of peroxide in oxytyrosinase from *Neurospora crassa* has been obtained by RR spectroscopy.[113] The spectrum produced by UV excitation is remarkably similar to those of the OxyHc's (Table 7). The peak at 755 cm^{-1} in oxytyrosinase again proves to be an O-O stretch of peroxide by its shift to 714 cm^{-1} when the protein is prepared with ^{18}O$_2$. The most intense spectral feature at 274 cm^{-1} can be assigned to ν(Cu-N, imidazole) as are the major peaks at 265 and 285 cm^{-1}, respectively, in mollusc and arthropod Hc's. Four histidine residues have been implicated as copper ligands by chemical studies of *N. crassa* tyrosinase in which the binuclear copper center is part of a single, 46,000 M_r polypeptide chain.[114]

The excellent matching of the UV/visible and RR spectra of oxytyrosinase and OxyHc indicates a near identity of copper ligation and oxygen binding in the two proteins. Spectroscopic and magnetic studies also point to the presence of the same endogenous bridging group, X, in Hc and tyrosinase.[115] A model for the active site of oxytyrosinase with a monophenolic substrate appropriately positioned for reaction with peroxide is shown in Figure 15. The major difference between the two proteins is in the far greater reactivity of oxytyrosinase towards phenolic substrates. This is viewed as being primarily due to the presence of a special substrate-binding cavity in tyrosinase, with both proteins having similar abilities to activate molecular oxygen.[113,114]

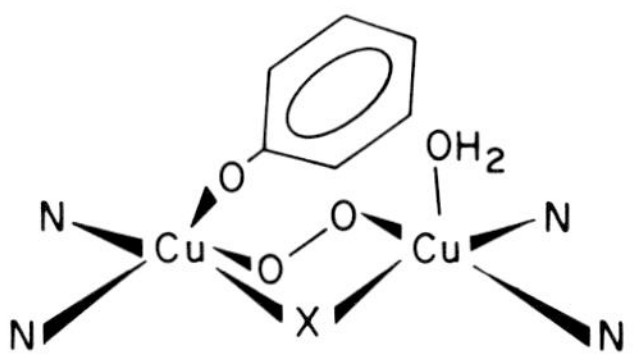

FIGURE 15.   Model of the binding site for oxygen and a monophenolic substrate in tyrosinase. (Adapted with permission from Himmelwright, R. S., Eickman, N. C., LuBien, C. D., Lerch, K., and Solomon, E. I., *J. Am. Chem. Soc.*, 102, 7339, 1980. Copyright 1980, American Chemical Society.)

C. Multicopper Oxidases

The enzymes ceruloplasmin (from blood serum), laccase (from the lacquer tree *Rhus vernicifera* or the fungus *Polyporus versicolor*), and ascorbate oxidase (from squash and cucumber) all contain a binuclear copper center in addition to type-1 blue and type-2 nonblue copper sites.[52,53] Each of these enzymes catalyses a net reaction in which four electrons are removed in the oxidation of two or more substrate molecules and O_2 is reduced to 2 H_2O. The binuclear copper site is believed to be the site of reduction of O_2,[88] and peroxy derivatives of laccase and ceruloplasmin have been characterized.[90] Although these bear some similarities to OxyHc and oxytyrosinase, they appear to have significantly less intense absorption at 330 nm ($\epsilon < 3,000\ M^{-1}cm^{-1}$) and antiferromagnetic interaction.[90] Conversely, the resting enzymes have a much greater absorbance at 330 nm ($\epsilon \simeq 3,000\ M^{-1}\ cm^{-1}$)[52] than the corresponding MetHc or resting tyrosinase. Since most spectroscopic investigations of the multicopper enzymes are complicated by contributions from three different types of copper, it would be extremely helpful to have a technique such as RR spectroscopy which could selectively enhance the vibrations of the binuclear copper center using ultraviolet excitation. Unfortunately, the photolability of these enzymes upon exposure to UV laser light has so far prevented the observation of RR spectra from the type-3 binuclear copper sites.

VI. TRIVALENT COPPER SITES

Trivalent copper has been proposed as an enzymatic intermediate in the reaction catalysed by galactose oxidase.[116] Oxidation of the Cu(II) form of the enzyme with hexacyanoferrate (III), hexachloroiridate, or Mn(III) results in an increase in enzymatic activity, a loss of Cu(II) EPR signal as expected for Cu(III), and an increase in absorbance at 440 nm ($\epsilon = 4,900\ M^{-1}\ cm^{-1}$) and 800 nm ($\epsilon = 2,800\ M^{-1}\ cm^{-1}$). However, more recently it has been shown that N-bromosuccinimide-inactivated galactose oxidase reacts with hexacyanoferrate (III) to produce a free radical and the previously observed visible absorption spectrum, but without loss of the Cu(II) EPR signal.[117] The loss of the Cu(II) signal in the native enzyme could thus be explained by spin coupling between Cu(II) and a free radical, which would also agree with the observation of a normal Cu(II) X-ray absorption edge from hexacyanoferrate(III)-treated galactose oxidase.[118] Although the proposal that a mononuclear copper enzyme can cycle between Cu(I) and Cu(III) is attractive because it would allow a two-electron reduction of oxygen without requiring the energetically unfavorable production of superoxide, it remains to be definitively proven.

Models for Cu(III) coordination in proteins are Cu(III)-tetraglycine, which is stable for several hours in aqueous solution,[119] and a Cu(III)-oxalodihydrazide macrocycle, which is completely stable in water.[120,121] In these compounds the copper is bound to deprotonated amide nitrogens, indicating that in proteins peptide nitrogens would be particularly good candidates for stabilizing Cu in the +3 oxidation state. The tetraglycine complex has the structure:

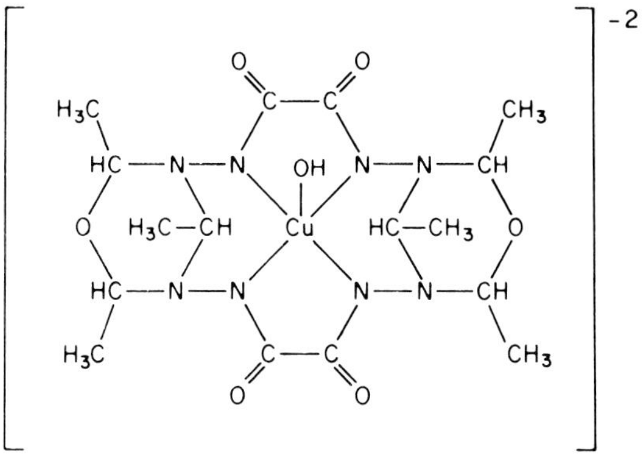

FIGURE 16. Structure of the Cu(III) macrocycle resulting from the condensation of oxalodihydrazide and acetaldehyde. Based on the X-ray crystallographic studies of Clark et al.[122]

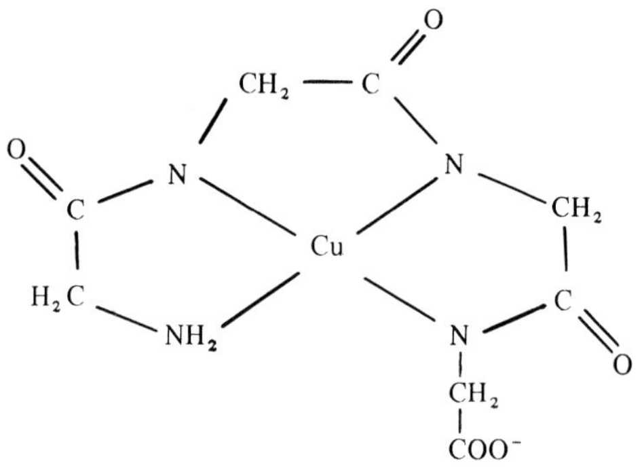

and the structure of the Cu(III)-oxalodihydrazide macrocycle is shown in Figure 16. (The macrocycle is formed from the condensation of Cu(II)-bisoxalodihydrazide with six acetaldehyde molecules and oxidized to Cu(III) by O_2, hexacyanoferrate(III), or peroxodisulfate.[120]) Both complexes exhibit intense CT bands in the visible region with absorption maxima at 365 nm ($\epsilon = 7,100\ M^{-1}\ cm^{-1}$) for Cu(III)-tetraglycine[119] and at 540 nm ($\epsilon = 29,000\ M^{-1}\ cm^{-1}$) for the Cu(III)-macrocycle.[120] Excitation within these CT bands leads to RR spectra[86,87] in which a number of vibrational modes are enhanced (Figures 17a and 18B).

Comparison with the Raman spectra of the simpler biuret and oxamide complexes, for which NCA is available,[28] shows that the wealth of spectral detail is due to the highly coupled nature of the vibrational modes. This is a consequence of the electron delocalization within the coordinated amide groups and the constrained rings formed by chelation.[28] Furthermore, most of the vibrational modes present in the nonresonance Raman spectra of Cu(III)-bisbiuret and Cu(III)-bisoxamide are enhanced by excitation within the CT bands of Cu(III)-tetraglycine and the Cu(III) macrocycle. In the case of the Cu(III) macrocycle, the major Raman spectral components in Figure 18B all fit an intensity enhancement profile commensurate with vibronic coupling to a single electronic transition at 540 nm (Figure 19). The RR behavior of these Cu(III) complexes points to the assignment of the intense absorption band as deprotonated (amide)N → Cu(III) CT with extended π orbital overlap between the N, C, and O atoms of the amide ligand. For both the Cu(III) macrocycles and Cu(III)-tetraglycine the strongest spectral features occur between 350 and 440 cm⁻¹. These are designated as predominantly Cu-N stretching vibrations. Other common spectral details include the 700- to 730-cm⁻¹ peak ascribed to a carbonyl deformation, the series of peaks

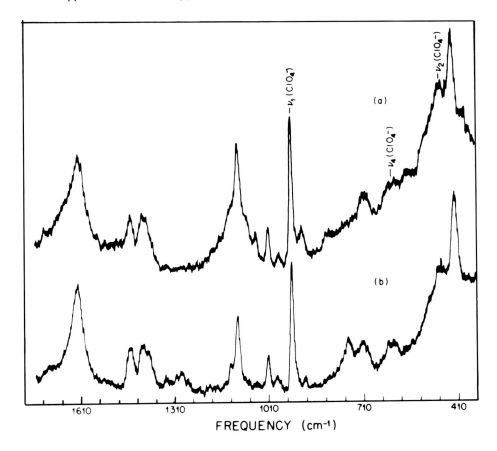

FIGURE 17.   Resonance Raman spectra of (a) Cu(III)-tetraglycine 3 m*M* and (b) Cu(III)-triglycine amide 2 m*M*, obtained with 363.8-nm excitation. (Reprinted with permission from Kincaid, J. R., Larrabee, J. A., and Spiro, Th. G., *J. Am. Chem. Soc.*, 100, 334, 1978. Copyright 1978, American Chemical Society.)

between 1,000 and 1,400 cm^{-1} due to coupled C-N stretching vibrations, and the relatively stronger peaks at 1,590 to 1,650 cm^{-1} assigned to carbonyl stretching.

A note of caution must be introduced at this point. The RR spectra of the Cu(III)-deprotonated amide complexes are surprisingly similar to the RR spectra of the blue copper proteins (Figures 6 and 8), particularly in the occurrence of several intense peaks in the 350- to 440-cm^{-1} region. Yet, X-ray crystal structure determinations on two blue copper proteins, Az[59] and Pc,[58] show no indication of amide ligation, and the RR spectrum of Az can be explained from a NCA treatment using only histidine, cysteine, and methionine as copper ligands (see Section III.A). Thus, it would be difficult to distinguish between these two sets of ligands in a copper-containing protein on the basis of a Raman spectrum alone.

VII. CONCLUSIONS

RR spectroscopy has proved to be a useful technique for obtaining direct information on the nature of copper ligands and coordination geometry in copper proteins. Spectral interpretations are aided by comparison with model complexes and conclusions are often testable by use of isotope, ligand, or metal substitutions and by NCA. Even though the resonance effect simplifies spectra by selective enhancement of metal chromophore vibrations, care must be exercised as many of the expected protein ligands exhibit quite similar RR spectra. The copper proteins most amenable to RR spectroscopy are those with intense LMCT bands: proteins containing blue (type-1) copper or binuclear (type-3) copper. The RR spectra of

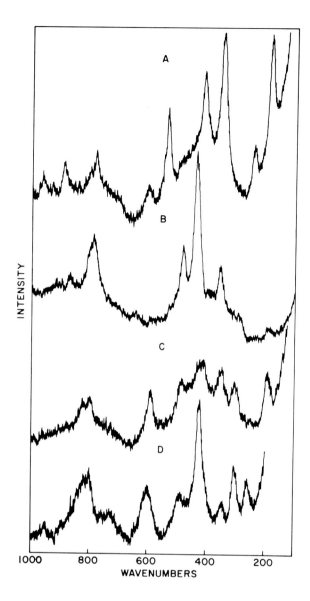

FIGURE 18. Resonance Raman spectra of Cu(III) macrocycles, approximately 0.1 mM in H$_2$O. Scan rate 50 cm^{-1}/min, slitwidth approximately 8 cm^{-1}. Cu(II) macrocycles were prepared by anaerobic reaction of copper(II) nitrate with 10 times excess aqueous oxalodihydrazide. Condensation was initiated by adjusting the pH to >7.5 with aqueous NH$_3$ and adding the appropriate aldehyde or ketone. Cu(III) macrocycles are produced by air oxidation. Macrocycles prepared with the following ratios of aldehyde or ketone to Cu(II) bis(oxalodihydrazide) had distinct absorption spectra, and Ar-Kr laser excitation used in each case was that closest to λ_{max}. (A) 10^5 times excess formaldehyde, λ_{max} = 545 nm (ϵ > 10,000 M^{-1} cm^{-1}), 530.9-nm excitation. (B) 10^4 times excess acetaldehyde, λ_{max} = 540 nm (ϵ = 29,500 M^{-1} cm^{-1}), 530.9-nm excitation. (C) 10 times acetaldehyde, λ_{max} = 608 nm (ϵ = 14,000 M^{-1} cm^{-1}), 647.1-nm excitation. (D) 10^4 times excess acetone, λ_{max} = 585 nm (ϵ = 12,000 M^{-1} cm^{-1}), 568.2-nm excitation.[87]

the blue copper proteins are well explained by a tetrahedral site containing histidine, cysteine, and possibly methionine as ligands. The binuclear copper site has thus far only been successfully studied in proteins which form an oxygenated species with intense O$_2^{2-}$ $\rightarrow$ Cu(II) CT in the visible region. In addition to the identification of the bound O$_2$ as peroxide, the

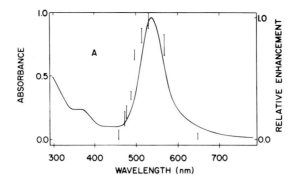

FIGURE 19.   Visible absorption spectrum (solid line, arbitrary ordinate) and Raman intensity enhancement profile (bars) for 40 μ*M* Cu(III)-macrocycle prepared from 10⁴ times excess acetaldehyde. Intensities of peaks at 352, 434, 477, and 785 cm⁻¹ relative to $v_1(NO_3^-)$ used as an internal standard are normalized to the relative intensity with 520.8-nm excitation.[87]

vibrational spectra also provide evidence for histidine ligation in the binuclear copper site. Future studies, particularly with the use of RR intensity enhancement profiles, will continue to aid in the detailed assignment of electronic transitions. Chromophoric enzymatic intermediates may also yield considerable structural information in future RR investigations.

ACKNOWLEDGMENTS

We wish to acknowledge the research contributions of our students and collaborators whose work is cited in this review. We thank Dr. Teresa B. Freedman (Syracuse University) for helpful discussions on the molecular orbital description of the hemocyanin copper center. Much of the work from the authors' laboratories discussed in this review was supported by a grant from the National Institutes of Health (GM 18865) whose sponsorship is gratefully acknowledged.

Note Added in Proof

Selenomethionine-substituted Az from *P. aeruginosa* has now been fully characterized.[123] The RR spectrum of the seleno-azurin derivative, however, was indistinguishable from that of native Az;[26] this observation shows that the purported methionine ligand does not contribute to the RR spectrum and may, in fact, not be ligated to the Cu(II) ion at all. Consistent with this interpretation are the crystallographically determined Cu-S(Met) distances of 2.90 Å in Pc[124] and ≈2.5 Å in Az,[125] as well as the single crystal EXAFS data from Pc which reveal no methionine contribution.[126] The resonance-enhanced vibration at ≈260 cm⁻¹ in all type-1 copper proteins is, therefore, more reasonably reassigned either to a Cu-S-C deformation mode of coordinated cysteine(s)[26] or to a Cu-N stretching mode of coordinated histidine(s).[127] The latter assignment is supported by observation of a 2 cm⁻¹ shift to lower frequency for a St sample prepared in D_2O. The RR spectra of Cu(II)-substituted liver alcohol dehydrogenase with one histidine and two cysteine ligands in a tetrahedral metal ion site are very similar to those of the blue copper proteins and lend additional support to the presence of strongly coupled metal-ligand vibrations in a constrained protein environment.[128] The appearance of a yet greater number of resonance-enhanced frequencies below 500 cm⁻¹ in type-1 proteins at cryogenic temperatures is being interpreted as evidence for an admixture of ligand deformation modes coupling with metal-ligand stretching and bending modes.[129] Further research is clearly required to fully understand the origin and meaning of the unprecedentedly large number of RR vibrational modes seen in the blue copper proteins.

REFERENCES

1. **Long, Th. V., II and Loehr, Th. M.,** The possible determination of iron coordination in nonheme iron proteins using laser Raman spectroscopy. Rubredoxin, *J. Am. Chem. Soc.,* 92, 6384, 1970.
2. **Spiro, Th. G. and Loehr, Th. M.,** Resonance Raman spectra of heme proteins and other biological systems, in *Advances in Infrared and Raman Spectroscopy,* Vol. 1, Clark, R. J. H. and Hester, R. E., Eds., Heyden, London, 1975, 98.
3. **Clark, R. J. H. and Stewart, B.,** The resonance Raman effect — review of the theory and of applications in inorganic chemistry, *Struct. Bonding (Berlin),* 36, 1, 1979.
4. **Raman, C. V. and Krishnan, K. S.,** A new type of secondary radiation, *Nature (London),* 121, 501, 1928.
5. **Tang, J. and Albrecht, A. C.,** Developments in the theories of vibrational Raman intensities, in *Raman Spectroscopy,* Vol. 2, Szymanski, H. A., Ed., Plenum Press, New York, 1970, 33.
6. **Behringer, J.,** Observed resonance Raman spectra, in *Raman Spectroscopy,* Vol. 1, Szymanski, H. A., Ed., Plenum Press, New York, 1967, 168.
7. **Spiro, Th. G. and Stein, P.,** Resonance effects in vibrational scattering from complex molecules, *Annu. Rev. Phys. Chem.,* 28, 501, 1977.
8. **Johnson, B. B. and Peticolas, W. L.,** The resonant Raman effect, *Annu. Rev. Phys. Chem.,* 27, 465, 1976.
9. **Long, D.A.,** *Raman Spectroscopy,* McGraw-Hill, New York, 1977.
10. **Druet, S. and Taran, J.-P.,** Coherent anti-Stokes Raman spectroscopy, in *Chemical and Biochemical Applications of Lasers,* Vol. 4, Moore, C. B., Ed., Academic Press, New York, 1979, 187.
11. **Hirakawa, A. Y. and Tsuboi, M.,** Molecular geometry in an excited electronic state and a preresonance Raman effect, *Science,* 188, 359, 1975.
12. **Ferraro, J. R.,** Advances in Raman instrumentation and sampling techniques, in *Raman Spectroscopy,* Vol. 1, Szymanski, H. A., Ed., Plenum Press, New York, 1967, 44.
13. **Downey, J. R., Jr. and Janz, G. J.,** Digital methods in Raman spectroscopy, in *Advances in Infrared and Raman Spectroscopy,* Vol. 1, Clark, R.J. H. and Hester, R. E., Eds., Heyden, London, 1975, 1.
14. **Loehr, Th. M., Keyes, W. E., and Pincus, P. A.,** A computer-controlled laser Raman spectrophotometer with interactive-graphics data analysis, *Anal. Biochem.,* 96, 456, 1979.
15. **Shriver, D. F.,** Raman spectroscopy, *Adv. Inorg. Biochem.,* 2, 117, 1980.
16. **Kiefer, W. and Bernstein, H. J.,** A cell for resonance Raman excitation with lasers in liquids, *Appl. Spectrosc.,* 25, 500, 1971.
17. **Kiefer, W., Schmid, W. J., and Topp, J. A.,** A universal rotating system for Raman spectroscopy, *Appl. Spectrosc.,* 29, 434, 1975.
18. **Keyes, W. E., Loehr, Th. M., Taylor, M. L., and Loehr, J. S.,** Protocatechuate 3,4-dioxygenase. Resonance Raman studies of the oxygenated intermediate, *Biochem. Biophys. Res. Commun.,* 89, 420, 1979.
19. **Woodruff, W. H. and Spiro, Th. G.,** Circulating sample cell for temperature control in resonance Raman spectroscopy, *Appl. Spectrosc.,* 28, 74, 1974.
20. **Mathies, R., Oseroff, A. R., and Stryer, L.,** Rapid-flow resonance Raman spectroscopy of photolabile molecules: rhodopsin and isorhodopsin, *Proc. Natl. Acad. Sci. U.S.A.,* 73, 1, 1976.
21. **Woodruff, W. H. and Atkinson, G.H.,** Vidicon detection of resonance Raman spectra—cytochrome c, *Anal. Chem.,* 48, 186, 1976.
22. **Sjöberg, B.-M., Gräslund, A., Loehr, J. S., and Loehr, Th. M.,** Ribonucleotide reductase: a structural study of the dimeric iron site, *Biochem. Biophys. Res. Commun.,* 94, 793, 1980.
23. **Freedman, T. B., Loehr, J. S., and Loehr, Th. M.,** A resonance Raman study of the copper protein, hemocyanin. New evidence for the structure of the oxygen-binding site, *J. Am. Chem. Soc.,* 98, 2809, 1976.
24. **Larrabee, J. A. and Spiro, Th. G.,** Structural studies of the hemocyanin active site. II. Resonance Raman spectroscopy, *J. Am. Chem. Soc.,* 102, 4217, 1980.
25. **Thompson, J. S., Marks, T. J., and Ibers, J. A.,** Blue copper proteins: synthesis, spectra, and structures of $Cu^I N_3(SR)$ and $Cu^{II} N_3(SR)$ active site analogues, *Proc. Natl. Acad. Sci. U.S.A.,* 74, 3114, 1977.
26. **Thamann, Th. J., Frank, P., Willis, L. J., and Loehr, Th. M.,** Normal coordinate analysis of the copper center of azurin and the assignment of its resonance Raman spectrum, *Proc. Natl. Acad. Sci. U.S.A.,* 79, 6396, 1982.
27. **Nakamoto, K.,** *Infrared Spectra of Inorganic and Coordination Compounds,* 2nd ed., Wiley-Interscience, New York, 1970.
28. **Thamann, Th. J. and Loehr, Th. M.,** Raman spectra and normal coordinate analysis of the copper(II) and copper(III) complexes of biuret and oxamide, *Spectrochim. Acta,* 36A, 751, 1980.

29. **Tosi, L. and Garnier, A.,** Circular dichroism and resonance Raman spectra of the Cu(II)-Cu(I) complex of D-penicillamine. The CuS(cys) stretching mode in blue copper proteins, *Biochem. Biophys. Res. Commun.*, 91, 1273, 1979.

30. **Siiman, O. and Carey, P. R.,** Resonance Raman spectra of some ferric and cupric thiolate complexes, *J. Inorg. Biochem.*, 12, 353, 1980.

31. **Ferris, N. S., Woodruff, W. H., Rorabacher, D. B., Jones, T. E., and Ochrymowycz, L. A.,** Resonance Raman spectra of copper-sulfur complexes and the blue copper protein question, *J. Am. Chem. Soc.*, 100, 5939, 1978.

32. **Herlinger, A. W., Wenhold, S. L., and Long, Th. V., II,** Infrared spectra of amino acids and their metal complexes. II. Geometrical isomerism in bis(amino acidato)copper(II) complexes, *J. Am. Chem. Soc.*, 92, 6474, 1970.

33. **Lever, A. B. P. and Ramaswamy, B. S.,** Isotopic studies of the metal-ligand bond. II. The far infrared spectra of some binuclear and polymeric copper carboxylate derivatives; variable temperature and isotopic studies of the copper-ligand vibrations, *Can. J. Chem.*, 51, 514, 1973.

34. **Percy, G. C. and Stenton, H. S.,** Infrared and electronic spectra of N-salicylideneglycinate complexes of cobalt and nickel, *Spectrochim. Acta*, 32A, 1615, 1976.

35. **Breslow, E.,** Metal-protein complexes, in *Inorganic Biochemistry*, Vol. 1, Eichhorn, G. L., Ed., Elsevier, Amsterdam, 1973, 227.

36. **Catterick, J. and Thornton, P.,** Structures and physical properties of polynuclear carboxylates, *Adv. Inorg. Chem. Radiochem.*, 20, 291, 1977.

37. **Stenkamp, R. E. and Jensen, L. H.,** Hemerythrin and myohemerythrin. A review of models based on X-ray crystallographic data, *Adv. Inorg. Biochem.*, 1, 219, 1979.

38. **Quiocho, F. A. and Lipscomb, W. N.,** Carboxypeptidase A: a protein and an enzyme, *Adv. Protein Chem.*, 25, 1, 1971.

39. **Tomimatsu, Y., Kint, S., and Scherer, J. R.,** Resonance Raman spectra of iron(III)-, copper(II)-, cobalt(III)-, and manganese(III)-transferrins and of bis(2,4,6-trichlorophenolato)diimidazolecopper(II) monohydrate, a possible model for copper(II) binding to transferrins, *Biochemistry*, 15, 4918, 1976.

40. **Ainscough, E. W., Brodie, A. M., Plowman, J. E., Bloor, S. J., Loehr, J. S., and Loehr, Th. M.,** Studies on human lactoferrin by electron paramagnetic resonance, fluorescence, and resonance Raman spectroscopy, *Biochemistry*, 19, 4072, 1980.

41. **Salama, S. and Spiro, Th. G.,** Resonance Raman spectra of cobalt(II)-imidazole complexes: analogues of the binding site of cobalt-substituted zinc proteins, *J. Am. Chem. Soc.*, 100, 1105, 1978.

42. **Loehr, Th. M., Keyes, W. E., and Loehr, J. S.,** Resonance Raman spectroscopic studies of non-heme-iron dioxygenases, in *Oxidases and Related Redox Systems*, King, T. E., Mason, H. S., and Morrison, M., Eds., Pergamon Press, Oxford, 1982, 463.

43. **Tang, S.-P. W., Spiro, Th. G., Mukai, K., and Kimura, T.,** Resonance Raman scattering and optical absorption of adrenodoxin and selena-adrenodoxin, *Biochem. Biophys. Res. Commun.*, 53, 869, 1973.

44. **Frank, P.,** The Biochemistry of Transition Metals: Copper, Vanadium, and Iron, Ph.D. thesis, Stanford University, Stanford, Calif., 1980.

45. **Tennent, D. L. and McMillin, D. R.,** A detailed analysis of the charge-transfer bands of a blue copper protein. Studies of the nickel(II), manganese(II), and cobalt(II) derivatives of azurin, *J. Am. Chem. Soc.*, 101, 2307, 1979.

46. **Ferris, N. S., Woodruff, W. H., Tennent, D.L., and McMillin, D. R.,** Native azurin and its Ni(II) derivative: a resonance Raman study, *Biochem. Biophys. Res. Commun.*, 88, 288, 1979.

47. **Albrecht, A. C. and Hutley, M. C.,** On the dependence of vibrational Raman intensity on the wavelength of incident light, *J. Chem Phys.*, 55, 4438, 1971.

48. **Wilson, E. B., Jr., Decius, J. C., and Cross, P. C.,** *Molecular Vibrations*, McGraw-Hill, New York, 1955.

49. **Califano, S.,** *Vibrational States*, John Wiley & Sons, Chichester, 1976.

50. **Abe, M., Kitagawa, T., and Kyogoku, Y.,** Resonance Raman spectra of octaethylporphyrinato-Ni(II) and *meso*-deuterated and ^{15}N substituted derivatives. II. A normal coordinate analysis, *J. Chem. Phys.*, 69, 4526, 1978.

51. **Warshel, A.,** Interpretation of resonance Raman spectra of biological molecules, *Annu. Rev. Biophys. Bioeng.*, 6, 273, 1977.

52. **Malkin, R. and Malmström, B. G.,** The state and function of copper in biological systems, *Adv. Enzymol.*, 33, 177, 1970.

53. **Fee, J. A.,** Copper proteins — systems containing the ''blue'' copper center, *Struct. Bonding (Berlin)*, 23, 1, 1975.

54. **Solomon, E. I., Hare, J. W., Dooley, D. M., Dawson, J. H., Stephens, P. J., and Gray, H. B.,** Spectroscopic studies of stellacyanin, plastocyanin, and azurin. Electronic structure of the blue copper sites, *J. Am. Chem. Soc.*, 102, 168, 1980.

55. **Gray, H. B. and Solomon, E. I.,** Electronic structures of blue copper centers in proteins, in *Copper Proteins, Metal Ions in Biology,* Vol. 3, Spiro, Th. G., Ed., John Wiley & Sons, New York, 1981, chap. 1.
56. **Markley, J. L., Ulrich, E. L., Berg, S.P., and Krogmann, D. W.,** Nuclear magnetic resonance studies of the copper binding sites of blue copper proteins: oxidized, reduced, and apoplastocyanin, *Biochemistry,* 14, 4428, 1975.
57. **Tullius, Th. D., Frank, P., and Hodgson, K. O.,** Characterization of the blue copper site in oxidized azurin by extended X-ray absorption fine structure: determination of a short Cu-S distance, *Proc. Natl. Acad. Sci. U.S.A.,* 75, 4069, 1978.
58. **Colman, P. M., Freeman, H. C., Guss, J. M., Murata, M., Norris, V. A., Ramshaw, J. A. M., and Venkatappa, M. P.,** X-ray crystal structure analysis of plastocyanin at 2.7 Å resolution, *Nature (London),* 272, 319, 1978.
59. **Adman, E. T., Stenkamp, R. E., Sieker, L. C., and Jensen, L. H.,** A crystallographic model for azurin at 3 Å resolution, *J. Mol. Biol.,* 123, 35, 1978.
60. **Peisach, J. and Blumberg, W. E.,** Structural implications derived from the analysis of electron paramagnetic resonance spectra of natural and artificial copper proteins, *Arch. Biochem. Biophys.,* 165, 691, 1974.
61. **McLendon, G. and Martell, A. E.,** Sequence structure relationships in metalloproteins. II. A proposed coordination model for plastocyanin, *J. Inorg. Nucl. Chem.,* 39, 191, 1977.
62. **Markley, J. L., Ulrich, E. L., and Krogmann, D. W.,** Spinach plastocyanin: comparison of reduced and oxidized forms by natural abundance carbon-13 nuclear magnetic resonance spectroscopy, *Biochem. Biophys. Res. Commun.,* 78, 106, 1977.
63. **Bergman, C., Gandvik, E.-K., Nyman, P. O., and Strid, L.,** The amino acid sequence of stellacyanin from the lacquer tree, *Biochem. Biophys. Res. Commun.,* 77, 1052, 1977.
64. **Miskowski, V., Tang, S.-P. W., Spiro, Th. G., Shapiro, E., and Moss, Th. H.,** The copper coordination group in ''blue'' copper proteins: evidence from resonance Raman spectra, *Biochemistry,* 14, 1244, 1975.
65. **Blumberg, W. E.,** Some aspects of models of copper complexes, in *The Biochemistry of Copper,* Peisach, J., Aisen, P., and Blumberg, W. E., Eds., Academic Press, New York, 1966, 49.
66. **Sugiura, Y., Hirayama, Y., Tanaka, H., and Ishizu, K.,** Copper(II) complex of sulfur-containing peptides. Characterization and similarity of electron spin resonance spectrum to the chromophore in blue copper proteins, *J. Am. Chem. Soc.,* 97, 5577, 1975.
67. **Hirayama, Y. and Sugiura, Y.,** An analogous ligand of blue copper active sites: synthesis, electron spin resonance characteristics of its copper(II) complex, and role of proline residue, *Biochem. Biophys. Res. Commun.,* 86, 40, 1979.
68. **Gould, D. C. and Ehrenberg, A.,** Cu^{2+} in non-axial field: a model for Cu^{2+} in copper enzymes, *Eur. J. Biochem.,* 5, 451, 1968.
69. **Sakaguchi, U. and Addison, A. W.,** Structural implications for blue protein copper centers from electron spin resonance spectra of $Cu^{II}S_4$ chromophores, *J. Am. Chem. Soc.,* 99, 5189, 1977.
70. **Bereman, R. D., Wang, F. T., Najdzionek, J., and Braitsch, D. M.,** Stereoelectronic properties of metalloenzymes. IV. Bis(imidotetraphenyldithiodiphosphino-S,S') copper(II) as a tetrahedral model for type I copper(II), *J. Am. Chem. Soc.,* 98, 7266, 1976.
71. **Jones, T. E., Rorabacher, D. B., and Ochrymowycz, L. A.,** Simple models for ''blue'' copper proteins. The copper-thiaether complexes, *J. Am. Chem. Soc.,* 97, 7485, 1975.
72. **Amundsen, A. R., Whelan, J., and Bosnich, B.,** Biological analogues. On the nature of the binding sites of copper-containing proteins, *J. Am. Chem. Soc.,* 99, 6730, 1977.
73. **Bernarducci, E., Schwindinger, W. F., Hughey, J. L., IV, Krogh-Jespersen, K., and Schugar, H. J.,** Electronic spectra of Cu(II)-imidazole and Cu(II)-pyrazole chromophores, *J. Am. Chem. Soc.,* 103, 1686, 1981.
74. **Birker, P. J. M. W. L. and Freeman, H. C.** Structure, properties, and function of a copper(I)-copper(II) complex of D-penicillamine: pentathallium(I) μ_8-chloro-dodeca(D-penicillaminato)octacuprate(I) hexacuprate(II) n-hydrate, *J. Am. Chem. Soc.,* 99, 6890, 1977.
75. **Czernuszewicz, R., Maslowsky, E., Jr., and Nakamoto, K.,** Infrared and Raman spectra of bis(imidotetraphenyldithiodiphosphino-S,S') complexes with Cu(II), Co(II) and Fe(II), *Inorg. Chim. Acta,* 40, 199, 1980.
76. **Thompson, J. S., Marks, T. J., and Ibers, J. A.,** Blue copper proteins. Synthesis, chemistry, and spectroscopy of $Cu^IN_3(SR)$ and $Cu^{II}N_3(SR)$ active site approximations. Crystal structure of potassium p-nitrobenzenethiolato(hydrotris(3,5-dimethyl-1-pyrazolyl)borato)cuprate(I)diacetone, K[Cu(HB(3,5-Me$_2$pz)$_3$)(SC$_6$H$_4$NO$_2$)]·2C$_3$H$_6$O, *J. Am. Chem. Soc.,* 101, 4180, 1979.
77. **Siiman, O., Young, N. M., and Carey, P. R.** Resonance Raman spectra of ''blue'' copper proteins and the nature of their copper sites, *J. Am. Chem. Soc.,* 98, 744, 1976.
78. **Siiman, O., Young, N. M., and Carey, P. R.,** Resonance Raman studies of ''blue'' copper proteins, *J. Am. Chem. Soc.,* 96, 5583, 1974.

79. **Tosi, L., Garnier, A., Hervé, M., and Steinbuch, M.,** Ceruloplasmin-anion interaction. A resonance Raman spectroscopic study, *Biochem. Biophys. Res. Commun.,* 65, 100, 1975.

80. **McCord, J. M. and Fridovich, I.,** Superoxide dismutase. An enzymic function for erythrocuprein (hemocuprein), *J. Biol. Chem.,* 244, 6049, 1969.

81. **Rotilio, G., Finazzi-Agrò, A., Calabrese, L., Bossa, F., Guerrieri, P., and Mondovì, B.** Studies of the metal sites of copper proteins. Ligands of copper in hemocuprein, *Biochemistry,* 10, 616, 1971.

82. **Haffner, P. H. and Coleman, J. E.,** Cu(II)-carbon bonding in cyanide complexes of copper enzymes. ^{13}C splitting of the Cu(II) electron spin resonance, *J. Biol. Chem.,* 248, 6626, 1973.

83. **Malmström, B. G., Reinhammar, B., and Vänngård, T.,** Two forms of copper(II) in fungal laccase, *Biochim. Biophys. Acta,* 156, 67, 1968.

84. **Richardson, J. S., Thomas, K. A., Rubin, B. H., and Richardson, D. C.,** Crystal structure of bovine Cu,Zn, superoxide dismutase at 3 Å resolution: chain tracing and metal ligands, *Proc. Natl. Acad. Sci. U.S.A.,* 72, 1349, 1975.

85. **Kato, M.,** Zur Komplexchemie der Biuretreaktion. V. Mitt. Spektrochemische Untersuchungen über die Biuret- und Proteinkomplexe des dreiwertigen Kobalts, *Z. Anorg. Allg. Chem.,* 300, 84, 1959.

86. **Kincaid, J. R., Larrabee, J. A., and Spiro, Th. G.,** Resonance Raman spectra of copper(III) peptide complexes, *J. Am. Chem. Soc.,* 100, 334, 1978.

87. **Dunn, J. B. R., Loehr, J. S., and Loehr, Th. M.,** unpublished results.

88. **Urbach, F. L.,** The properties of binuclear copper centers in model and natural compounds, in *Metal Ions in Biological Systems,* Vol. 13, Sigel, H., Ed., Marcel Dekker, New York, 1981, 73.

89. **Lontie, R. and Witters, R.,** Hemocyanin, in *Inorganic Biochemistry,* Vol. 1, Eichhorn, G. L., Ed., Elsevier, Amsterdam, 1973, 344.

90. **Solomon, E. I.,** Binuclear copper active site: hemocyanin, tyrosinase, and type 3 copper oxidases, in *Copper Proteins, Metal Ions in Biology,* Vol. 3, Spiro, Th. G., Ed., John Wiley & Sons, New York, 1981, chap. 2.

91. **Loehr, J. S., Freedman, T. B., and Loehr, Th. M.,** Oxygen binding to hemocyanin: a resonance Raman spectroscopic study, *Biochem. Biophys. Res. Commun.,* 56, 510, 1974.

92. **Freedman, T. B., Yoshida, C. M., and Loehr, Th. M.,** Resonance Raman spectra of μ-peroxo binuclear cobalt(III) complexes, *J. Chem. Soc., Chem. Commun.,* 1016, 1974.

93. **Kozuka, M., Suzuki, M., Nishida, Y., Kida, S., and Nakamoto, K.,** Resonance Raman spectra of molecular oxygen adducts of Co(II) chelates in solution equilibria, *Inorg. Chim. Acta,* 45, L111, 1980.

94. **Evans. J. C.,** The peroxide-ion fundamental frequency, *Chem. Commun.,* 682, 1969.

95. **Shibahara, T.,** Laser-Raman and infrared spectra of μ-superoxo-dicobalt(III) complexes, *J. Chem. Soc., Chem. Commun.,* 864, 1973.

96. **Dooley, D. M., Scott, R. A., Ellinghaus, J., Solomon, E. I., and Gray, H. B.,** Magnetic susceptibility studies of laccase and oxyhemocyanin, *Proc. Natl. Acad. Sci. U.S.A.,* 75, 3019, 1978.

97. **Thamann, Th. J., Loehr, J. S., and Loehr, Th. M.,** Resonance Raman study of oxyhemocyanin with unsymmetrically labeled oxygen, *J. Am. Chem. Soc.,* 99, 4187, 1977.

98. **Klotz, I. M., Duff, L. L., Kurtz, D. M., Jr., and Shriver, D. F.,** Oxidation state and structural disposition of dioxygen in oxygen-carrying proteins, in *Invertebrate Oxygen-Binding Proteins: Structure, Active Site, and Function,* Lamy, J. and Lamy, J., Eds., Marcel Dekker, New York, 1981, 469.

99. **Salvato, B., Ghiretti-Magaldi, A., and Ghiretti, F.,** Acid-base titration of hemocyanin from *Octopus vulgaris,* Lam., *Biochemistry,* 13, 4778, 1974.

100. **Larrabee, J. A., Spiro, Th. G., Ferris, N. S., Woodruff, W. H., Maltese, W. A., and Kerr, M. S.,** Resonance Raman study of mollusc and arthropod hemocyanins using ultraviolet excitation: copper environment and subunit inhomogeneity, *J. Am. Chem. Soc.,* 99, 1979, 1977.

101. **Eickman, N. C., Himmelwright, R. S., and Solomon, E. I.,** Geometric and electronic structure of oxyhemocyanin. Spectral and chemical correlations to met apo, half met, met, and dimer active sites, *Proc. Natl. Acad. Sci. U.S.A.,* 76, 2094, 1979.

102. **Wing, R. M. and Callahan, K. P.,** The characterization of metal-oxygen bridge systems, *Inorg. Chem.* 8, 871, 1969.

103. **Summerville, D. A., Jones, R. D., Hoffman, B. M., and Basolo, F.** Assigning oxidation states to some metal dioxygen complexes of biological interest, *J. Chem. Educ.,* 56, 157, 1979.

104. **Hay, P. J., Thibeault, J. C., and Hoffmann, R.,** Orbital interactions in metal dimer complexes, *J. Am. Chem. Soc.,* 97, 4884, 1975.

105. **Brown, J. M., Powers, L., Kincaid, B., Larrabee, J. A., and Spiro, Th. G.,** Structural studies of the hemocyanin active site. I. Extended X-ray absorption fine structure (EXAFS) analysis, *J. Am. Chem. Soc.,* 102, 4210, 1980.

106. **Chen, J. T., Shen, S. T., Chung, C. S., Chang, H., Wang, S. M., and Li, N. C.,** *Achatina fulica* hemocyanin and its interactions with imidazole, potassium cyanide, and fluoride as studied by spectrophotometry and nuclear magnetic resonance and resonance Raman spectrometry, *Biochemistry,* 18, 3097, 1979.

107. **Gielens, C., Maes, G., Zeegers-Huyskens, Th., and Lontie, R.,** Raman resonance studies of functional fragments of *Helix pomatia* β$_c$-haemocyanin, *J. Inorg. Biochem.,* 13, 41, 1980.

108. **Mason, H. S.,** Oxidases, *Annu. Rev. Biochem.,* 34, 595, 1965.

109. **Makino, N., McMahill, P., Mason, H. S., and Moss, Th. H.,** The oxidation state of copper in resting tyrosinase, *J. Biol. Chem.,* 249, 6062, 1974.

110. **Jolley, R. L., Jr., Evans, L. H., and Mason, H. S.,** Reversible oxygenation of tyrosinase, *Biochem. Biophys. Res., Commun.,* 46, 878, 1972.

111. **Makino, N. and Mason, H. S.,** Reactivity of oxytyrosinase toward substrates, *J. Biol. Chem.,* 248, 5731, 1973.

112. **Mason, H. S.** Structures and functions of the phenolase complex, *Nature (London),* 177, 79, 1956.

113. **Eickman, N. C., Solomon, E. I., Larrabee, J. A., Spiro, Th. G., and Lerch, K.,** Ultraviolet resonance Raman study of oxytyrosinase. Comparison with oxyhemocyanins, *J. Am. Chem. Soc.,* 100, 6529, 1978.

114. **Lerch, K.,** Amino acid sequence of tyrosinase from *Neurospora crassa, Proc. Natl. Acad. Sci. U.S.A.,* 75, 3635, 1978.

115. **Himmelwright, R. S., Eickman, N. C., LuBien, C. D., Lerch, K., and Solomon, E. I.,** Chemical and spectroscopic studies of the binuclear copper active site of *Neurospora* tyrosinase: comparison to hemocyanins, *J. Am. Chem. Soc.,* 102, 7339, 1980.

116. **Hamilton, G. A., Adolf, P. K., de Jersey, J., DuBois, G. C., Dyrkacz, G. R., and Libby, R. D.,** Trivalent copper, superoxide, and galactose oxidase, *J. Am. Chem. Soc.,* 100, 1899, 1978.

117. **Winkler, M. E. and Bereman, R. D.,** Stereoelectronic properties of metalloenzymes. VI. Effects of anions and ferricyanide on the copper(II) site of the histidine and the tryptophan modified forms of galactose oxidase, *J. Am. Chem. Soc.,* 102, 6244, 1980.

118. **Blumberg, W. E., Peisach, J., Kosman, D. J., and Mason, H. S.,** Is trivalent copper a viable oxidation state in the enzymatic turnover of copper proteins?, in *Oxidases and Related Redox Systems,* King, T. E., Mason, H. S., and Morrison, M., Eds., Pergamon Press, Oxford, 1982, 207.

119. **Margerum, D. W., Chellappa, K. L., Bossu, F. P., and Burce, G. L.,** Characterization of a readily accessible copper(III)-peptide complex, *J. Am. Chem. Soc.,* 97, 6894, 1975.

120. **Keyes, W. E., Dunn, J. B. R., and Loehr, Th. M.,** A water-stable Cu(III) complex, *J. Am. Chem. Soc.,* 99, 4527, 1977.

121. **Keyes, W. E., Swartz, W. E., Jr., and Loehr, Th. M.,** X-ray photoelectron spectroscopy of a copper(III) macrocyclic complex, *Inorg. Chem.,* 17, 3316, 1978.

122. **Clark, G. R., Skelton, B. W., and Waters, T. N.,** An investigation of the interaction between copper(II) ion, oxalyldihydrazide, and molecular oxygen: crystal and molecular structures of products resulting from the addition of an excess of acetaldehyde, *J. Chem. Soc. Dalton,* 1528, 1976.

123. **Frank, P., Licht, A., Tullius, Th. D., Hodgson, K. O., and Pecht, I.,** A selenomethionine containing auxotroph of *Pseudomonas aeruginosa* azurin, *Biochemistry,* in press.

124. **Freeman, H. C.,** Electron transfer in 'blue' copper proteins, in *Coordination Chemistry,* Vol. 21, Laurent, J. P., Ed., Pergamon Press, Oxford, 1981, 29.

125. **Adman, E. T. and Jensen, L. H.,** Structural features of azurin at 2.7 Å resolution, *Isr. J. Chem.,* 21, 8, 1981.

126. **Scott, R. A., Hahn, J. E., Doniach, S., Freeman, H. C., and Hodgson, K. O.,** Polarized x-ray absorption spectra of oriented plastocyanin single crystals. Investigation of methionine-copper coordination, *J. Am. Chem. Soc.,* 104, 5364, 1982.

127. **Nestor, L., Larrabee, J. A., Woolery, G., Reinhammar, B., and Spiro, Th. G.,** unpublished results.

128. **Maret, W., Zeppezauer, M., Sanders-Loehr, J., and Loehr, Th. M.,** Resonance Raman spectra of copper(II)-substituted liver alcohol dehydrogenase: a type 1 copper analogue, *Biochemistry,* 22, 3202, 1983.

129. **Woodruff, W. H., Norton, K. A., Swanson, B. I., and Fry, H. A.,** Cryoresonance Raman spectroscopy of plastocyanin, azurin, and stellacyanin. A reevaluation of the identity of the resonance-enhanced modes, *J. Am. Chem. Soc.,* 105, 657, 1983.

Chapter 6

STRUCTURE AND EVOLUTION OF THE SMALL BLUE PROTEINS

Lars Rydén

TABLE OF CONTENTS

I. INTRODUCTION

The blue proteins are conspicuous because of their color and in several instances they have first been detected and isolated with the color as guidance. This is particularly true for the small blue proteins for which copper content and color are the only easily measurable properties. The higher plant blue proteins were observed in conjunction with enzyme purifications. Thus the *Rhus* blue protein,[1] later renamed stellacyanin (St),[2] was obtained from the lacquer of the Japanese lacquer tree as a "by-product" in the isolation of laccase in 1961, and others were found in connection with the isolation of peroxidase from horseradish roots[3] and of ascorbate oxidase from cucumber.[4]

In contrast, the bacterial blue proteins have been obtained by investigators of bacterial redox systems. The first to be reported was the *Pseudomonas* blue protein in 1956,[5] later renamed azurin (Az),[6] followed by similar proteins from *Alcaligenes, Paracoccus,* and *Thiobacillus* to mention a few. Only in the latter case has a definite role in the bacterial redox systems been ascribed to one of them. The *Thiobacillus ferrooxidans* blue protein seems to be involved in the iron-oxidation system in these bacteria.[7]

The isolation of a blue protein from a green alga, *Chlorella,* and from chloroplasts of higher plants, reported in 1960,[8] followed upon investigations of the role of copper in photosynthesis. This protein, which was given the name plastocyanin (Pc),[9] has later been assigned a precise role in the photosynthetic redox chain in photosystem II. It has also been found in the prokaryotic cyanobacteria, also called blue-green algae, and is thus an interesting link between prokaryotic and eukaryotic cells. It is today the best known of the small blue proteins both from a structural and a functional point of view.

The small blue proteins have invariably been found to contain copper(II) with spectral properties of what generally is referred to as type 1; they have an absorption maximum close to 600 nm with an absorption coefficient in the range of 3,000 to 5,000 M^{-1} cm^{-1} and a unique EPR spectrum with a small hyperfine coupling constant. A large number of spectroscopic studies have been published over the years. It is only more recently that these have been paralleled by investigations on the primary and tertiary structures. The relatively small size of this group of proteins — they all fall in the range 10,000 to 25,000 in M_r, the larger ones containing an appreciable amount of carbohydrate — makes them suitable objects for such studies. During the 1970s a number of amino-acid sequences have become available, mainly from laboratories where the primary field of interest was molecular evolution. Deductions about copper-liganding amino acids were possible from the sequence data. The breakthrough in the area came, however, when the high resolution X-ray structures of poplar Pc[10] and of *Pseudomonas* Az[11] were published in 1978.

Precise knowledge of the copper ligands (His, His, Cys, Met) and the geometry of the copper site in these two proteins have been of crucial importance to later work. In particular it has been argued that the same copper coordination exists in the type-1 sites in the large multicopper oxidases and that these in fact evolved from the small blue proteins. These results are critically reviewed below.

Most of the early data on the blue proteins, including the small blue proteins, are found in earlier reviews, see, e.g., the one by Fee,[12] and will only occasionally be quoted here. In addition, the detailed review on Pc structure, evolution, and function by Boulter and co-workers[13] should be mentioned.

II. OCCURRENCE AND PHYSICOCHEMICAL PROPERTIES

A. Bacterial Blue Proteins

A blue pigment in *Pseudomonas aeruginosa,* first observed by Verhoeven and Takeda,[5] was later isolated by Horio,[14,15] who showed it to be a low M_r copper protein. A similar protein was later obtained by Sutherland and Wilkinson[6] from several strains of *Bordetella*. They introduced the name azurin for this protein. Their studies were later enlarged to include several species of *Pseudomonas* and *Alcaligenes*.[16] The proteins produced by these bacteria have been shown to be closely homologous by amino-acid sequencing (Section III.B.). More recently a similar protein has been reported to be present in *Paracoccus denitrificans*[17] and also in *Th. ferrooxidans*.[7] Apart from the cyanobacterial Pcs, which will be mentioned below, these seem to be the only prokaryotic blue proteins reported up to date.

The bacterial species mentioned fall in a rather close group. *Pseudomonas*, *Alcaligenes*, and *Bordetella* are found in the Pseudomonaceae, a family of free-living Gram-negative aerobic bacteria. *Paracoccus* belongs to the Enterobacteriaceae, but resembles several species of the genera *Alcaligenes* and *Pseudomonas*, the main difference being the coccoid shape during exponential growth and the absence of flagellation.[18] *Paracoccus denitrificans* and some species of *Thiobacillus*, in particular the denitrifying strains, are in turn quite similar and were earlier confused.[19] Doudoroff[20] summarizes the situation: "On many grounds, it seems likely that the genera *Paracoccus*, *Pseudomonas*, and *Alcaligenes* are very closely related to each other and that at least the facultatively organotrophic species of the genus *Thiobacillus* belong in the same generic cluster."

The relatedness of the proteins isolated from *Paracoccus* and *Thiobacillus* to the *Pseudomonas* blue protein needs to be proved by amino-acid sequencing. Given their very similar physical properties (Table 1), it seems at present reasonable to assume that they all form a close family, which might be called azurins. This name is also given the *Paracoccus* blue protein, while the discoverers of the *Thiobacillus* protein preferred the name rusticyanin to stress its involvement in iron oxidation.

The M_r values of the Az's are all close to 16,000 (Table 1). The precise figure given by amino-acid sequencing is 14,600. A single chain is present. The copper content indicates the presence of a single copper atom per mole of protein. No carbohydrate has been reported.

B. Higher Plant Nonphotosynthetic Blue Proteins

The first small blue protein from plant tissue was isolated by Omura from the lacquer of the Japanese lacquer tree, *Rhus vernicifera*.[1] The protein was later characterized in more detail by Peisach and co-workers[2] who gave it the name stellacyanin. Later similar proteins have been reported to be present in mung bean seedlings,[25] rice bran,[24] horseradish roots,[3] cucumber,[27] and squash.[26] Those from horseradish and squash were named umecyanin and mavicyanin, respectively. The sources represent a number of angiospermic plants with no apparent generic grouping. As reported in Table 1, the proteins all have M_r values around 15,000 to 23,000 and contain carbohydrate and a single copper atom. The higher M_r values are associated with a higher content of carbohydrate so that the peptide-chain length does not exceed 150 residues in any of them. The proteins have been isolated from nonphotosynthetic tissue and are thus not to be confused with Pc, present in the green parts of the plants (see below). Two of the proteins for which amino-acid sequence data are available (St and umecyanin, see below) have been found to be closely homologous. In view of the rather similar physicochemical properties of the rest of them, it seems reasonable to assume that they represent a single family of plant proteins. If this is confirmed by future sequence data it would be desirable to introduce one name for the blue plant glycoproteins. In this review I will occasionally use the name phytocyanin.

The presence of a small blue plant protein, that differed markedly from those discussed above, in cucumber seedlings was reported by Vickery and Purves[28] in 1972. The same protein was described by Markossian and co-workers[4] 2 years later. Crystallographic data were finally provided by Colman and co-workers[29] in 1977. This protein was named cusacyanin or plantacyanin.[4] The latter name, which has gained some popularity, will be used here. Its M_r from crystallographic data was 10,100.[29] Gel chromatography gives a value

Table 1
PHYSICOCHEMICAL PROPERTIES OF THE SMALL BLUE PROTEINS (1 Cu/mol)

Protein source	Commonly used name	M_r	Method	Peptide-chain length	Carbohydrate (%)	pI	Ref.
Bacterial proteins							
Pseudomonas aeruginosa	Azurin	14,600	Amino-acid sequence	129	Absent	5.4	14,15,21
Pseudomonas fluorescens	Azurin	14,600	Amino-acid sequence	128	Absent		14,15,21
Pseudomonas denitrificans	Azurin	14,600	Amino-acid sequence	128	Absent		14,15,21
Bordetella bronchiseptica	Azurin	14,600	Amino-acid sequence	129	Absent		6
Alcaligenes faecalis	Azurin	14,600	Amino-acid sequence	128	Absent		16
Alcaligenes denitrificans	Azurin	14,600	Amino-acid sequence	129	Absent		16
Paracoccus denitrificans	Azurin	13,800	Gel chromatography	(124)[a]	ND[b]		17
Thiobacillus ferrooxidans	Rusticyanin	16,500	Sedimentation equilibrium	(159)	ND		7,22
Nonphotosynthetic plant glycoproteins							
Lacquer tree latex	Stellacyanin	19,000	Composition	107	40	9.9	1,2
Horseradish root	Umecyanin	14,600	Composition	(125)	3.7	5.9	3,23
Rice bran		18,300	Sedimentation equilibrium	(150)	9.5		24
Mung bean seedlings		23,000	Cu content	(145)	(30)[c]	7.0	25
Green squash fruits	Mavicyanin	18,000	SDS gel electrophoresis	(150)	7	8.9	26
Cucumber peelings		20,000	SDS gel electrophoresis		ND	7.5	27
Nonphotosynthetic plant small proteins							
Cucumber seedlings	Plantacyanin	10,100	Crystal structure	(93)	ND	10.6	4,28,29
Spinach leaves	Plantacyanin	9,000	SDS gel electrophoresis	(81)	ND		30
Photosynthetic proteins							
Green plant chloroplasts	Plastocyanin	11,000	Amino-acid sequence	99	Absent	4.1	9
Green algae (*Chlorella*)	Plastocyanin	11,000	Amino-acid sequence	98	Absent		31,32
Cyanobacteria (*Anabaena*)	Plastocyanin	11,500	Amino-acid sequence	105	Absent		33

a   Values in parentheses are calculated from the M_r.
b   ND = not determined.
c   Only identified as nonprotein material.

close to 8,000[4] while SDS gel electrophoresis and copper content indicate 9,000 to 9,700.[30] No carbohydrate has been reported. This would then correspond to a peptide chain of $\approx$90 amino-acid residues. It contains a single copper atom per molecule. It differs notably from the other blue proteins in its isoelectric point: plantacyanin is a basic protein (pI 10.6), while St is slightly basic and Pc markedly acidic (Table 1). The notion that plantacyanin represents a new protein, different from those discussed above, was also indicated by the results from a study of the copper proteins in cucumber peelings.[27] Four blue proteins — ascorbate oxidase, Pc, plantacyanin, and a St-like protein (phytocyanin) — were separated. Together these accounted for 70 to 80% of the soluble copper in the initial extract. The occurrence of plantacyanin is not well known. The only additional report describes its presence in spinach.[30].

No physiological role has been ascribed to either plantacyanin or the phytocyanins.

C. Plastocyanins: The Photosynthetic Blue Proteins

The isolation of a small blue protein from a green alga, *Chlorella*, was reported by Katoh in 1960.[8] This protein was later shown to be present also in spinach chloroplasts and given the name plastocyanin by Katoh and Takamiya.[9] Its association with chloroplasts has later been substantiated by EPR measurements on isolated chloroplasts.[34] Pc has today been isolated from a large number of vascular plants (angiosperms, gymnosperms, and ferns)[13] as well as from the eukaryotic green algae *Chlorella*, *Enteromorpha*, and *Scenedesmus*.[31,32] Its presence in the blue-green alga, cyanobacterium, *Anabaena*,[33] enlarges this to include prokaryotic oxygenic photosynthetic cells (see also Section IV.E).

A number of Pc's has today been sequenced and the peptide chain is in all cases close to 100 residues long (Section III.B). No carbohydrate has been found, and claims to the contrary have not been confirmed. A single copper atom is present per molecule. Most M_r measurements give values close to 10,000, indicating that a single peptide chain is present per molecule. A value of 40,000 has, however, been reported for *Scenedesmus* Pc using SDS gel electrophoresis.[35,36] Only after heat denaturation did bands at 20,000 and 10,000 appear, and reduction was necessary to obtain a single component at 10,000. Apparently di- and tetramerization were involved here. If the protein is a tetramer its biologically active form seems less clear since similar observations have not been reported for other Pc's. If it is, then certainly the EPR measurements indicate that the coppers are magnetically isolated, i.e., no functional association seems to occur.[13]

Pc is well-studied functionally.[13] Its likely position in the photosynthetic redox chain is at the end of photosystem II receiving electrons from cytochrome *f* for the reduction of P-700 in photosystem I. The close similarity of the redox potential of cytochrome *f* (365 mV) and Pc (370 mV) makes the exact position uncertain. The likely localization of Pc is the inner surface of the chloroplast membrane in contact with the thylakoid space.

III. STRUCTURE

A. Spectroscopic and Redox Properties of the Copper Chromophore

All of the small blue proteins have spectroscopic properties that fall in a narrow range (Table 2) due to the type-1 copper chromophore.[37] The main feature in the optical spectrum is a maximum close to 600 nm (16,500 cm^{-1}) with a molar absorption coefficient in the region 3,000 to 5,000 M^{-1} cm^{-1}. The EPR spectrum is characterized by an unusually small hyperfine coupling constant; $A_\parallel$ is close to 0.006 cm^{-1}. The symmetry of the spectrum varies, but this does not necessarily imply very different geometry at the copper site (see below). The redox potentials are high but comparatively variable (Table 2).

Plantacyanin is the least studied and at the same time the least typical of the four main types of proteins. It has a distinctly green color due to a secondary maximum in the optical

Table 2
SPECTROSCOPIC AND REDOX PROPERTIES OF THE SMALL BLUE PROTEINS

Protein source (cf. Table 1)	Redox potential (mV)	(pH)	λ_{max} (nm)	ϵ_M ($M^{-1}\,cm^{-1}$)	g_z	g_y	g_x	A_z	A_y	A_x	Ref.
Azurins (bacterial proteins)											
Pseudomonas aeruginosa	300	(7.0)	631	3,800	2.260	2.052		0.006		0	38—40
Pseudomonas denitrificans	230	(6.8)	620		2.26	2.055		0.006			39,41
Pseudomonas fluorescens			625	3,500	2.261	2.052		0.0058		0	39
Paracoccus denitrificans	230	(7.0)	595,448^b	1,530	2.290	2.052		0.0077			17
Thiobacillus ferrooxidans (rusticyanin)	680	(2.0)	597	2,200	2.229	2.064	2.019	0.0045	0.0020	0.0065	42
Phytocyanins (nonphotosynthetic plant glycoproteins)											
Lacquer tree latex (stellacyanin)	184	(7.1)	617	3,550	2.287	2.077	2.025	0.0035	0.0029	0.0057	38,43,44
Horseradish roots (umecyanin)	283	(7.0)	610	3,500	2.317	2.05			0.0035		23,45,46
Rice bran	275	(7.4)	600	4,300				<0.002			24
Green squash fruits (mavicyanin)	285	(7.0)	600	5,000	2.287	2.077	2.025	0.0035	0.0029	0.0057	26
Plantacyanins (nonphotosynthetic plant small proteins)											
Cucumber seedlings			597,443^b	3,500	900^b 2.207	2.08	2.02	0.0055	0.001	0.006	27,29
Spinach leaves			593	800							30
Plastocyanins (photosynthetic proteins)											
French bean	347		606	4,360	2.226	2.053		0.0063		<0.0017	38,40,47
Chlorella	390		597	4,700							8

EPR parameters^a: $g_\parallel$ $g_\perp$; $A_\parallel$ $A_\perp$

^{a} A values in cm^{-1}.
^{b} A secondary maximum.

Table 3
CHARGE-TRANSFER SPECTRAL DATA FOR THREE SMALL BLUE PROTEINS

| Protein | Absorption | | Circular dichroism | | Charge-transfer assignment |
	λ_{max} (nm)	ϵ_M ($M^{-1}cm^{-1}$)	λ (nm)	$\Delta\epsilon$ ($M^{-1}\ cm^{-1}$)	$\rightarrow d_{x2-y2}$
Plastocyanin	752	1,289	781	3.78	πS
	606	4,364	606	4.08	σS
	552	1,163	526	0.4	σS*
	450[a]	300	472	−1.32	πN
	428[a]	100	417	1.26	πN
Azurin	779	686	800	−5.9	πS
	631	3,798	621	6.5	σS
	567	504	526	1.2	σS*
	481	198	467	−1.8	πN
Stellacyanin	789	341	781	−5.0	πS
	617	3,549	606	3.6	σS
	676	1,542	526	0.75	σS**[b]
	450	942	446	−7.35	πN[b]

Note: The proteins are plastocyanin from French bean, azurin from Pseudomonas aeruginosa, and stellacyanin from Japanese lacquer tree latex (data from Reference 38). Values were obtained at 270 K (optical spectra) and 295 K (CD spectra) unless otherwise stated, as shown in Figure 1. S denotes thiol sulfur, S* thioether sulfur, and S** disulfide sulfur.

[a] Data recorded at 35 K.
[b] Assignments based on discussion in Reference 38.

spectrum close to 440 nm.[27,29,30] The molar absorption coefficient at 595 nm has been reported to be 800 M^{-1} cm^{-1} from one laboratory[30] and 3,500 M^{-1} cm^{-1} from another.[29] The former value is outside the range typical of type-1 copper sites. The EPR spectrum[27,30] shows strong rhombic distortion, similar to that of St (see below).

Solomon and co-workers[38] have reported a detailed analysis of the optical and circular dichroic spectra of three well-characterized blue proteins: the Az from *P. aeruginosa*, the Pc from French bean *(Phaseolus vulgaris)*, and St, a phytocyanin, plant glycoprotein from the Japanese lacquer tree *(R. vernicifera)* (Table 3). The corresponding spectra are shown in Figure 1. The electronic assignments are similar in all of the three proteins and assume two nitrogenous and two sulfur ligands. In all cases the electronic properties can be rationalized in terms of a tetragonally flattened tetrahedral geometry. If the distortion from a tetrahedron to a square is expressed by an angle β, where the value 90° represents the square-planar limit (see Reference 38), this angle is calculated to be 60, 61, and 60° for the three proteins. This is clear evidence for close similarity in coordination geometry.

Also the EPR characteristics of these three proteins are consistent with a flattened tetra-hedral ligand structure.[39,43,47] St deviates from the other two proteins in having a rhombic EPR spectrum in contrast to an axial one. In both cases, however, calculated values of $g_\parallel$ and $g_\perp$ as well as of $A_\parallel$ and $A_\perp$ could be accommodated with a D_{2d} symmetry, although for St a further splitting was obtained.[38] The calculated values are in fair agreement with experimental data. It is to be noted in this context that St and umecyanin, which are closely homologous (Section III.B), and thus expected to have similar copper sites, have quite different EPR spectra. The former is rhombic,[43] while the latter has axial symmetry.[45] A similar situation might exist within the bacterial small blue proteins since *Pseudomonas* Az[39] and *Thiobacillus* rusticyanin[42] exhibit axial and rhombic symmetry, respectively. Possibly

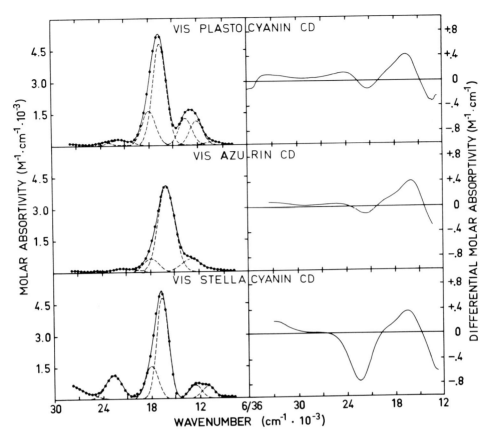

FIGURE 1. Optical and circular dichroic spectra of three blue proteins: plastocyanin from French bean, azurin from *Pseudomonas aeruginosa*, and stellacyanin from Japanese lacquer tree latex.[38] The spectra were obtained at 35 K. The solid curves are the experimental spectra and the dashed curves Gaussian resolutions of these. (Reprinted with permission from Solomon, E. I., Hare, J. W., Dooley, D. M., Dawson, J. H., Stephens, P. J., and Gray H. B., *J. Am. Chem. Soc.*, 102, 168, 1980. Copyright 1980, American Chemical Society.)

the symmetry of the EPR spectrum reflects details of the ligand field and not its main geometry.

The resonance Raman spectra of the type-1 copper proteins are characterized by a number of strong bands in the region of 350 to 470 cm^{-1} and a weak band near 265 cm^{-1}.[48,49] The first set of bands has been assigned to Cu-S(Cys) and Cu-N(His) stretching vibrations.[50] The weaker band has been shown to be typical of Cu-S(thioether) stretchings.[50] This is certainly caused by the methionine ligand in Pc and Az, but not in St, which lacks methionine. The presence of such a band in the spectrum of this protein has therefore been suggested to be due to a Cu-S(disulfide) bond.[50]

For St a ^{1}H, ^{14}N, and ^{63,65}Cu ENDOR study is also available.[51] The unusual copper hyperfine splitting tensor could be explained in terms of a flattened tetrahedral geometry confirming the interpretation of the optical spectra. No assumption had to be made as to the nature of the ligands, pointing to the possibility that a blue type-1 site might exist with other ligands than those known to occur in Pc and Az. Preliminary results from ENDOR studies of Az from *P. aeruginosa* are comparable to those of St.[51] The assignments made of the ^{14}N ENDOR spectrum indicate that the St copper has two or possibly three nitrogenous ligands (see also Section IV.C).

The redox potentials of the type-1 sites are all higher than the value of a tetragonal Cu(II) aquo ion (153 mV), but quite variable (Table 2). The lowest value (180 mV) is found for

St[44] and the highest (345 to 390 mV) for the Pc's.[40] Since evidence exists for a nearly identical geometry of the copper sites in these proteins, the variation observed in redox potential is possibly due to environmental effects and differences in ligands. This assumption is consistent with the absence of a methionine in St. It should, however, be noted that the Az's and Pc's with identical ligands still have quite different redox potentials (Table 2).

B. Amino-Acid Sequences

The amino-acid sequences of 9 Az's, 14 Pc's, and 2 nonphotosynthetic plant glycoproteins (phytocyanins) are shown in Figure 2. Each of these three groups represents a family of closely homologous proteins. For Pc, a composite sequence based on a set of 69 sequences, 3 of which were from species of algae, is shown in Figure 3. Of these, 14 were complete, but most were sequenced up to approximately residue 40 (span: 30 to 46). In the Az family 48 out of 130 positions (37%) are invariant in all 9 sequences. Deletions occur only at the two termini. In the Pc family 26% of the residues is invariant in all sequences. If only the higher plant Pc's are included, this figure increases to 48%. Eight deletions or insertions are found when the higher plant and algal sequences are compared. Most of the conserved residues in the Az and Pc family can be rationalized in terms either of copper binding or chain folding (hydrophobics, glycine, and proline) as discussed in Section III.C.

In the nonphotosynthetic plant glycoprotein family, 1 complete sequence, St of 107 residues,[66] and 1 partial sequence, umecyanin with 88 residues out of a total of $\approx$125,[67] are available. In the 90 positions where these two align, 37% identities were found and only two deletions had to be introduced. The alignment shows that the extra 20 residues of umecyanin appear as a C-terminal elongation. The homology between the two proteins is close enough to permit the assumption that they have a very similar chain folding and identical copper ligands. The fact that the two proteins have different EPR characteristics and redox potentials (Section III.A) indicates, however, that details of the copper binding differ. The close homology would also permit one to expect that the single disulfide bridge in St is conserved in umecyanin. The amino-acid composition of the latter also accounts for three residues of cysteine.

The carbohydrate of St is found at three positions: 27, 60, and 102. In all cases the carbohydrate is bound by an asparagine in an Asn-X-Thr sequence in accordance with what is generally valid for glucosamine-containing glycoproteins. A possible carbohydrate-binding site in umecyanin is position 78 in the sequence X-Thr-Thr. This position was reported to be a "hole" in the Edman degradation, as expected for a carbohydrate-binding asparagine, and the corresponding peptide contained carbohydrate. It is thus evident that the carbohydrate sites are not conserved in the two proteins.

For some other small blue proteins amino-acid composition data are available. Mavicyanin is reported to contain four cysteine residues. The *Paracoccus denitrificans* Az contains a single cysteine, making it likely that the disulfide bridge is not present in this protein. Rusticyanin, the Az from *Thiobacillus*, also contains a single residue of cysteine (the analytical data are, however, inconsistent) and is devoid of arginine.[22]

C. Three-Dimensional Structures

Several groups have reported preliminary crystallographic data for different small blue proteins. Az was crystallized already by Horio.[15] X-ray data were published by Strahs[70] and Norris and co-workers[71] for the *Pseudomonas denitrificans* protein. Finally a crystallographic structure at 3 Å resolution was reported by Adman et al.[11] for *P. aeruginosa* Az. The map was derived from an averaging of the four molecules in the asymmetric unit and used two heavy atom derivatives. Preliminary data for poplar Pc were reported from Freeman's laboratory in Sydney after a long series of failing trials to crystallize Pc from other sources.[72] A well-resolved structure at 2.7 Å resolution was described a year later, based on ortho-

Azurins

```
                      1         10        20        30        40        50        60        70        80        90       '00       110       120
P. aeruginosa         AECSVDIQGNDQMQFNTNAITTVDKSCKQFTVNLSHPGNLPKNVMGHNWVLSTAADMOGVVTDGMASGLDKQYLKPDDSRVIAHTKLIGSGEKDSVTFDVSKLKEGEQYMFFCTFPGHSALMKGTLTLK-
P. denitrificans      E  SVDIQGN Q QFSTNAITVD S     VN S P SLPKNV  W LTTAA MOGVVT   MAA LDKN V DG T VI  KII S    V F   I S   KAGDA AF S    SAM K TLT K-
P. fluorescens B-93   E  KTTIDST Q SFNTKAIEID S  T  VE S T SLPKNV  L ISKQA MOPIAT   LSA IDKN L EG T VI  KVI A    K L   I S   NAAEK GF S    ISM K TVT K-
P. fluorescens C-18   E  KVTVDST Q SFDTKAIEID S  T  VD K S NLPKNV  W LTTQA MOPTAV   MAA IDKN L EG T II  KII A    T V   F S   KADGK MF S    IAM K TVT K-
P. fluorescens D-35   E  KVDVDST Q SFNTKEITID S     VN T S SLPKNV  W LSKSA MAGIAT   MAA IDKD L PG S VI  KII S    V F   F S   TAGES EF S    NSM K AVV K-
Bordetella           E  SVDIAGT Q QFDKKAIEVS S     VN K T KLPRNV  W LTKTA MQAVEK   IAA LDNQ L AG T VL  KVL G    S V   F A   AAGDD TF S    GAL K TLK VD
  bronchiseptica
Alcaligenes           Q  EATIESN A QYDLKEMVVD S     VH K Y KMAKAV  W LTKEA KEGVAT   MNA LAQD V AG T VI  KVI G    V V   F S   TPGEA AY S    WAM K TLK SN
  denitrificans
Alcaligenes faecalis  -  DVSIEGN S QFNTKSIVVD T  E  IN K T KLPAAA  V VSKKS ESAVAT   MKA LNND V AG E VI  SVI G    T V   F S   KEGED AF S    WSI T EIK GS
Alcaligenes sp        E  SVDIAGN G QFDKKEITVS S     VN K P KLAKNV  W LTKQA MOGAVN   MAA LDNN V KD A VI  KVI G    T V   F S   AAGED AY S    FAL K VLK VD
```

Plastocyanins

```
                              1         10        20        30        40        50        60        70        80        90       100
Anabaena (cyanobacterium)     ETYTVKLGSDKGLLVFEPAKLTIKPGDTVEFLNNKVPPHNVVFDAALNPAKSADLAKSLSHKQLLMSPGQSTSTTFPADAPAGEYTFYCEPHRGAGMVGKITVAG
Chlorella (green alga)        -DVTVK ADS A V E SSVTIKA ETVTWV AGF I EDEV SAGNAEAL     --DT-A T GYF E Q    KTI Q-
Rumex (dock)                  --IEIK GDD A A V GSFTVAA EKIVFK AGF V EDEV AGVDASKI MSEEDL NAP ETYAVTL--SE-K T SFY S Q    V KV Q-
Sambucus (elder)              VEIL GE S A Y SN SVPS EK T           EIS SA     S DD     P TYS T   TE S T K     S                  N
Spinacia (spinach)            VEVL GG S A L GD SVAS EE V           EIS AG     N ED     P TYK T   TE K T K     S                  K
Cucurbita (marrow)            IEVL GD S A I ND SVAA EK V           EIS AS     D ND     P VYK N   TE A T S     A                  N
Capsella (shepherd's purse)   IEVL GG S A V ND SIAK EK V           EIA AS     A ED     A TYE A   TE A T A     S                  K
Solanum (potato)              LDVL GD S A I GN SVAA EK T           EIA AS     P ED     P TYS T   SE K T S     S                  K
Solanum crispum               IEVL SD G A V GN SISA EK T           EIA AS     P ED     P TYS T   SE K T S     A                  Q
Vicia (broad bean)            VEVL AS G A V NS EVSA DT V           EIA AA     P EE     P TYS K   DA K T K     S                  N
Phaseolus (French bean)       LEVL SG S V V SE SVPS EK V           EIA AV     P EE     P TYV T   DT K T S     S                  K
Mercurialis (dog's mercury)   LDVL SD E A V NN SVPS EK T           EIS AS     D AD     P TYA T   TE K S S     A                  N
Lactuca (lettuce)             AEVL SS G V E ST SVAS EK V           EIA AS     S ED     P TYA T   TE K T S     S                  K
Populus (poplar)              IDVL AD S A V SE SISP EK V           SIS AS     S ED     K TFE A   SN K E S     S
```

Plant glycoproteins

```
                              1         10        20        30        40        50        60        70        80        90       100
Lacquer tree (stellacyanin)   TVYTVGDSAGWKVPFFGDVDYDWKWASNKTFHIGDVLVFKYDRRFHNVDKVTQKNYQSCNDTTPIASYNTGNDRINLKTVGQKYYICGVPKHCDLGQKVHINVTVRS
Horseradish (umecyanin)       ED D GDME  R --S PKFYIT  TG   RV  E  E DFAAGM D AV  KDAFDN KKEN   SHMT  PPVK M X T PQ    T G......
```

FIGURE 2. Amino-acid sequences of small blue proteins. In each of the three families a space in sequences below the top most one means a residue identical to the one in the first line. In higher plant plastocyanins identity refers to the first of these in the third line. Disulfide bridges are indicated by horizontal bars. carbohydrate-binding residues by *. and deletions by — . The umecyanin sequence has been only partially determined. Azurins;[52-54] plastocyanins: *Anabaena*,[33] *Chlorella*,[31] dock,[55] elder,[56] spinach,[57] vegetable marrow,[58] shepherd's purse,[59] potato,[60] *Solanum crispum*,[61] broad bean,[62] French bean,[63] dog's mercury,[64] lettuce,[64] poplar;[65] plant glycoproteins: stellacyanin,[66] umecyanin.[67] A leucine reported in position 97 in the dock plastocyanin sequence has been disregarded for obvious reasons.

```
    1         10        20        30        40        50        60        70        80        90        100
                                      +                                              +         +
EAQDIELGADKGLLAFEPADVEIKAGDSVEWLNNKVPPHNIIFDAAANPAKAADDAKSLAHKLLLMSAGQSTATAFPADAPAGEYGVFCEPHAGAGMK-TITVAG
-DYELKM DES A V S DEFNLAK  EEIIFV AAK   VV EDLI SGSNAEKL MDEAQY NAP ETYEAKL--SE-K S KYY A R VGKV N-
 TALVL GGD E I GKIQVDP K T K SGF       EV VDSL I -N DD     K VFKVN TT ST SF S Q      Q
 -FT N  SNE G  L KNLS NS T V I        Y      S      A -P EE        ST   N      T
  I     SG N  V NQ T PV                           G   S N      V   S T
  L     N  S  QS   S                             S   -             V
  M     T     ST   T.                            V
  V     V
```

FIGURE 3. Composite plastocyanin sequence, where each position contains all residues observed there. Underlined residues have only been observed in algal plastocyanins. The copper-liganding residues are indicated by a + sign; — indicates a deletion. Data are from Figure 2 and References 13, 68, and 69 (see also Section IV.B).

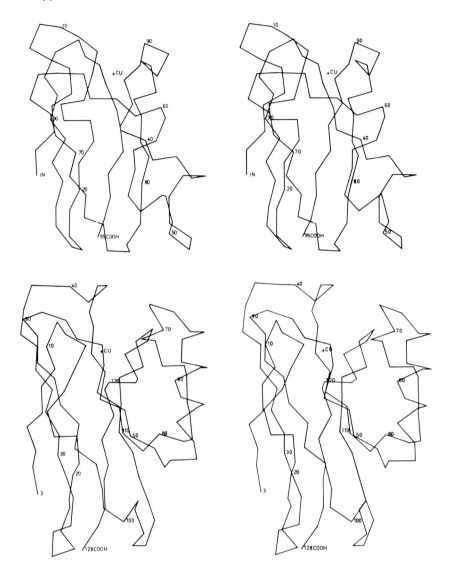

FIGURE 4.   Stereo pairs showing α-carbon plots of poplar plastocyanin (top) and *Pseudomonas aeruginosa* azurin (bottom). The original structures[10,11] were refined to an *R*-factor of 0.17 for plastocyanin and 0.35 for azurin. In the latter case data were collected to 2.7 Å resolution. The coordinates have been obtained from the Protein Data Bank, Brookhaven National Laboratory, Upton, N.Y., where they were reported in 1980.

rhombic crystals with a single molecule in the asymmetric unit and four heavy atom derivatives.[10] Finally crystallographic data for plantacyanin from cucumber seedlings have been published.[29] This smallest known of the small copper proteins crystallizes in the $P2_12_12_1$ space group and contains a single molecule in the asymmetric unit. No structure has yet been reported, however. The descriptions of the three-dimensional structures below of Az and Pc are based on the refined data reported to the Protein Data Bank in 1980 for the two proteins.

A stereo-pair showing the poplar Pc structure is given in Figure 4. The molecule forms a slightly flattened barrel 4.0 nm high and 2.8 × 3.2 nm wide. The peptide chain folds into eight strands roughly parallel to the long axis. Seven of these are in correct conformation for β-structure, while the fifth middle strand is irregular and even contains a turn of approximate helix. The core of the molecule is hydrophobic and made up of six of the seven phenylalanine side chains. On the surface of the molecule a negative patch of six carboxylates and a hydrophobic patch on the top of the molecule formed by seven residues widely spread in the linear structure are noteworthy. The copper site at the top of the molecule consists of three liganding side-chains — Cys-89, His-92, and Met-97 — from the turn between strands 7 and 8, and a histidine side-chain (His-39) from strand 4.

The described characteristics account for several of the invariant amino acids in the Pc family. Of the core residues, two (Phe-14 and Phe-41) are invariant in all Pc's and two (Phe-29 and Phe-82) are invariant in the higher plants and conservatively substituted in the algal sequences. The remaining two, Phe-19 and Phe-70, are conservatively replaced. It is interesting to note that the negative patch is conserved in all higher plant sequences, but almost completely lost in the algal proteins. It is thus quite possible that it has a unique role for Pc in the chloroplast membrane. The hydrophobic patch residues are the most invariant: six of them — Gly-10, Leu-12, Pro-36, Gly-89, Ala-90, and Gly-91 — are completely conserved in all sequences. Finally this is also valid for the four copper ligands. Three more of the conserved glycines — Gly-24, Gly-67, and Gly-78 — occur at the N-terminal ends of strands 3, 6, and 7, while three of the conserved prolines occur at or close to the turns, where they probably cannot easily be replaced by other side chains.

The Az molecule (Figure 4) forms a more pear-like body 4.5 nm high and slightly flattened, 2.5×3.0 nm wide. The copper site is at the small end at the top of the molecule. The peptide chain again is folded into eight strands of roughly pleated sheet conformation. Between strands 4 and 5 there is an added external flap containing a short piece of helix. The flap includes residues 53 to 80 in the linear sequence. Also this molecule has an aromatic core. In its center is located the single tryptophan, Trp-48, flanked by Phe-29, Phe-97, and Phe-110. The copper site is again composed of Cys-112, His-117, and Met-121 from the turn between strands 7 and 8 and a histidine side-chain, His-46, from the beginning of strand 4.

The copper sites in the two molecules are shown in stereo in Figure 5. It is clear that the geometry is highly irregular with large deviations from tetrahedral angles. Presumably it is the accessibility to two oxidation states that has promoted the evolution of such a structure and increased the redox potential, when compared to the Cu(I)/Cu(II) pair in aqueous solution. This question will be further discussed in Chapter 7 on the reactivity of the small blue proteins.

There is a large body of data on side-chain reactivities in both Pc and Az. Most of these are superseded by the crystallographic work, but some data are still of interest. A detailed study by Ugurbil and co-workers[73-75] on the tyrosines and tryptophans of Az proved these to be in a hydrophobic environment as confirmed by the structure. Two histidine side-chains did not titrate between pH 4 and 11, which is explained by the presence of two imidazole ligands to the copper.[73,76] A single-carbon amide carbonyl resonance with an unusual chemical shift is not rationalized in terms of the structure.[73] It might, however, be noted that the asparagine following the Cu-liganding His-46 is conserved in all Az's and Pc's as well (Section IV.A).

IV. MOLECULAR EVOLUTION

A. General Comments

The blue proteins have in common a copper(II) site with specific and unusual properties which suggests that the copper environment is similar in all of them. This situation — if true — can be the result of either convergent evolution, where similar requirements have promoted the evolution of the same structure from different origins (= analogy), or divergent evolution, where several or all of these proteins have evolved from a common ancestor (=

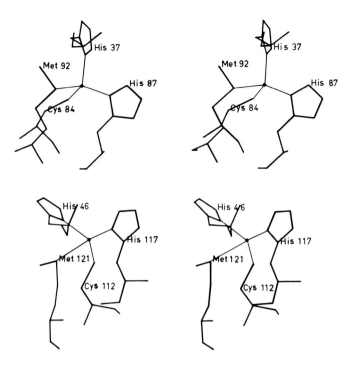

FIGURE 5.   Stereo pairs showing the copper sites of poplar plastocyanin
(top) and *Pseudomonas aeruginosa* azurin (bottom). The coordinates have
been obtained as mentioned in Figure 4. Copper is shown by a +. The
histidine from strand 4 (His-37 in plastocyanin and His-46 in azurin) is at
the top and the three ligands from strands 7 and 8 are at the bottom (Cys-
84, His-87, and Met-92 in plastocyanin and Cys-112, His-117, and Met-
121 in azurin).

homology). If this latter possibility is valid, the evolutionary distances involved are tre-
mendous: the common ancestor of say a pseudomonad bacterium and a higher plant inhabited
our planet not later than some 3,000 million years ago. Nevertheless, molecular evolution
aims at clarifying such relationships, many of which are inaccessible to traditional methods
relying on morphology and the study of fossils.

Comparing two amino-acid sequences which are not closely related runs into the problem
of the best positioning of deletions. This problem has, however, been solved in a mathematically strict way.[77] The method uses a table of scores (also called weights) for the 220
possible amino-acid pairs and aligns the sequences in such a way that a maximum total score
is obtained. The table used is constructed for detecting distant relationships.[78] Constraints
are used such that a small number of deletions are introduced into the longer sequence and/
or that the deletions are collected into a minimal number of gaps. The deviation of the total
score from the mean of a distribution of total scores obtained for a number of alignments
where the sequences have the same composition but have been scrambled (randomized) is
a measure of the similarity of the two sequences. This deviation is expressed in number of
standard deviations of the random distribution (SD units) and is thus directly convertible to
the probability that the pair of real sequences is a part of the random distribution. Thus 3,
4, and 5 SD units correspond to probabilities of $1.4 \cdot 10^{-3}$, $3.2 \cdot 10^{-5}$, and $2.9 \cdot 10^{-6}$,
respectively. The generally accepted limit for significant similarity is 3 SD units.[78]

A high alignment score or, for that matter, an obvious similarity such as 30% identities
with few deletions, is considered a proof of homology, i.e., common ancestry. The aligned
sequences can then be combined into a phylogenetic tree for the examination of cladistic

relationships, evolutionary rates, and construction of the amino-acid sequences of the ancestors. The question of homology can, however, be tested directly in the tree.[79] If the ancestors are growing more similar towards the root this indicates divergence, while the opposite trend is a sign of convergent evolution. One of the few such tests that have been applied refers to the small blue proteins (see below). A note of caution should be added in connection with these similarity measures. A nonsignificant alignment score does not imply a nonhomologous relationship. Rather that if such a relationship exists it is too distant to be detected. As more data come into the picture, or a ''missing link'' is found, the homology might be ascertained. The negative assertion, that a sequence similarity is due to convergence, is in principle possible with the above-mentioned test, but has so far not been made. A classic case of convergent structures of *active sites* is afforded by chymotrypsin and subtilisin.[80] The similarity does not, however, apply for the rest of the structure; e.g., the active-site residues do not occupy corresponding sites in the linear sequence. As to the case of the blue proteins, a corresponding positioning of copper-binding residues in linear sequences is consistent with the hypothesis of divergence, to be proved or disproved when enough data are available.

Several new questions have been approached in connection with the study of the evolution of the small blue proteins. One is the comparison of whole families of sequences rather than individual sequences. This was solved by the use of composite sequences (Figure 3).[81] Another difficulty arose in the comparison of the blue oxidases with the small blue proteins. These two groups of proteins have peptide chains of very different lengths. The normal alignment procedure was questionable since the shorter sequence was spread out over the entire length of the longer one. A method was thus developed to find out which part of the longer sequence should be used to find a maximum alignment score in the comparison.[82] A third problem refers to the quantitative evaluation of conserved copper-binding residues. The table of scores for amino-acid pairs is based on the assumption that the amino acids have a ''normal'' role to play in the structure of the protein, such as being inside or at the surface. A conserved methionine and a methionine converted to a leucine do not have very different scores. If the methionine is copper-binding the difference is, however, essential and should be reflected in the parameters used. The effect of changes of the scores used was therefore studied.[83]

It is well-known that peptide-chain folding is more conserved than amino-acid sequence. The three-dimensional structures available for Az and Pc have been of great value in the study of the evolution of these proteins. Apart from confirming conclusions based mainly on the comparison of linear sequences, these have permitted new approaches also for the evaluation of possible homologies with proteins for which the tertiary structure is not known. This again touches on the question of a better definition of the evolutionary value of individual residues and the possible location of gaps in the three-dimensional fold.

B. Plastocyanins and Azurins

A large number of sequence data has been collected by the Durham group for the Pc's,[13] especially from flowering plants, to aid in finding a phylogeny of these. Pc was chosen because of its higher evolutionary rate as compared to the previously studied cytochrome *c*. The data have permitted the construction of a taxonomy mainly on the level of families and genera.[84,85] The results agree in a general sense with traditional taxonomy but deviations occur, especially as to the accumulation of families in higher taxa. The evaluation of the results from these studies has to await more protein data sets.

Also the Az sequence data were collected with an aim at studying phylogeny.[52,86] It appears that firm conclusions were difficult, possibly due to gene transfer. Such events would indeed destroy the underlying assumptions for the student of molecular evolution. A more fruitful line of investigation seems to have been the study of ribosomal RNA sequences,

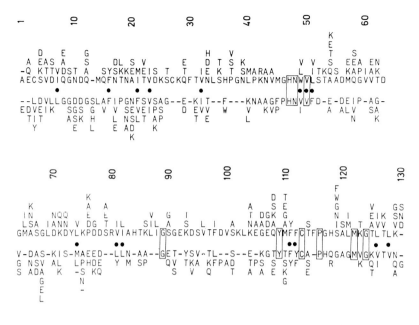

FIGURE 6.   Alignment of a composite azurin sequence (top) with a composite plastocyanin sequence (bottom), based on Reference 81. Residues conserved in both sequences are boxed, while those with conserved aromatic or hydrophobic character are indicated by a dot. (From Rydén, L. and Lundgren, J.-O., *Nature (London)*, 261, 344, 1976. With permission.)

suitable for detecting distant relationships.[87] The results on bacterial phylogeny will be of great benefit also for understanding the evolution of prokaryotic blue proteins (Section IV.E).

Az's and Pc's show some similarities in the C-terminal end of their amino-acid sequences.[52] Similarities towards the N-terminus were, however, not evident by visual inspection. Rydén and Lundgren made an alignment of the composite sequences of the two families of proteins, using average weights for each position in the calculations.[81] They found that only 7% of the positions had conserved residues (Figure 6). The alignment score obtained was nevertheless 7.3 SD units which is overwhelming evidence for the similarity not owing to chance. The direct test for divergence/convergence also strongly suggested divergent evolution (+6.9 SD units).[79] This should be compared to the weak evidence obtained when individual members of the two families were matched.[81]

The three-dimensional structures of Az and Pc published later fully confirmed the conclusions (Section III.C). The two structures have the same general folding of the peptide chain. The four copper-binding residues in both proteins are found among the conserved residues (positions His-47, Cys-113, Met-122, and His-117/118), with the exception that a histidine was displaced one position in the two sequences. The greater length of the Az's, about 30 residues, is accounted for by insertions essentially at the surface of the protein. An inspection of the two structures indicates that the insertion pattern of Az's should not be taken as detailed three-dimensional evidence. Rather it is suggested that the alignment breaks down in areas with different folding. It is particularly true for the C-terminal end of the fourth strand from the N-terminus, where Az has added an extra loop, and the fifth strand where the structure is rather irregular with several kinks, and a single turn of helix is present in both proteins. A similar breakdown occurs for the end of strand 3 where Az's are longer. The presence of an insertion right close to the copper-binding cysteine is more surprising but apparently possible in the turn from strands 7 to 8.

C. Higher Plant Glycoproteins (Stellacyanin)

The question if the other fully sequenced small blue protein, the nonphotosynthetic plant

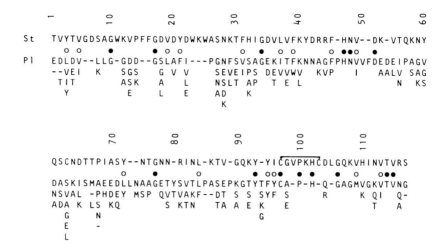

FIGURE 7.   Alignment of stellacyanin (top) with a composite plastocyanin sequence (bottom), based on Reference 83. An extra weight was given for a conserved copper-binding residue. Conserved residues are indicated by filled circles and conserved aromatic or hydrophobic character by open circles. The single disulfide bridge in stellacyanin is shown by the horizontal bar.

glycoprotein St, is related to the Az and the Pc families has also been studied.[83] In the comparisons with the two composite sequences, alignment scores of 1 and 3.8 SD units were obtained. The latter value, for the Pc-St match, is significant on a level of 10^{-4}. Of the known copper ligands in Pc, methionine is missing and His-43 did not line up with a histidine in the St sequence, although there was one close by, a situation which also occurred in the Pc-Az match (see above). To overcome this, an extra weight was given for conservation of copper-binding residues. The addition necessary for the alignment of the two histidines was equivalent to the weight of two average pairs and thus produced two additional deletions in the new match (Figure 7). The alignment score for this comparison increased to 5.3 SD units, expressing the fact that three out of four copper-ligating residues were conserved in the real sequences, but only occasionally so in the randomized ones.

The copper-binding cysteine in Pc lines up with a cysteine known to be part of a disulfide bridge in St, while the single thiol in St lines up with a residue in the irregular fifth strand in Pc relatively far from the copper site. If the two proteins are homologous, one would expect their peptide-chain folding to be similar and the copper sites to be in corresponding places in the molecules. This would imply that the single cysteine in St is not a copper ligand, while the disulfide bridge is one. The latter statement has some experimental support such as the Raman spectrum of St, which suggests that a thioether-copper bond, possibly from a disulfide, is present (Section III.A).

An interesting model of a type-1 copper site with two cysteines has been constructed by Maret et al.[88] They incubated alcohol dehydrogenase devoid of the Zn^{2+} ion, present at the active site in the native enzyme, in a solution of *bis*(ethylenediamine) copper(II) to obtain a deep blue solution. The resulting complex had a 30% occupancy of the catalytic site with copper ions. Its molar absorption coefficient of 2,000 M^{-1} cm^{-1} at 620 nm and an $A_\parallel$ of 0.0030 cm^{-1} in the EPR spectrum were similar to those of the type-1 copper proteins. The catalytic site in alcohol dehydrogenase is known from X-ray studies to contain one histidine, two cysteines, and one water ligand in a distorted tetrahedral geometry. In contact with air the copper-substituted enzyme lost its blue color, possibly under formation of a disulfide bridge between the two sulfur ligands and reduction of copper. The color returned when hexacyanoferrate(III) was added. The redox processes in these changes seem unclear, but it is nevertheless evident that a blue site can be created also with two nearby cysteines in the absence of methionine, as might be the case in St. It is also interesting that the probable geometry of the site is similar to what is found in Pc and Az.

Further evidence as to the structure of the St copper site comes from a study by Hill and Lee[89] on the proton NMR of the protein. Of the four histidine residues in St, two were freely titrating in the holoprotein while two additional sharp resonances were observed in the apoprotein, suggesting that these correspond to two copper ligands. The suggestion is supported by the ENDOR spectra.[51] Two additional ligands thus remain to be assigned. These authors suggest, on the basis of the lower redox potential of St, that these are cysteines and that a disulfide is not formed until the copper is removed, somewhat analogous to the situation in copper-alcohol dehydrogenase. This suggestion is, however, not consistent with spectral evidence (Section III.A), nor is it consistent with the finding that the alkylation of Cys-59 makes impossible the reconstitution of St from the apoprotein.[90] If in fact St has a copper site in a place in the molecule corresponding to the sites in Pc and Az, the four ligands would be the two cysteines, Cys-87 and Cys-93, His-46, and either His-92 or His-100.

The umecyanin sequence is so similar to that of St, that it is evident that they have the same copper ligation. It thus will serve as a check of the above ideas. To the extent of the available information, His-46 is still a candidate since it is present in both proteins. The C-terminal end of the sequence is awaited with suspense to see if any of the other two possible histidines has been replaced.

The pattern of deletions in the St-Pc alignment indicates that the extra residues in St occur in the first two strands and in the turn between these. A deletion in St close to residue 70 is found in the irregular fifth strand. The three carbohydrate sites would be found at the second and eighth strand on the front side of the molecule and where strand five makes a small kink at the backside of the molecule as shown in Figure 4. In summary, the additional data available are compatible with a Pc-like peptide-chain fold also for St/umecyanin and with the hypothesis that all small blue proteins studied in any detail up to date share a common ancestry.

D. Relationship to the Multicopper Oxidases and Cytochrome *c* Oxidase

Recently, amino-acid sequence data have become available for some of the multicopper oxidases. A total of 650 residues out of the ≈1,050 residues of human ceruloplasmin has been published.[91-93] A basic repeat unit of 340 residues was found.[93] Some sequence data from fungal laccase show this protein to be homologous to ceruloplasmin.[94] No data are available for the third of the three most studied blue oxidases, ascorbate oxidase.

Cytochrome *c* oxidase from bovine mitochondria contains seven distinct subunits which now have been sequenced.[95] The primary structure of three subunits of the human and mouse mitochondrial enzyme was deduced from the mitochondrial DNA sequence.[96,97]

A visual inspection of the sequences suggests that part of the ceruloplasmin and laccase contains a copper-binding site similar to the Az/Pc one with the residues cysteine, histidine, and methionine close to each other in the proper order. The same is observed towards the C-terminus of subunit II of the mitochondrial cytochrome *c* oxidase. For the critical evaluation of these similarities a new method, the score-loss method, was developed to find out which part of the longer sequence should be compared with the shorter one to obtain a maximal alignment score.[82] Significant similarities were obtained for the Pc/ceruloplasmin comparison, 4.5 SD units,[93] and the Az/cytochrome *c* oxidase subunit II comparison, 3.5 SD units.[82] These two matches were also the only ones that showed well-defined points for the truncation of the longer sequence.

The two alignments obtained are shown in Figures 8 and 9. In both cases all four of the copper-binding residues are conserved. In ceruloplasmin weak evidence, 2.2 SD units, for a duplication in the 340-residues long repeat unit was obtained. The data tentatively suggest

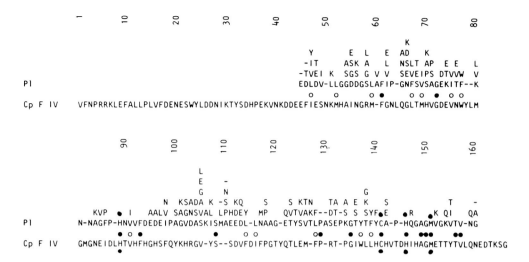

```
       —      10     20     30     40     50     60     70     80
                                                        K
                                           Y     E  L    E  AD   K
                                           -IT    ASK A   L  NSLT AP  E E    L
                                           -TVEI K SGS G V V  SEVEIPS DTVVW   V
P1                                         EDLDV-LLGGDDGSLAFIP-GNFSVSAGEKITF--K
                                            o  o     o •     o o  •  o o
Cp F IV  VFNPRRKLEFALLPLVFDENESWYLDDNIKTYSDHPEKVNKDDEEFIESNKMHAINGRM-FGNLQGLTMHVGDEVNWYLM


              90     100    110    120    130.    140    150    160
                L
                E     -
                G     N
             N KSADA K -S KQ    S    S KTN   TA A E K   S              T   -
          KVP • I  AALV SAGNSVAL LPHDEY  MP  QVTVAKF--DT-S S SYF•E  •R  •K QI   QA
P1        N-NAGFP-HNVVFDEDEIPAGVDASKISMAEEDL-LNAAG-ETYSVTLPASEPKGTYTFYCA-P-HQGAGMVGKVTV-NG
            • o •          •    o o      •• •  • o o •  •  •••      ••
Cp F IV   GMGNEIDLHTVHFHGHSFQYKHRGV-YS--SDVFDIFPGTYQTLEM-FP-RT-PGIWLLHCHVTDHIHAGMETTYTVLQNEDTKSG
            •                                    •      •        •
```

FIGURE 8. Alignment of a composite plastocyanin sequence (top) with the 159-residues long C-terminal fragment of human ceruloplasmin (bottom), based on Reference 82. The part of the ceruloplasmin-fragment sequence that gave a maximum alignment score in the alignment was found by the score-loss method. Meanings of the circles are as in Figure 7.

that the origin of the blue oxidases is an ancestral oxidase formed by duplication of a Pc unit elongated by 50 residues on its N-terminus. The matter is further discussed in Chapter 2 on ceruloplasmin in Volume III.

The possibility that the Az's and mitochondrial cytochrome *c* oxidase subunit II are related is supported by circumstantial evidence. The mitochondrion is believed to be a descendant of a *Paracoccus*-like bacterium which entered a symbiotic relationship with a larger eukaryotic cell.[98] The cytochrome oxidase of *Paracoccus denitrificans* today consists of two subunits,[99] similar to the two larger subunits in the mitochondrial enzyme, I and II, precisely those that are encoded by the mitochondrial genome and contain the prosthetic copper and heme.[100,101] Some immunochemical cross-reactivity between the two oxidases has been observed, which implies similarities in structure.[102] The data thus suggest that mitochondrial and *Paracoccus* cytochrome oxidase are related. *Paracoccus* is also one of the few bacteria that have been reported to contain an Az (Section II.A). Indeed its presence in *Paracoccus* was suggested previous to the publication of this finding on theoretical grounds.[83] The amino-acid sequence of *Paracoccus* Az would thus be expected to be more similar to the mitochondrial subunit and would be valuable for further illuminating the relationships discussed. It should be borne in mind that the well-studied cytochrome oxidase of *P. aeruginosa*, a bacterium that also contains Az, is distinct from the *Paracoccus* enzyme, and that there is no evidence that this enzyme contains copper.

E. On the Origin of the Blue Proteins

In bacterial phylogeny, based on ribosomal RNA sequences, the Az-containing genera *Pseudomonas, Alcaligenes,* and *Paracoccus* are found in three of the four branches of the purple photosynthetic bacteria.[87] These constitute in turn one of the four major groupings of the eubacteria, one of the three kingdoms of living cells (the other two being the archaebacteria and the urkaryotes or ancestral eukaryotes). The origin of the Az's would thus be located in an ancestral purple photosynthetic bacterium. The origin of the Pc's would be expected to be found among the cyanobacteria, which form one of the other major groupings of the eubacteria. The relationship of the cyanobacteria to higher plant chloroplasts remains unclear, especially with regard to the Pc's.[103] These have so far only been isolated from heterocyst-forming filamentous cyanobacteria, while none of the investigated unicellular

```
     10        20        30        40        50        60        70        80        90        100       110

                                                      D    E     G       S D L S V          E       H  V
                                                    - EA S A    A SYS KKEME     IS T    T    IE K T SMARAA    L   V
                                            Q KT TVDST                                                            IT
Az                                          AECSV-DIQGND-QMQFN-TNAIT---VDKSC-KQ-FTVNLSHPGNLPKNVMGHNW--VLS-

Cox  MAYPMQLGFQDATSPIMEELLHFHDHTLMIVFLISSLVLYIISLMLTTKLTHTSTMDAQEVETIWTILPAIILILIALPSLRILYMMDEINNPSLTVKTMGHQWYWSYEY


     120       130       140       150       160       170       180       190       200       210       220

     K                           K    A    A                          D  T              F
     E             S EA  AEN  K    N QQ     A   E       V G  S         S  E              W
     T             S  K AP  IAK LSA    IA NN V DG T IL   SILA  T   L I   A NAA    GK  G    N I    EIK SS    V  G-
     KQ                                                              TPD    DA  A        S    ISM T AVV VN
Az   TA-AD-M-QG--VVTDGMASG----LD-KDYLK-PDD-S-RVIAHTK-LIGS-GEKD-SVTFD-V-SKLKEG--EQY-M-FF--CT-F-PG-HSALMKGTLTLKD

Cox  TDYEDLSFDSYMIPTSELKPGELRLLEVDNRVVLPMEMTIRMLVSSEDVLHSWAVPSLGLKTDAIPGRLNQTTLMSSRPGLYYGQCSEICGSNH-SFMPIVLELVPLKYFEKWSASML
```

FIGURE 9.   Alignment of a composite azurin sequence (top) with the 227-residues long subunit II of mitochondrial cytochrome *c* oxidase (bottom), based on Reference 82. The score-loss method was used in the alignment (see Figure 8). Meanings of the circles are as in Figure 7.

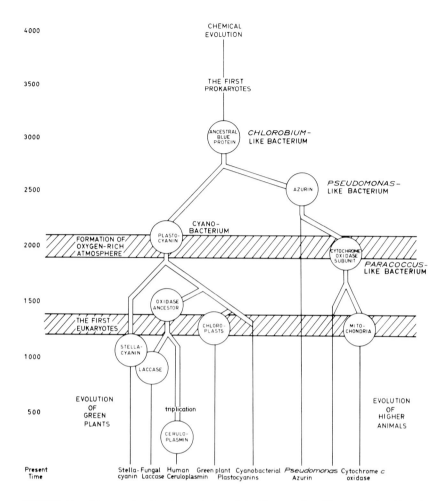

FIGURE 10. Diagram showing the evolution of the blue proteins. The ancestral oxidase is assumed to be the immediate precursor of the multicopper oxidases laccase and ceruloplasmin of today. Time scale in million years.

species contained the protein.[103] Still it is the latter which are assumed to be the ancestors of the chloroplasts via an endosymbiotic relationship with a eukaryotic cell.[104] The data are at present inconclusive. Fossil evidence suggests that the major cyanobacterial lines were established at least 3,000 million years ago.[105] The common ancestor of the blue copper-containing proteins would then have existed before this date. It is clear that the blue proteins represent an ancient line of metalloproteins that later has evolved into a great diversity of forms participating both in the evolution of oxygen — in oxygenic photosynthesis — and in the reduction of oxygen for efficient energy production — in mitochondrial and bacterial redox chains — as well as in many other metabolic roles, only some of which are known. The relationships between these proteins are delineated in Figure 10.

ACKNOWLEDGMENTS

The stereo pairs of the Pc and Az structures were obtained with a graphic display system of the Department of Molecular Biology, Protein Crystallography Group, at the University of Uppsala, with the aid of Dr. Alwyn Jones and Dr. Hans Eklund, which is gratefully acknowledged.

Note Added in Proof

The complete amino-acid sequence of the 96-residues long basic small blue protein (plantacyanin) from cucumber seedlings has been reported.[106] A 45% identity with the St sequence is observed. It thus appears that the nonphotosynthetic plant small blue proteins form a single family of homologous proteins (the phytocyanins). In the cucumber protein a Met residue is found in the potential copper-binding position where it is lacking in St and umecyanin. The 124-residues long sequence of a small blue protein from the bacterium *Achromobacter cycloclastes*[107] is not closely homologous to the Az's or any other of the known blue proteins. The single cysteine is found in position 78 and is followed by a His (position 81) and two Met residues closely thereafter. A His-Asn-Val sequence is found in positions 40-42. It thus constitutes a new class of small blue proteins.

REFERENCES

1. **Omura, T.,** Studies on laccases of lacquer trees. IV. Purification and properties of a blue protein obtained from latex of *Rhus vernicifera, J. Biochem. (Tokyo),* 50, 395, 1961.
2. **Peisach, J., Levine, W. G., and Blumberg, W. E.,** Structural properties of stellacyanin, a copper mucoprotein from *Rhus vernicifera,* the Japanese lac tree, *J. Biol. Chem.,* 242, 2847, 1967.
3. **Paul, K.-G. and Stigbrand, T.,** Umecyanin, a novel intensely blue copper protein from horseradish root, *Biochim. Biophys. Acta,* 221, 255, 1970.
4. **Markossian, K. A., Aikazyan, V. Ts., and Nalbandyan, R. M.,** Two copper-containing proteins from cucumber *(Cucumis sativus), Biochim. Biophys. Acta,* 359, 47, 1974.
5. **Verhoeven, W. and Takeda, Y.,** The participation of cytochrome *c* in nitrate reduction, in *Inorganic Nitrogen Metabolism,* McElroy, W. D. and Glass, B., Eds., The Johns Hopkins Press, Baltimore, 1956, 159.
6. **Sutherland, I. W. and Wilkinson, J. F.,** Azurin: a copper protein found in *Bordetella, J. Gen. Microbiol.,* 30, 105, 1963.
7. **Cobley, J. G. and Haddock, B. A.,** The respiratory chain of *Thiobacillus ferro-oxidans:* the reduction of cytochromes by Fe^{2+} and the preliminary characterization of rusticyanin, a novel "blue" copper protein, *FEBS Lett.,* 60, 29, 1975.
8. **Katoh, S.,** A new copper protein from *Chlorella ellipsoidea, Nature (London),* 186, 533, 1960.
9. **Katoh, S. and Takamiya, A.,** A new leaf copper protein "plastocyanin", a natural Hill oxidant, *Nature (London),* 189, 665, 1961.
10. **Colman, P. M., Freeman, H. C., Guss, J. M., Murata, M., Norris, V. A., Ramshaw, J. A. M., and Venkatappa, M. P.,** X-ray crystal structure analysis of plastocyanin at 2.7 Å resolution, *Nature (London),* 272, 319, 1978.
11. **Adman, E. T., Stenkamp, R. E., Sieker, L. C., and Jensen, L. H.,** A crystallographic model for azurin at 3 Å resolution, *J. Mol. Biol.,* 123, 35, 1978.
12. **Fee, J. A.,** Copper proteins. Systems containing the "blue" copper center, *Struct. Bonding (Berlin),* 23, 1, 1975.
13. **Boulter, D., Haslett, B. G., Peacock, D., Ramshaw, J. A. M., and Scawen, M. D.,** Chemistry, function, and evolution of plastocyanin, *International Review of Biochemistry,* Vol. 13, Northcote, D. H., Ed., 1977, 1.
14. **Horio, T.,** Terminal oxidation system in bacteria. I. Purification of cytochromes from *Pseudomonas aeruginosa, J. Biochem. (Tokyo),* 45, 195, 1958.
15. **Horio, T.,** Terminal oxidation system in bacteria. II. Some physical and physiological properties of purified cytochromes of *Pseudomonas aeruginosa, J. Biochem. (Tokyo),* 45, 267, 1958.
16. **Sutherland, I. W.,** The production of azurin and similar proteins, *Arch. Mikrobiol.,* 54, 350, 1966.
17. **Martinkus, K., Kennelly, P. J., Rea, T., and Timkovich, R.,** Purification and properties of *Paracoccus denitrificans* azurin, *Arch. Biochem. Biophys.,* 199, 465, 1980.
18. **Davis, D. H., Stanier, R. Y., Doudoroff, M., and Mandel, M.,** Taxonomic studies on some Gram-negative polarly flagellated "hydrogen bacteria" and related species, *Arch. Mikrobiol.,* 70, 1, 1970.
19. **Taylor, B. F., and Hoare, D. S.,** New facultative *Thiobacillus* and a reevaluation of the heterotrophic potential of *Thiobacillus novellus, J. Bacteriol.,* 100, 487, 1969.

20. **Doudoroff, M.,** Genus *Paracoccus,* in *Bergey's Manual of Determinative Bacteriology,* 8th ed., Buchanan, R. E. and Gibbons, N. E., Eds., Williams & Wilkins, Baltimore, 1974, 438.
21. **Horio, T., Sekuzu, I., Higashi, T., and Okunuki, K.,** Cytochrome oxidase of *Pseudononas aeruginosa* and ox-heart muscle, and their related respiratory components, in *Haematin Enzymes,* Falk, J. E., Lemberg, R., and Morton, R. K., Eds., Pergamon Press, Oxford, 1961, 302.
22. **Cox, J. C. and Boxer, D. H.,** The purification and some properties of rusticyanin, a blue copper protein involved in iron(II) oxidation from *Thiobacillus ferrooxidans, Biochem. J.,* 174, 497, 1978.
23. **Stigbrand, T.,** Structural properties of umecyanin, a copper protein from horseradish roots, *Biochim. Biophys. Acta,* 236, 246, 1971.
24. **Morita, Y., Wadano, A., and Ida, S.,** Studies on respiratory enzymes in rice kernel. VI. Characterization of a blue protein of rice bran, *Agric. Biol. Chem.,* 35, 255, 1971.
25. **Shichi, H. and Hackett, D. P.,** Purification and properties of a blue protein from etiolated mung bean seedlings, *Arch. Biochem. Biophys.,* 100, 185, 1963.
26. **Marchesini, A., Minelli, M., Merkle, H., and Kroneck, P. M. H.,** Mavicyanin, a blue copper protein from *Cucurbita pepo medullosa.* Purification and characterization, *Eur. J. Biochem.,* 101, 77, 1979.
27. **Aikazyan, V. Ts. and Nalbandyan, R. M.,** Copper-containing proteins from *Cucumis sativus, FEBS Lett.,* 104, 127, 1979.
28. **Vickery, L. E. and Purves, W. K.,** Isolation of indole-3-ethanol oxidase from cucumber seedlings, *Plant Physiol.,* 49, 716, 1972.
29. **Colman, P. M., Freeman, H. C., Guss, J. M., Murata, M., Norris, V. A., Ramshaw, J. A. M., Venkatappa, M. P., and Vickery, L. E.** Preliminary crystallographic data for a basic copper-containing protein from cucumber seedlings, *J. Mol. Biol.,* 112, 649, 1977.
30. **Aikazyan, V. Ts. and Nalbandyan, R. M.,** Plantacyanin from spinach, *FEBS Lett.,* 55, 272, 1975.
31. **Kelly, J. and Ambler, R. P.,** The amino acid sequence of plastocyanin from *Chlorella fusca, Biochem. J.,* 143, 681, 1974.
32. **Kunert, K.-J. and Böger, P.,** Absence of plastocyanin in the alga *Bumilleriopsis* and its replacement by cytochrome 553, *Z. Naturforsch.,* 30c, 190, 1975.
33. **Aitken, A.,** A Prokaryote-eukaryote relationships and the amino acid sequence of plastocyanin from *Anabaena variabilis, Biochem. J.,* 149, 675, 1975.
34. **Malkin, R. and Bearden, A. J.,** Light-induced changes of bound chloroplasts plastocyanin as studied by EPR spectroscopy: the role of plastocyanin in noncyclic photosynthetic electron transport, *Biochim. Biophys. Acta,* 292, 169, 1973.
35. **Siegelman, M. H., Rasched, I., and Böger, P.,** Evidence for a 40,000 dalton species of plastocyanin as the active component, *Biochem. Biophys. Res. Commun.,* 65, 1456, 1975.
36. **Siegelman, M. H., Rasched, I. R., Kunert, K.-J., Kroneck, P., and Böger, P.,** Plastocyanin: possible significance of quaternary structure, *Eur. J. Biochem.,* 64, 131, 1976.
37. **Vänngård, T.,** cited in Reference 12.
38. **Solomon, E. I., Hare, J. W., Dooley, D. M., Dawson, J. H., Stephens, P. J., and Gray, H. B.,** Spectroscopic studies of stellacyanin, plastocyanin, and azurin. Electronic structure of the blue copper sites, *J. Am. Chem. Soc.,* 102, 168, 1980.
39. **Brill, A. S., Bryce, G. F., and Maria, H. J.,** Optical and magnetic properties of *Pseudomonas* azurins, *Biochim. Biophys. Acta,* 154, 342, 1968.
40. **Sailasuta, N., Anson, F. C., and Gray, H. B.,** Studies of the thermodynamics of electron transfer reactions of blue copper proteins, *J. Am. Chem. Soc.,* 101, 455, 1979.
41. **Suzuki, H. and Iwasaki, H.,** Studies on denitrification. VI. Preparation and properties of crystalline blue protein and cryptocytochrome c, and role of copper in denitrifying enzyme from a denitrifying bacterium, *J. Biochem. (Tokyo),* 52, 193, 1962.
42. **Cox, J. C., Aasa, R., and Malmström, B. G.,** EPR studies on the blue copper protein, rusticyanin. A protein involved in Fe^{2+} oxidation at pH 2.0 in *Thiobacillus ferro-oxidans, FEBS Lett.,* 93, 157, 1978.
43. **Malmström, B. G., Reinhammar, B., and Vänngård, T.,** The state of copper in stellacyanin and laccase from the lacquer tree *Rhus vernicifera, Biochim. Biophys. Acta,* 205, 48, 1970.
44. **Reinhammar, B. R. M.,** Oxidation-reduction potentials of the electron acceptors in laccases and stellacyanin, *Biochim. Biophys. Acta,* 275, 245, 1972.
45. **Stigbrand, T., Malmström, B. G., and Vänngård, T.,** On the state of copper in the blue protein umecyanin, *FEBS Lett.,* 12, 260, 1971.
46. **Stigbrand, T.,** Oxidation-reduction potential of umecyanin, *FEBS Lett.,* 23, 41, 1972.
47. **Blumberg, W. E. and Peisach, J.,** The optical and magnetic properties of copper in *Chenopodium album* plastocyanin, *Biochim. Biophys. Acta,* 126, 269, 1966.
48. **Miskowski, V., Tang, S.-P. W., Spiro, T. G., Shapiro, E., and Moss, T. H.,** The copper coordination group in "blue" copper proteins: evidence from resonance Raman spectra, *Biochemistry,* 14, 1244, 1975.
49. **Siiman, O., Young, N. M., and Carey, P. R.,** Resonance Raman spectra of "blue" copper proteins and the nature of their copper sites, *J. Am. Chem. Soc.,* 98, 744, 1976.

50. **Ferris, N. S., Woodruff, W. H., Rorabacher, D. B., Jones, T. E., and Ochrymowycz, L. A.** Resonance Raman spectra of copper-sulfur complexes and the blue copper protein question, *J. Am. Chem. Soc.*, 100, 5939, 1978.

51. **Roberts, J. E., Brown, T. G., Hoffman, B. M., and Peisach, J.,** Electron nuclear double resonance spectra of stellacyanin, a blue copper protein, *J. Am. Chem. Soc.*, 102, 825, 1980.

52. **Ambler, R. P.,** Sequence data acquisition for the study of phylogeny, in *Recent Developments in the Chemical Study of Protein Structures,* Previero, A., Pechère, J.-F., and Coletti-Previero, M.-A., Eds., Inserm, Paris, 1971, 289.

53. **Ambler, R. P. and Brown, L. H.,** The amino acid sequence of *Pseudomonas fluorescens* azurin, *Biochem. J.*, 104, 784, 1967.

54. **Dayhoff, M. O., Ed.,** *Atlas of Protein Sequence and Structure,* Vol. 5, Suppl. 3, National Biomedical Research Foundation, Washington, D.C., 1978.

55. **Haslett, B. G., Bailey, C. J., Ramshaw, J. A. M., Scawen, M. D., and Boulter, D.,** The amino acid sequence of plastocyanin from *Rumex obtusifolius, Phytochemistry,* 17, 615, 1978.

56. **Scawen, M. D., Ramshaw, J. A. M., Brown, R. H., and Boulter, D.,** The amino-acid sequence of plastocyanin from *Sambucus nigra* L. (elder), *Eur. J. Biochem.*, 44, 299, 1974.

57. **Scawen, M. D., Ramshaw, J. A. M., and Boulter, D.,** The amino acid sequence of plastocyanin from spinach (*Spinacia oleracea* L.), *Biochem. J.*, 147, 343, 1975.

58. **Scawen, M. D. and Boulter, D.,** The amino acid sequence of plastocyanin from *Cucurbita pepo* L. (vegetable marrow), *Biochem. J.*, 143, 257, 1974.

59. **Scawen, M. D., Ramshaw, J. A. M., Brown, R. H., and Boulter, D.,** The amino acid sequences of plastocyanin from *Mercurialis perennis* and *Capsella bursa-pastoris, Phytochemistry,* 17, 901, 1978.

60. **Ramshaw, J. A. M., Scawen, M. D., Bailey, C. J., and Boulter, D.,** The amino acid sequence of plastocyanin from *Solanum tuberosum* L. (potato), *Biochem. J.*, 139, 583, 1974.

61. **Haslett, B. G., Evans, I. M., and Boulter, D.,** Amino acid sequence of plastocyanin from *Solanum crispum* using automatic methods, *Phytochemistry,* 17, 735, 1978.

62. **Ramshaw, J. A. M., Scawen, M. D., and Boulter, D.,** The amino acid sequence of plastocyanin from *Vicia faba* L. (broad bean), *Biochem. J.*, 141, 835, 1974.

63. **Milne, P. R., Wells, J. R. E., and Ambler, R. P.,** The amino acid sequence of plastocyanin from French bean *(Phaseolus vulgaris), Biochem. J.*, 143, 691, 1974.

64. **Ramshaw, J. A. M., Scawen, M. D., Jones, E. A., Brown, R. H., and Boulter, D.,** The amino acid sequence of plastocyanin from *Lactuca sativa* (lettuce), *Phytochemistry,* 15, 1199, 1976.

65. **Ambler, R. P.,** in *Poplar Plastocyanin Structure,* Guss, J. M. and Freeman, H. C., Eds., Protein Data Bank, Brookhaven National Laboratory, Upton, N.Y., 1980.

66. **Bergman, C., Gandvik, E.-K., Nyman, P. O., and Strid, L.,** The amino acid sequence of stellacyanin from the lacquer tree, *Biochem. Biophys. Res. Commun.,* 77, 1052, 1977.

67. **Bergman, C.,** Amino Acid Sequence Studies of Two Small Blue Proteins: Stellacyanin and Umecyanin, Ph.D. thesis, University of Göteborg, Sweden, 1980.

68. **Haslett, B. G., Gleaves, T., and Boulter, D.,** N-terminal amino acid sequences of plastocyanins from various members of the Compositae, *Phytochemistry,* 16, 363, 1977.

69. **Haslett, B. G. and Boulter, D.,** The N-terminal amino acid sequence of plastocyanin from *Stellaria media.* An exercise to establish criteria for the idetification of residues from a sequenator, *Biochem. J.*, 153, 33, 1976.

70. **Strahs, G.,** Azurin: X-ray data for crystals from *Pseudomonas denitrificans, Science,* 165, 60, 1969.

71. **Norris, G. E., Anderson, B. F., Baker, E. N., and Rumball, S. V.,** Purification and preliminary crystallographic studies on azurin and cytochrome c' from *Alcaligenes denitrificans* and *Alcaligenes* sp. NCIB 11015, *J. Mol. Biol.*, 135, 309, 1979.

72. **Chapman, G. V., Colman, P. M., Freeman, H. C., Guss, J. M., Murata, M., Norris, V. A., Ramshaw, J. A. M., and Venkatappa, M. P.,** Preliminary crystallographic data for a copper-containing protein, plastocyanin, *J. Mol. Biol.*, 110, 187, 1977.

73. **Ugurbil, K., Norton, R. S., Allerhand, A., and Bersohn, R.,** Studies of individual carbon sites of azurin from *Pseudomonas aeruginosa* by natural-abundance carbon-13 nuclear magnetic resonance spectroscopy, *Biochemistry,* 16, 886, 1977.

74. **Ugurbil, K. and Bersohn, R.,** Tyrosine emission in the tryptophanless azurin from *Pseudomonas fluorescens, Biochemistry,* 16, 895, 1977.

75. **Ugurbil, K., Maki, A. H., and Bersohn, R.,** Study of the triplet state properties of tyrosines and tryptophan in azurins using optically detected magnetic resonance, *Biochemistry,* 16, 901, 1977.

76. **Hill, H. A. O., Leer, J. C., Smith, B. E., Storm, C. B., and Ambler, R. P.,** A possible approach to the investigation of the structures of copper proteins: ^{1}H n.m.r. spectra of azurin, *Biochem. Biophys. Res. Commun.*, 70, 331, 1976.

77. **Needleman, S. B. and Wünsch, C. D.,** A general method applicable to the search for similarities in the amino acid sequence of two proteins, *J. Mol. Biol.*, 48, 443, 1970.

78. **Schwartz, R. M. and Dayhoff, M. O.,** Matrices for detecting distant relationships, in *Atlas of Protein Sequence and Structure,* Vol. 5, Suppl. 3, Dayhoff, M. O., Ed., National Biomedical Research Foundation, Washington, D.C., 1978, 353.

79. **Fitch, W. M.,** Distinguishing homologous from analogous proteins, *System. Zool.,* 19, 99, 1970.

80. **Wright, C. S., Alden, R. A., and Kraut, J.,** Structure of subtilisin BPN′ at 2.5 Å resolution, *Nature (London),* 221, 235, 1969.

81. **Rydén, L. and Lundgren, J.-O.,** Homology relationships among the small blue proteins, *Nature (London),* 261, 344, 1976.

82. **Rydén, L. and Lundgren, J.-O.,** The relationship of the small blue proteins with the copper-containing oxidases, *Protides Biol. Fluids,* Peeters, H., Ed., 28, 87, 1980.

83. **Rydén, L. and Lundgren, J.-O.,** On the evolution of blue proteins, *Biochimie,* 61, 781, 1979.

84. **Boulter, D., Gleaves, J. T., Haslett, B. G., Peacock, D., and Jensen, U.,** The relationships of 8 tribes of the Compositae as suggested by plastocyanin amino acid sequence data, *Phytochemistry,* 17, 1585, 1978.

85. **Boulter, D., Peacock, D., Guise, A., Gleaves, J. T., and Estabrook, G.,** Relationships between the partial amino acid sequences of plastocyanin from members of ten families of flowering plants, *Phytochemistry,* 18, 603, 1979.

86. **Ambler, R. P.,** Cytochrome *c* and copper protein evolution in procaryotes, in *The Evolution of Metalloenzymes, Metalloproteins and Related Materials,* Leigh, G. J., Ed., Symposium Press, London, 1977, 100.

87. **Fox, G. E., Stackebrandt, E., Hespell, R. B., Gibson, J., Maniloff, J., Dyer, T. A., Wolfe, R. S., Balch, W. E., Tanner, R. S., Magrum, L. J., Zablen, L. B., Blakemore, R., Gupta, R., Bonen, L., Lewis, B. J., Stahl, D. A., Luehrsen, K. R., Chen, K. N., and Woese, C. R.,** The phylogeny of prokaryotes, *Science,* 209, 457, 1980.

88. **Maret, W., Dietrich, H., Ruf, H.-H., and Zeppezauer, M.,** Active site-specific reconstituted copper(II) horse liver alcohol dehydrogenase: a biological model for type 1 Cu^{2+} and its changes upon ligand binding and conformational transitions, *J. Inorg. Biochem.,* 12, 241, 1980.

89. **Hill, H. A. O. and Lee, W. K.,** Investigation of the structure of the blue copper protein from *Rhus vernicifera* stellacyanin by ¹H nuclear magnetic resonance spectroscopy, *J. Inorg. Biochem.,* 11, 101, 1979.

90. **Morpurgo, L., Finazzi-Agrò, A., Rotilio, G., and Mondovì, B.,** Studies on the metal sites of copper proteins. IV. Stellacyanin: preparation of apoprotein and involvement of sulfhydryl and tryptophan in the copper chromophore, *Biochim. Biophys. Acta,* 271, 292, 1972.

91. **Kingston, I. B., Kingston, B. L., and Putnam, F. W.,** Complete amino acid sequence of a histidine-rich proteolytic fragment of human ceruloplasmin, *Proc. Natl. Acad. Sci. U.S.A.,* 76, 1668, 1979.

92. **Dwulet, F. E. and Putnam, F. W.,** Complete amino acid sequence of a 50,000-dalton fragment of human ceruloplasmin, *Proc. Natl. Acad. Sci. U.S.A.,* 78, 790, 1981.

93. **Rydén, L.,** Model of the active site in the blue oxidases based on the ceruloplasmin-plastocyanin homology, *Proc. Natl. Acad. Sci. U.S.A.,* 79, 6767, 1982.

94. **Briving, C., Gandvik, E.-K., and Nyman, P. O.,** Structural studies around cysteine and cystine residues in the "blue" oxidase fungal laccase B. Similarity in amino acid sequence with ceruloplasmin, *Biochem. Biophys. Res. Commun.,* 93, 454, 1980.

95. **Steffens, G. J. and Buse, G.,** Studies on cytochrome *c* oxidase. IV. Primary structure and subunit function of polypeptide II, *Hoppe-Seyler's Z. Physiol. Chem.,* 360, 613, 1979.

96. **Anderson, S., Bankier, A. T., Barrell, B. G., De Bruijn, M. H. L., Coulson, A. R., Drouin, J., Eperon, I. C., Nierlich, D. P., Roe, B. A., Sanger, F., Schreier, P. H., Smith, A. J. H., Staden, R., and Young, I. G.,** Sequence and organization of the human mitochondrial genome, *Nature (London),* 290, 457, 1981.

97. **Bibb, M. J., Van Etten, R. A., Wright, C. T., Walberg, M. W., and Clayton, D. A.,** Sequence and gene organization of mouse mitochondrial DNA, *Cell,* 26, 167, 1981.

98. **John, P. and Whatley, F. R.,** *Paracoccus denitrificans* and the evolutionary origin of the mitochondrion, *Nature (London),* 254, 495, 1975.

99. **Ludwig, B. and Schatz, G.,** A two-subunit cytochrome *c* oxidase (cytochrome *aa₃*) from *Paracoccus denitrificans, Proc. Natl. Acad. Sci. U.S.A.,* 77, 196, 1980.

100. **Winter, D. B., Bruyninckx, W. J., Foulke, F. G., Grinich, N. P., and Mason, H. S.,** Location of heme *a* on subunits I and II and copper on subunit II of cytochrome *c* oxidase, *J. Biol. Chem.,* 255, 11408, 1980.

101. **Saraste, M., Penttilä, T., and Wikström, M.,** Quaternary structure of bovine cytochrome oxidase, *Eur. J. Biochem.,* 115, 261, 1981.

102. **Poole, R. K.,** The oxygen reactions of bacterial cytochrome oxidases, *Trends Biochem. Sci.,* 7, 32, 1982.

103. **Aitken, A.,** Protein evolution in cyanobacteria, *Nature (London),* 263, 793, 1976.

104. **Taylor, F. J. R.,** Implications and extensions of the serial endosymbiosis theory of the origin of eukaryotes, *Taxon,* 23, 229, 1974.

105. **Schopf, J. W.,** The development and diversification of precambrian life, *Origins Life,* 5, 119, 1974.
106. **Murata, M., Begg, G. S., Lambrou, F., Leslie, B., Simpson, R. J., Freeman, H. C., and Morgan, F. J.,** Amino acid sequence of a basic blue protein from cucumber seedlings, *Proc. Natl. Acad. Sci. U.S.A.,* 79, 6434, 1982.
107. **Ambler, R. P.,** Reported in the Atlas of Protein Sequence and Structure, National Biomedical, Washington, D.C., 1982.

Chapter 7

THE REACTIVITY OF COPPER SITES IN THE "BLUE" COPPER PROTEINS

Ole Farver and Israel Pecht

TABLE OF CONTENTS

I. INTRODUCTION

The biological role of copper ions is in mediation and catalysis of redox reactions. The advent of knowledge of the structure of the blue single copper proteins has caused a marked increase in the interest in that group of proteins which serve as electron mediators.[1-4] The so-called type-1 copper site in these blue proteins gives rise, in the Cu(II) state, to the characteristic intense absorption band near 600 nm, as well as to a very small hyperfine splitting in the $g_{\parallel}$ region of the EPR spectrum.[5] None of these spectral properties has been successfully reproduced together in low M_r Cu(II) complexes. Finally, the redox potential of the type-1 sites is exceptionally high compared with the Cu(II)/Cu(I) couple in aqueous solution (Appendix 1).

Three-dimensional structures are now available for two of the blue copper proteins. Crystallographic X-ray studies by Freeman and co-workers,[6] recently refined to 1.6 Å resolution, have established the structures of both oxidized and reduced forms of plastocyanin (Pc) from the poplar tree *(Populus nigra)*.[7] The molecular structure is a β-barrel with the copper ion placed near the top of the molecule (Figure 1). The metal ion is coordinated by the side chains of His-37, Cys-84, His-87, and Met-92 in a highly distorted tetrahedral geometry and is further surrounded by conserved hydrophobic residues. Only one of the ligating residues, the imidazole of His-87, has an edge exposed to the solvent boundary of the molecule. The Cu(II)-ligand bond lengths in oxidized Pc were found to be 2.04 (His-37), 2.10 (His-87), 2.13 (Cys-84), and 2.90 Å (Met-92). It is important to notice that in the unprotonated form of reduced Pc, the Cu-ligand distances coincide within 0.2 Å with those in the oxidized protein.[7]

The interior of the molecule consists mainly of hydrophobic side chains, but some are directed into the solvent such as, e.g., Tyr-83. Another striking feature is a surface concentration of conserved acidic residues close to Tyr-83 suggesting some kind of functional significance. This aspect will be discussed later (cf. Section II.A). Another interesting observation is that while the structure of Cu(II)-Pc is pH-independent, the reduced protein seems to exist in two conformers, a redox-active and a redox-inactive protonated form.[7] At low pH the Cu(I)-N(His-87) distance becomes sufficiently long to accommodate a proton between the copper atom and the N_δ-atom. This could stabilize the Cu(I) ion in a trigonal planar geometry, making the protonated form of Cu(I)-Pc redox-inactive. This aspect will be discussed further below (Section II.A).

The second blue protein whose structure has been determined is azurin (Az) from the bacterium *Pseudomonas aeruginosa*. The X-ray analysis has been performed at 3 Å resolution[8] for the oxidized protein and recently refined to 2.7 Å.[9] The overall molecular structure of Az (Figure 2) is very similar to that of Pc, as was also suggested from sequence analogies. The Cu-atom is coordinated to the analogs of the ligands in Pc: His-46, Cys-112, His-117, and Met-121. The metal ion is again buried in a highly conserved hydrophobic core and may be somewhat less accessible than in Pc, since in Az the Cu-site is further shielded by hydrophobic side chains.[10] Another important difference between the two protein structures is that there is no cluster of surface charges on Az. Instead, the invariant acidic and basic groups occur in pairs.[9] These differences naturally have implications for the electron-transfer mechanisms which will be discussed below.

The above observation that the metal ion in the blue copper proteins is substantially buried in the solvent-inaccessible interior of the protein is most probably a reflection of the functional evolution of these proteins, allowing for the required control and specificity of the electron-transfer processes mediated by them. In these reactions the redox centers are widely separated, hence a direct orbital overlap between the metal ions is not possible without a major rearrangement of the surrounding amino-acid side chains. With the flattened tetrahedral geometry of the copper center, we have a situation where coordination-sphere reorganization associated

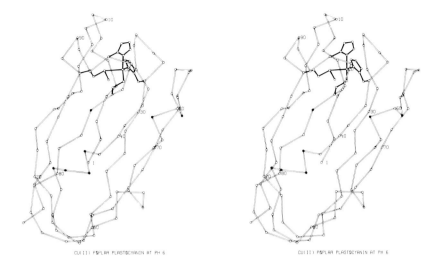

CU(II) POPLAR PLASTOCYANIN AT PH 6   CU(II) POPLAR PLASTOCYANIN AT PH 6

FIGURE 1. Stereoscopic view of the poplar Cu(II)-plastocyanin molecule. The amino-acid residues (other than those which coordinate the Cu atom) are represented by their C_α atoms. Every tenth C_α atom is numbered. Black C_α atoms belong to residues 42 to 45, 59, and 61 of the "acidic patch", and to the invariant Tyr-83. Residue 45 is Ser in poplar plastocyanin, while Glu in French bean plastocyanin. (Courtesy of Dr. H. C. Freeman.)

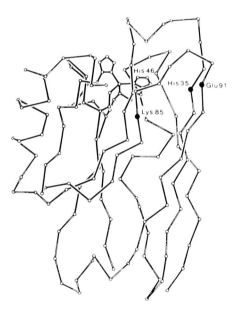

FIGURE 2. View of the polypeptide backbone of *Pseudomonas aeruginosa* azurin. Only C_α atoms are shown, and the view is from the side which becomes labeled by $Cr(II)_{aq}$ ions. The four amino-acid side chains binding the copper are indicated. (Drawn by Dr. H. C. Freeman, from coordinates of the model kindly provided by Dr. E. Adman.)

with the electron transfer is minimized. This is attained by the particular ligand geometry which accommodates both redox states of the copper ion (cf. above).

Experimental evidence for this notion has recently been presented.[11] By subpicosecond irradiation of Az solutions at 625 nm, the solution was transiently bleached, but recovered its original color with a time constant of 1.6 ± 0.2 psec due to reverse charge transfer. The two processes were described[11] by (1) charge transfer from the coordinated thiolate to the metal center:

$$R-S^-Cu^{2+} \xrightarrow{h\nu} R-S^{\bullet}Cu^+ \qquad (1)$$

followed by (2) a radiationless relaxation process:

$$R-S^{\bullet}Cu^+ \to R-S^-Cu^{2+} + \text{phonons} \qquad (2)$$

The very fast electronic relaxation implies that the rearrangement of the copper ligand sphere is not rate determining for the electron transfer.[11]

Structural considerations greatly favor an outer sphere electron-transfer mechanism in reactions involving blue copper proteins. Outer-sphere reactions are those in which no ligands are shared between the two redox centers in the transition state.[12] Here electron transfer can take place via molecular orbitals weakly delocalized between the redox centers of the reaction partners. Since no bonds are formed or broken in an outer-sphere reaction it is the simplest to treat theoretically. The most useful theory is due to Marcus,[13] and the kinetics of electron transfer in protein systems have been extensively analysed within this framework. A detailed review has recently been published on this subject,[1] therefore only a brief summary will be given here.

The cross-reaction rate constant, k_{12}, for an outer-sphere electron-transfer reaction between two species is related to the equilibrium constant of that reaction, K_{12}, and to the self-exchange rates of the two reactants, k_{11} and k_{22}. This yields the Marcus equation:[13]

$$k_{12} = (k_{11} k_{22} K_{12} f)^{1/2} \qquad (3)$$

The last term f is defined by:

$$\log f = (\log K_{12})^2 / 4 \log (k_{11} k_{22}/z^2) \qquad (4)$$

where z is the collision frequency for neutral molecules in solution ($\approx 10^{11} M^{-1} \sec^{-1}$). For reactions with small driving forces (i.e., small difference in redox potential), $f \approx 1$. Non-specific electrostatic contributions to the rate constants can be corrected for by adjusting these to a certain ionic strength by means of Debye-Hückel theory. The Marcus theory has been extensively used for discussing differences in kinetic parameters of both protein-protein and protein-small molecule electron-exchange reactions.[1,4] Thus, if the same electron-transfer pathway and mechanism are employed by a certain protein in all its redox reactions, and there is no pronounced association between the reactants, the calculated self-exchange rate should be constant. In contrast, obtaining different values for self-exchange rates of the cross-reactions between a given copper protein and various redox partners, implies differences in the mechanisms of electron transfer.[1]

Quantum mechanical tunneling has also been invoked to explain biochemical electron-transfer processes.[14-16] (For a comprehensive presentation and discussion of the present state of tunneling theory as applied to biochemistry, see Reference 16.) In principle, at least short-range electron tunneling may occur whenever energy barriers are involved in electron transfer. The electron transfer rate constant, k, is related to a "tunneling" matrix element, $\mathbf{T}_{ab}$, the magnitude of which depends on the extent of electronic overlap of the donor and acceptor wave functions. $\mathbf{T}_{ab}$ is small because of the exponential decrease of the wave functions in the region of the energy barrier between the two redox sites. In calculating the rate of tunneling through the barrier it is assumed that the translational energy of the reactants is conserved as internal vibrational energy of the activated complex. This energy is coupled to the electronic state of the donor molecule and thus serves to reduce the height of the tunneling barrier.

Significant in the present context is that the tunneling matrix element can be expressed as an exponential function of the intersite distance, which means that the rate constant, k, of electron flow through the barrier will depend on this distance, i.e., $k \sim e^{-r}$. It is also important to realize that r is the edge-to-edge distance between the two electronic systems rather than strictly distance between the metal centers.

A critical factor in biological electron-transfer mechanisms is the necessity of the reacting molecules to attain the proper relative orientation before electron transfer. Generally, the detailed structural interactions provide the necessary basis for both specificity and control of biological processes. Besides yielding optimal orientations, interactions between electron-transport proteins may promote structural changes which can serve to facilitate the electron-transfer reaction. A basic, if somewhat metaphysical problem associated with the tunneling process is that the protein has been relegated to a role which is in essence to provide a continuum between the metal ions. It remains to be shown whether the design of these complex macromolecules, involving extensive evolutionary processing, was only to provide a certain distance between redox centers and their appropriate potentials.

The single blue copper proteins mediate electrons between supramolecular assemblies in the biochemical energy-conversion systems e.g., Pc in the photosynthetic apparatus. In analysing these protein-protein electron-transfer reactions one has to consider the complexity of interactions between the macromolecules including specific multipoint binding along with potential conformational transitions in them. One major approach in the field involved, therefore, the examination of electron-transfer reactions between a protein and a low M_r redox partner, usually a transition metal-ion complex. This provides a way of dissecting the problem by using the better understood low M_r complexes as a reference. The latter compounds have several important advantages. They provide a wide range of redox potentials, i.e., the driving force of the redox reaction can be controlled. They are all well-characterized single electron, outer-sphere reactants. Their size and hydrophobic/hydrophilic balance can also be modulated; since these reagents often have special spectroscopic features, they are convenient for kinetic studies. However, it is only by studying the physiologically significant reactions that the role of specificity and control can be examined.

In the following paragraphs we will focus primarily on those aspects of the mechanism of electron transfer of single copper proteins which can be interpreted in terms of presently available structural information.[6-10]

II. SINGLE BLUE COPPER PROTEINS

A. Plastocyanins

These are the best characterized among the single blue copper proteins. Their M_r values are $\approx 10,000$ (Appendix 1) and the amino-acid sequences of a very large number of Pc's have been determined and show a remarkably high degree of conservation.[17] The great similarity of PMR spectra of Pc(I) from a variety of higher plants[18] led to the suggestion that the structure of poplar Pc referred to in Section I is maintained by this whole group. They function as electron carriers in all higher plants, in many green algae, and some blue-green ones. The site of their action is commonly believed to be located between the membrane-bound cytochrome f and the P-700 chlorophyll pigment of photosystem I. The redox potential of Pc at pH 7 is ≈ 370 mV.[5] This value is between those of Cyt f and P-700, in agreement with the position of Pc in the electron-transfer chain. The orientation of Pc on the thylakoid membrane is still a matter of controversy. It is not clear whether it is bound to the inner or outer thylakoid surface. Cyt f and P-700 are probably located close to the interior thylakoid surface, whereas Pc is apparently bound loosely to the membrane as it can be solubilized readily.

Since Pc is relatively easy to prepare, is well-characterized, and its structure is available at high resolution in both oxidation states, it offers an interesting system for correlating the electron-transfer activity with its structure. Indeed, several independent lines of research have been pursued in order to resolve the electron-transfer routes to and from the copper site in that protein. Recently the electron-transfer reactions between Pc and a large number of transition metal-ion complexes have been studied in great detail.[19-25] Parameters such as pH of the medium, electrostatic charges, and hydrophobicity of the reagents were varied. Results were interpreted in terms of the active sites on Pc[24,25] and distances over which electrons are transferred.[22] Independent examination of the potential electron-transfer loci has been made by the use of redox-inert paramagnetic complexes perturbing the PMR lines of Pc.[26-28] More recently an affinity-labeling procedure has been employed in order to localize the sites of electron transfer on Pc.[29] In the following paragraphs, accounts of the results obtained will first be given and later, these will be discussed in an effort to integrate them into a possibly unified scheme of reaction routes.

The pH dependence of Pc oxidation and reduction by low M_r reagents have been extensively studied by Sykes and co-workers.[23-25] From the concentration dependence of the observed reaction rates it was inferred that the electron-transfer step is preceded, in many cases, by the formation of a relatively stable, outer-sphere precursor complex between Pc and the redox agent. This leads to the following scheme for Pc oxidation:

$$Pc(I) + R \overset{K}{\rightleftharpoons} Pc(I) \cdot R \tag{5}$$

$$Pc(I) \cdot R \overset{k}{\rightarrow} products \tag{6}$$

and analogously for Pc(II) reduction. The observed rate constants for Pc(I) oxidation by $Co(phen)_3^{3-}$ and $Fe(CN)_6^{3-}$ show a marked dependence on pH. In both cases the rate decreases with decreasing pH, corresponding to the rapid formation of a redox-inactive protonated form of the protein with a $pK \approx 6$.[24,25] The question as to which of the above steps is affected by the pH changes has also been examined. It was found that for the reaction of Pc(I) with $Co(phen)_3^{3+}$, the association constant decreased with decreasing pH, while the values of the electron-transfer rates remained essentially constant. With $Fe(CN)_6^{3-}$, the observed pH dependence was also attributed to inhibition of complex formation. It is noteworthy that the "switch off" observed in the reactivity of Pc at low pH can be correlated with the formation of a conformer of reduced Pc, having the imidazole of His-87 protonated (cf. Section I) as resolved by X-ray data.[7] This form is most probably the "redox-inactive" one, which is also in line with pH dependence of the redox potential of Pc determined as early as 1962.[30] In view of the structural and thermodynamic evidence for a drastic change taking place at the redox center,[7] the assignment of the decreased reactivity to a failure in forming a precursor complex is not satisfactory.

Whether the same binding site in Pc is employed by both $Co(phen)_3^{3+}$ and by $Fe(CN)_6^{3-}$ was not resolved by these studies.[24,25] However, the recently adopted and very informative PMR study has illuminated much of this problem.[26-28] The high resolution PMR spectrum of Pc(I) is well-characterized.[18] By studying the broadening caused to proton resonances of certain residues by redox-inactive paramagnetic complexes, two distinct regions of interaction were revealed. With $Cr(CN)_6^{3-}$ as a probe, resonance lines of three of the copper ligands (His-37, His-87, and Met-92) were perturbed.[26-28] Considering the close similarity between the hexacyano complexes of iron and chromium, this finding implies that negatively charged oxidants like $Fe(CN)_6^{3-}$ associate with the protein surface at an area close to the copper coordination sphere before electron transfer. The most probable locus is near the exposed edge of imidazole-87 at the "north" end of the molecule. In contrast, the site identified by positively charged complexes is near Tyr-83 as judged from the pronounced resonance line broadening of this residue on interaction of Pc(I) with $Cr(NH_3)_6^{3+}$ [27,28] or $Cr(phen)_3^{3+}$.[26-28] Since it was found that addition of $Cr(phen)_3^{3+}$ inhibits the oxidation of Pc(I) by $Co(phen)_3^{3+}$,[25] a distinct site of interaction for positively charged oxidants, different from that used by the negatively charged ones, is strongly indicated. Tyr-83 is exposed to the solvent and situated next to a cluster of carboxylate side chains. Consequently, binding near this residue is clearly promoted by electrostatic interactions with the array of negative charges, which select and facilitate the complex formation with positively charged reagents.

A more direct identification of an electron-transfer site at the above-described region on French bean Pc has recently been achieved.[29] This was done by employing $Cr(II)_{aq}$ ions as an affinity-labeling agent. The rationale of the approach is that while Cr(II) ions are strong reductants and exhibit exceptionally fast ligand exchange, their oxidation product, Cr(III) complexes, are highly substitution inert.[12] Hence, the coordination sphere of the Cr(II) ion at the transition state of electron transfer is expected to be the one found in the reaction products. A stoichiometric binding of one Cr(III) per mole protein was found upon reduction by Cr(II) ions. Enzymatic cleavage of this derivative, followed by analysis of the proteolytic fragments, led to the identification of the Cr(III)-binding peptide.[29] This peptide contains four potential ligands for Cr(III): Asp-42, Glu-43, Asp-44, and Glu-45, which then most likely mark the region from which the electron is transferred to Cu(II) in Pc (cf. Figure 1). This patch is very close to Tyr-83, the solvent-exposed aromatic residue mentioned above. The proximity of this side chain to the Cr(III) binding site was further supported by the tyrosine emission properties of the labeled protein, which markedly differ from those of the native protein. The distance between the copper center and the proposed Cr(III) binding site is ≈ 12 Å, and the intervening region contains an array of highly conserved aromatic residues. Based on these observations an electron-transfer route from the solvent-protein interface to the copper ion has been proposed which involves electronic delocalization through a weakly coupled π-system. Such "through space" influence with direct interaction between π-systems of intervening aromatic rings instead of a "through bond" system has been suggested by Taube[31] for the electron transfer in model systems of mixed valence complexes.

A systematic examination of the electron-transfer reactions between Pc and a selected group of transition metal complexes has been carried out by Gray and co-workers.[19-22] The data obtained were analysed within the Marcus formalism framework, and calculated self-exchange rate constants, corrected for electrostatic effects, were presented. Some of these results are shown in Table 1, and it is seen that the self-exchange rates for each of the proteins span a wide range. Comparing these apparent self-exchange constants provides a means to assess other important factors that influence the reactivity of Pc, in addition to the driving force and electrostatic interaction. The most important among these seems to be the extent of orbital overlap between the redox orbitals of the reagents, and this has led to the definition of a kinetic accessibility scale for redox metalloproteins.[1] Complexes having hydrophobic ligands with good π-conducting molecular orbitals such as 1,10-phenanthroline (phen) or pyridine (py) yield much higher self-exchange constants than, e.g., EDTA or oxalato complexes. The latter are quite hydrophilic and lack extended π-orbitals. The noted difference has been attributed to penetration of the hydrophobic ligands into the protein interior, leading to closer contact with the aromatic ligands of the metal centers. An isokinetic relationship was reported for the reduction by Pc(I) of a variety of $Co(phen)_3^{3+}$ complexes having substituted phen ligands. This strongly suggests that the mechanism of electron transfer does not alter substantially with the derivatization of the aromatic ring system.[19] In addition, both activation enthalpy and entropy decrease with increasing size of the ring system. This relation is consistent with the reaction becoming increasingly nonadiabatic concomitantly with a poorer orbital overlap between the redox centers.

Table 1

RATE CONSTANTS FOR PROTEIN-SMALL MOLECULE AND PROTEIN-PROTEIN ELECTRON TRANSFER (k_{12}) TOGETHER WITH THOSE CALCULATED FOR PROTEIN SELF-EXCHANGE (k_{11}) ($M^{-1}sec^{-1}$)[a]

Reagent	Azurin (Pseudomonas aeruginosa)		Plastocyanin (higher plants)		Stellacyanin (Rhus vernicifera)	
	k_{12}	k_{11}	k_{12}	k_{11}	k_{12}	k_{11}
$Co(OX)_3^{3-}$[b]	$2.9 \cdot 10^{-2}$	$2.6 \cdot 10^{-3}$	$2.4 \cdot 10^{-1}$	$5.4 \cdot 10^{-1}$	$7.3 \cdot 10^2$	$2.5 \cdot 10^4$
$Fe(EDTA)^{2-}$[c]	$1.3 \cdot 10^3$	$1.2 \cdot 10^{-2}$	$8.2 \cdot 10^4$	$3.4 \cdot 10^1$	$4.3 \cdot 10^5$	$2.3 \cdot 10^5$
$Fe(CN)_6^{3-}$[c]	$2.7 \cdot 10^4$	$2.0 \cdot 10^1$	$7.0 \cdot 10^4$	$7.6 \cdot 10^3$		
$Co(phen)_3^{3+}$[c]	$3.2 \cdot 10^3$	$1.6 \cdot 10^4$	$4.9 \cdot 10^3$	$2.6 \cdot 10^3$	$1.8 \cdot 10^5$	$3.0 \cdot 10^5$
$Co(dm-phen)_3^{3+}$[d]	$1.5 \cdot 10^3$	$6.3 \cdot 10^5$	$8.0 \cdot 10^2$	$1.5 \cdot 10^4$	$1.9 \cdot 10^4$	$5.8 \cdot 10^5$
$Ru(NH_3)_5py^{3+}$[c]	$2.0 \cdot 10^3$	$2.7 \cdot 10^3$	$7.1 \cdot 10^3$	$1.2 \cdot 10^4$	$1.9 \cdot 10^5$	$1.7 \cdot 10^5$
Cyt 553	$1.4 \cdot 10^{7f}$	$7.3 \cdot 10^{6g}$	$4.0 \cdot 10^{5f}$	$4.1 \cdot 10^{2g}$		
Cyt c_{551}	$6.1 \cdot 10^{6h}$	$1.0 \cdot 10^{6g}$	$7.5 \cdot 10^{5i}$	$6.0 \cdot 10^{2g}$	$1.6 \cdot 10^{5j}$	$6.1 \cdot 10^{3g}$
Cyt c	$3.0 \cdot 10^{3i}$	$1.1 \cdot 10^{4g}$	$1.0 \cdot 10^{6i}$	$1.3 \cdot 10^{8g}$	$3.5 \cdot 10^{2j}$	$4.5 \cdot 10^{3g}$
Cyt f	$1.0 \cdot 10^{6i}$	$2.4 \cdot 10^{8k}$	$3.6 \cdot 10^{7i}$	$8.6 \cdot 10^{9k}$		

[a]   Protein-small molecule reactions are, with the exception of Fe(EDTA)$^{2-}$ as reagent, all protein oxidations. All rate constants for protein-protein reactions are given with the blue proteins as oxidants.

[b]   Reference 21.

[c]   Reference 1.

[d]   *Tris* (5,6-dimethyl-1,10-phenanthroline)cobalt(III), Reference 1.

[e]   Reference 20.

[f]   Reference 41.

[g]   Our calculations, based on cytochrome self-exchange rate constants given in Reference 41.

[h]   Reference 55.

[i]   Reference 39.

[j]   Reference 69.

[k]   Our calculation based on an estimated k_{11} for Cyt f = $2.5 \cdot 10^4$ M^{-1} sec^{-1}, determined from the Fe(CN)$_6^{3-}$ - Cyt f reaction.[39]

Electron transfer from Fe(EDTA)$^{2-}$ to Pc(II) is also characterized by a very small activation enthalpy and a large negative activation entropy ($\Delta H^{\ddagger}$ = 8.8 kJ mol^{-1}, $\Delta S^{\ddagger}$ = -121 J K^{-1} mol^{-1}). It is interesting that the activation parameters for the above reactions where the overlap of the redox orbitals is expected to be small also fit the tunneling mechanism. Hence a very small reorganizational activation requirement is expected as well as a large negative activation entropy, as a result of a small transmission coefficient.

The possible relation existing between the calculated self-exchange rate constants and the distance over which the electron is transferred has been further pursued. Using the tunneling hypothesis, an equation was derived which relates the electrostatic-corrected self-exchange rates to an electron-transfer distance.[22] Based on the reaction with Fe(EDTA)$^{2-}$, the shortest distance from the copper site in Pc to the surface was calculated as 2.6 Å. This is in disagreement with the crystallographic result even involving the imidazole ring of His-87 as the way of closest approach. The distances reported there[22] also do not conform to the suggested binding site for Co(phen)$_3^{3+}$ as implied from other studies *(vide supra)*.[26-29] Since two different electron-transfer sites are present on the protein, this should be a primary consideration in the analysis of the data. As all the negatively charged redox agents used are hydrophilic while the positively charged ones are hydrophobic, it is difficult to differentiate between the roles of separate electron-transfer routes and the chemical properties of the agents. Another feature which makes the distance calculations problematic is the employed values of the overall second-order rate constants, whereas in many cases, precursor complex formation precedes an *intra*molecular electron-transfer step. That is, the electron transfer cannot be handled as a simple bimolecular reaction. An inherent problem is also that the tunneling theory treats the protein interior as a homogeneous entity without, e.g., specific electron-conducting pathways.

The electron-transfer pathways outlined above on the basis of the Cr(II) affinity labeling[29] provide an attractive rationale for the kinetic accessibility scale of Gray and co-workers.[1] Redox agents, such as $Co(phen)_3^{3+}$ with hydrophobic π-conducting ligands, were found to exhibit much higher reactivities than the more hydrophilic ones.[21] This is understood in terms of the above-mentioned proposal since the former residues will provide an excellent overlap with the π-symmetric orbitals of Tyr-83. It is also noteworthy that the proposed binding site of Cr(III) in the inner-sphere reduction of Pc(II) by Cr(II) coincides with the one which was suggested for interaction between cationic inorganic outer-sphere complexes and the native reduced protein, based upon the specific broadening of the PMR signals of the Tyr-83 protons.[26-28] This raises the question of whether this electron-transfer locus is relevant to the sites of interaction between Pc and its physiological partners in the photosynthetic system. This has recently been studied,[32] and the results will be discussed below.

The amount of data on protein-protein reactions involving Pc is more limited than the corresponding small molecule-protein reactions. However, it is primarily the former type of studies which provides information about the possible role of specificity between physiological partners.

There is a relatively large body of data dealing with the role played by Pc in the photosynthetic process. The *in situ* turnover of Pc has been monitored and kinetically analysed by flash spectroscopy in broken chloroplasts and algal cells.[33-35] For the electron transfer from Pc to isolated photosystem I reaction centers of spinach leaves, it was found that the observed rate was proportional to the Pc concentration, and a second-order constant of $1.5 \cdot 10^8 \ M^{-1} \ sec^{-1}$ at room temperature has been reported.[34] It was further demonstrated that low Mg^{2+} concentrations stimulated the electron-transfer rate by about one order of magnitude.[34] The same reaction was also studied at low temperature ($-20°C$) and it was reported to be qualitatively similar to that observed at room temperature but 100-fold slower.[35] As previously noted, the conclusion reached was that Pc donates the electron directly to P-700 in a bimolecular reaction.

For the reaction of Pc(II) with the *f*-type cytochrome 553 of green algae it has, however, been shown that besides the reaction sequence Cyt 553 → Pc → photosystem I, solubilized Cyt 553 can be photooxidized at 700 nm without Pc being present.[36] There are significant differences between cytochromes of algae and higher plants. Deficiency of copper in certain algae,[37] yielding low Pc levels, has been demonstrated to cause a situation where the role of Pc is taken over by Cyt 553.

Wood and Bendall[38] have shown that Pc reacts two orders of magnitude faster with P-700 from disrupted pea chloroplasts than does Az ($k = 8 \cdot 10^7$ vs. $3 \cdot 10^5 \ M^{-1} \ sec^{-1}$, respectively). This was taken as a demonstration of reaction specificity. It was also shown that Cyt *f* was a poor electron donor for P-700, which supports the notion that in higher plants the reaction order is Cyt *f* → Pc → P-700. Wood[39] has also studied electron transfer between Pc and cytochromes from various sources and found that reduction of Pc(II) by reduced Cyt *f* is at least 30 times faster ($k = 3.6 \cdot 10^7 \ M^{-1} \ sec^{-1}$ at 25°C, pH 7.0) (Table 1) when compared to Pc(II) reduction by other cytochromes. Again, this result was interpreted as reflecting biological discrimination. Thus the variation in rates of reduction of Pc by the different cytochromes was taken as evidence for some degree of specificity.[39] However, high reaction rates between proteins alone cannot be taken as evidence for physiological specificity just as a slow rate cannot eliminate this possibility. Factors such as driving force of the electron-transfer reaction (difference in redox potentials) and the self-exchange rate constant of each individual protein will have a decisive influence on the protein-protein electron-transfer rate. The activation parameters for the Cyt *f*-Pc reaction ($\Delta H^{\ddagger} = 44$ kJ mol^{-1}, $\Delta S^{\ddagger} = +46$ J K^{-1} mol^{-1} at pH 7.0) show that the high rate is due to the very favorable entropy term.[39] Since both proteins are highly negatively charged at this pH, it was claimed that electrostatic interactions play a minor role here compared with other contributions. This, however, seems to oversimplify the situation as it is not the overall charge that determines the interactions but rather the particular distribution on the proteins surface. As pointed out recently by San Pietro and co-workers,[40] changes in the reactivity of different Pc's may be traced back to differences in the patch of negative residues.

Wherland and Pecht[41] have carried out an extensive survey of protein-protein electron-transfer reactions between blue copper proteins and *c*-type cytochromes using the chemical relaxation temperature-jump method. Pc from both algae and bean leaves were found to exhibit only a single equilibrium step in reactions with algal Cyt 553. These data, along with earlier published results, were analysed according to the Marcus theory formalism, and are presented in Table 1. Using the self-exchange rate constants for the cytochromes reported in this paper,[41] we have calculated the self-exchange rate constants, k_{11}, for the blue copper proteins. These are shown (Table 1) in order to allow comparing them with the corresponding calculated self-exchange rate constants determined on the basis of protein-small molecule reactions. It is seen that k_{11} calculated for Pc from reactions with the small, prokaryotic cytochromes (Cyt 553 and Cyt c_{551}) is much lower than that calculated on the basis of its reactivity with several of the inorganic reagents. Also the value of k_{11} is virtually independent of these two cytochromes, although Cyt 553 is assumed to be the natural partner for algal Pc. These observations strongly suggest that there is no kinetic selectivity in these reactions. With the two high M_r cytochromes, the calculated self-exchange rate constants are also comparable, although the mammalian Cyt *c* from horse heart can hardly be considered a natural partner for parsley Pc. It is difficult to reconcile the lack of specificity between any of the protein pairs examined as implied by the Marcus analysis with their expected functional characteristics. At least a few of the pairs examined should have shown some degree of specificity and it could well be that while the self-exchange reactions occur via one route, the electron transfer proceeds through another. Hence caution should be used in employing that type of analysis without considering the particular details of the reactions.

The recent identification of an electron-transfer route in Pc by affinity labeling with Cr(II) ions (cf. above)[29] raised the question of the relevance of the labeled region near Tyr-83 to the interaction between Pc and its physiological partners in the photosynthetic system.[32] In order to examine that, photooxidation of native and Cr-labeled Pc(I) by isolated photosystem I reaction centers was kinetically investigated. It was found that with the Cr(III)-labeled protein, the rate was significantly slower than with native Pc ($k = 6.9 \cdot 10^7$ and $1.8 \cdot 10^8$ M^{-1} sec^{-1}, respectively). This was contrasted by the finding that the rate of photoreduction of the Cr-labeled Pc(II) by Cyt *f* in broken chloroplasts was the same as for native Pc(II). From the rate dependence on the Pc concentrations it was concluded that in both cases the rate-determining steps were indeed those involving Pc. These observations indicate that the area of Cr coordination is involved in the reactions with P-700, whereas another region is used for Cyt *f* electron transfer to Pc(II); i.e., there are at least two sites of physiological significance for electron transfer. This suggestion was made earlier, based on the X-ray structural analysis,[6] and was further confirmed by NMR studies with inorganic outer-sphere reagents mentioned earlier.[26-28] The present study using the physiological partners now lends this direct experimental support.[32] Interestingly, the presence of Cr(III) does not block the electron transport from the copper center to P-700, but only attenuates the rate to a third of its value. This may reflect some perturbation of the interaction between the two proteins by Cr(III) preventing an optimal contact between them. As described earlier in detail, the area where Cr(III) is found to bind on Pc is very hydrophilic due to several negatively charged

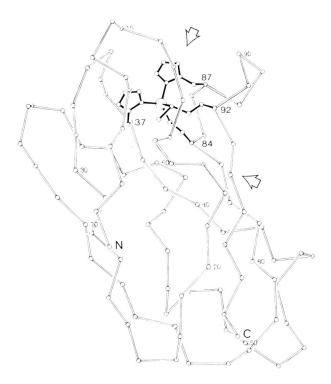

FIGURE 3.   The polypeptide backbone of poplar plastocyanin represented by C$_\alpha$ atoms with arrows indicating regions potentially involved in electron transfer. For further details see text. (Adapted from Freeman, H. C., *Coordination Chemistry*, Vol. 21, Laurent, J. P., Ed., Pergamon Press, Oxford, 1981, 29. With permission.)

carboxylates around Tyr-83.[6,7] It was therefore proposed that this highly conserved patch of acidic side chains serves as the recognition site on Pc for the reaction centers (Figure 3).[32]

A second potential electron-transfer site in Pc is at the partially exposed imidazole of His-87, which is one of the Cu-ligands (cf. Section I). As already mentioned, an edge of this aromatic ring is exposed to solvent and closely surrounded by several invariant hydrophobic amino-acid residues.[6,7] Since the rate of photoreduction of Pc(II) was found to be independent of the presence of the Cr(III) label, it was suggested that electron transfer between Cyt *f* and Pc proceeds by this path (Figure 3).[32] These conclusions describing the hydrophilic electrostatic interactions of Pc with P-700 and hydrophobic interactions with its membranous electron donor, Cyt *f*, are fully compatible with recent reports where the efficiency of electron transfer to P-700 has been studied. The stimulating effect of divalent cations on the electron transfer from Pc(I) to P-700 and a comparison to the efficiencies of electron donors from a variety of eukaryotic and prokaryotic photosynthetic organisms all support the notion of an electrostatic hydrophilic nature of this interaction.[32]

The proposed electron-transfer sites on Pc for its biological partners are shown in Figure 3. The possibility that only a single electron-transfer site is available, namely the exposed imidazole of His-87, and that the negative array next to Tyr-83 is only a directing-binding site, can now be dismissed in view of the affinity-labeling results which showed that the attachment point of the electron-donating Cr(II) ion is indeed at the negative patch.

B. Azurins

These single blue copper proteins have a M_r of $\approx14,000$ (Appendix 1). The three-dimensional structure is available for the Cu(II) state[8-10] and is less well-resolved (2.7 Å only) in comparison with Pc. The higher M_r and the presence of four molecules per asymmetric unit of the crystal made the X-ray crystallographic study of Az more cumbersome. Az's occur in bacteria of the *Pseudomonas* and *Alcaligenes* strains serving as electron mediators, most probably between Cyt c_{551} and cytochrome oxidase:[42]

$$\text{Cyt } c_{551} \rightarrow \text{Az} \rightarrow \text{cytochrome oxidase} \rightarrow O_2 \tag{7}$$

This sequence is supported by the serial redox potentials, e.g., for Az from *P. aeruginosa*, a value of 304 mV at pH 7 has been reported,[43] which is between the potentials of Cyt c_{551} and cytochrome oxidase. Sequence homologies between Pc and Az indicate an interesting evolutionary relationship.[44]

Gray and co-workers[19-21] have carried out a kinetic study of electron transfer between Az and inorganic complexes. The data have been analysed according to the Marcus formalism for outer-sphere reactions (cf. Section I). In analogy with Pc results, it was concluded that the most important factor which governs the reactivity of the low M_r reagents other than driving force and self-exchange rates was the extent of orbital overlap between the redox centers.[1] Thus, the kinetic accessibility scale defined for metalloproteins applies equally well for Az (cf. Table 1). Comparing the activation parameters for electron transfer from $Fe(EDTA^2)^-$ to Pc(II) and Az(II) showed that both reactions are characterized by very small activation enthalpies and the difference in reactivity is due to different activation entropies [$\Delta H^\ddagger = 8.4$ kJ mol^{-1} and $\Delta S^\ddagger = -155$ J K^{-1} mol^{-1} for Az(II) reduction by $Fe(EDTA)^{2-}$].[1] This was explained as reflecting differences in the degree of exposure of the redox site in the protein, with the copper center of Az being less accessible than that of Pc.[1] As already mentioned, this is in agreement with the X-ray crystallographic data.[10] The same isokinetic relationship found with Pc was also demonstrated for Az in its reactions with substituted 1,10-phenanthroline Co(III) complexes.[19] Using the tunneling hypothesis described in Section I, the shortest distance from the active site in Az to the surface was found to be 5.5 Å. This value was determined using the self-exchange rate constant for Az, calculated from its reaction with the "nonpenetrating" reagent, $Fe(EDTA)^{2-}$. Since the molecular structure of Az shows that one of the ligating imidazole rings is partly exposed at the protein surface,[8-10] this distance seems far too large, at least if this histidine (His-117) is involved in the electron-transfer process. The same problem raised in the context of analysing the data of Pc applies also here, namely the Az-small molecule electron-transfer is not a simple bimolecular reaction. In many cases, a rapid precursor-complex formation has been detected followed by a rate-limiting electron-transfer step *(vide infra)*.[43,45]

One of the earliest demonstrations of precursor-complex formation between a blue copper protein and an inorganic redox agent was in the electron-transfer reaction of Az with $Fe(CN)_6^{3-,4-}$ as studied by the chemical relaxation temperature-jump technique.[43] Only a single relaxation time was found over the wide concentration range examined. When Az(II) was titrated with increasing concentrations of $Fe(CN)_6^{4-}$, no saturation of the rate was found. However, when Az(I) was titrated with $Fe(CN)_6^{3-}$, a leveling off of the reciprocal relaxation time with increasing oxidant concentration was observed. This concentration dependence, together with an analysis of the relaxation amplitudes, led to the proposition of the following reaction scheme:

$$\text{Az(II)} + \text{Fe(II)} \underset{}{\overset{K_1}{\rightleftharpoons}} \text{Az(II)} \cdot \text{Fe(II)} \underset{k_{-3}}{\overset{k_3}{\rightleftharpoons}} \text{Az(I)} \cdot \text{Fe(III)} \overset{K_2}{\rightleftharpoons} \text{Az(I)} + \text{Fe(III)} \tag{8}$$

At 25°C the rate constants were $k_3 = 6.4$ sec^{-1} and $k_{-3} = 45$ sec^{-1}, the association constants $K_1 = 54$ M^{-1} and $K_2^{-1} = 610$ M^{-1}. Combining all the kinetic and thermodynamic data, a self-consistent thermodynamic profile was constructed which showed that the free-energy profile was symmetrical around the transition state, and that the intramolecular electron-transfer step within the complex was rate determining for the reaction.[43] In the three-dimensional structure of Az(II) one does not find any cluster of positively charged residues,[9] so that the formation of the relatively stable precursor complex cannot be due primarily to electrostatic interactions. As mentioned above, however, one of the copper ligands (His-117) is partly exposed and at the same time surrounded by a large number of hydrophobic amino-acid side chains.[9] These could serve as a site for nonelectrostatic interactions with the cyanide ligands of $Fe(CN)_6^{3-/4-}$.

Sykes and co-workers have carried out extensive stopped-flow kinetic studies of electron-transfer reactions between Az and inorganic complexes.[45] In these studies the pH dependence of the reaction rates and their inhibition by redox inert complexes were examined. As with Pc,[23-25] and in agreement with the above temperature-jump study,[43] it was found that the electron-transfer step is preceded by complex formation between the copper protein and the low M_r redox partner. Using $Co(phen)_3^{3+}$ as oxidant, the same type of pH dependence as with Pc(I) oxidation was reported,[45] namely increasing reactivity with rising pH. However, with $Fe(CN)_6^{3-}$ as oxidant, the opposite effect was observed, since the rate becomes faster with increasing protonation of Az(I).[45] Both results are consistent with electrostatic interactions playing a decisive role in precursor-complex formation. The role of a protonated, redox-inactive form of Az(I) will be discussed below.

The pK values for the pH-dependent oxidation rates of Az(I) with $Co(phen)_3^{3+}$ (7.6) and with $Fe(CN)_6^{3-}$ (7.1) were assigned to the two nonligating histidines.[45] These pK's were determined earlier by PMR studies,[46,47] which therefore led to the conclusion that the two redox agents make use of different sites on Az(I), with $Co(phen)_3^{3+}$ interacting close to His-83 while $Fe(CN)_6^{3-}$ uses a site that influences His-35. The uncertainties of the pk values and the results of affinity labeling with Cr(II) *(vide infra)* make these assignments of the active sites questionable. Nevertheless, the different pK values obtained with the two oxidants clearly suggest that two independent routes are involved.

The electron-transfer routes to the redox centers of copper proteins have also been studied by the pulse-radiolysis technique. Interaction of short pulses (≈ 0.1 μsec) of 2 to 15 MeV-accelerated electrons with water produces hydrated electrons and hydroxyl radicals together with smaller amounts of H_2, O_2, superoxide, and peroxide.[48] This method therefore provides the possibility to study very fast electron-transfer processes between redox proteins and hydrated electrons or hydroxyl radicals, because its time scale extends into the nanoseconds range. Moreover, because of the extreme redox potentials of these agents, -2.7 V for the hydrated electron and 1.8 V for the hydroxyl radical at pH 7,[49] one may derive from them a wide range of milder reductants and oxidants. These secondary reagents may be applied as well. Since the hydrated electron has an intense and broad absorption band in the visible region, it is convenient to monitor its reactions directly.[49]

Az was found to be reduced by hydrated electrons in a biphasic manner dominated by a direct bimolecular process at an essentially diffusion-controlled rate ($1.0 \cdot 10^{11}$ M^{-1} sec^{-1}).[50] It was found that the absorption band of the blue Cu(II) ion disappears with the same rate as that of e_{aq}^-. However, a considerable amount of the reducing equivalents was not utilized for the direct reduction of Az(II). The specific reduction yield amounted to only 20% of the produced electrons. Therefore a large fraction of the reducing equivalents decays by non-specific reaction pathways. The low yield may be a reflection of the relatively lower accessibility of the redox site in Az as compared, e.g., with the cytochrome heme group. Thus, also, for the reduction of Cyt c a diffusion-controlled reduction rate was observed,[51] but in this case, an almost full conversion of e_{aq}^- into reduced protein is obtained. Another interesting feature of the reaction between e_{aq}^- and Az is the formation of a transient with absorption maximum at 410 nm which is produced concomitantly with the decay of e_{aq}^-.

This transient, which is produced in small yield, is most probably due to electron attachment to the disulfide bridge in Az forming an $RSSR^-$ radical ion.[48,50] Further, the decay process of this transient showed no clear dependence on protein concentration suggesting an intra-molecular process. This was also substantiated by the observation that the slow reduction phase of the blue copper chromophore occurs at a rate similar to that of the transient decay $(1.4 \cdot 10^3 \ sec^{-1})$.[50]

In order to delineate the electron-transfer pathway to the redox center in *Pseudomonas* Az, the Cr(II) affinity-labeling procedure (Section II.A) has also been applied to Az(II).[52] As described in Section II.A for Pc, the rationale of the method is that Cr(III) complexes, in contrast to the corresponding Cr(II) complexes, generally are substitution inert. Therefore one would expect that the coordination sphere of the transition state of the electron-transfer step will remain intact in the Cr(III) product.[12] It was found that $Cr(II)_{aq}$ ions reduced Az(II) stoichiometrically and that Cr(III) remained bound to the reduced protein. Proteolytic diges-tion of the labeled Az with either trypsin or chymotrypsin led in each case to single, different peptides with an overlapping amino-acid sequence containing all the Cr(III) label. The respective peptides were uniquely identified by amino-acid analysis.[52] It was concluded that chromium was coordinated to the copper protein via the carboxylate of Glu-91 and the amine of Lys-85. This is then assigned to be the site from which Cr(II) transfers the electron. The characteristic patch of negative charges, present on Pc, is missing on Az, so there is no analogous "electrostatic" reaction site.[10] The Cr(III) attachment site seems to be governed primarily by the availability of an effective electron path leading to the copper redox center.[52] From the crystallographic data[8-10] one derives a distance of ≈ 10 Å between the Cr and Cu metal centers. The peptide loop from Lys-85 to Glu-91 defines an opening into the interior of the protein, exposing N_δ of His-35 (Figure 4). From analysis of the atomic coordinates, there is sufficient room to accommodate a water molecule within this loop.[52] Further, the imidazole rings of His-35 and the copper ligating His-46 are virtually parallel, with the N_δ - C_ϵ edge of His-35 overlapping with the C_δ-N_ϵ edge of His-46, and with an interplane distance of 3.8 Å. This means that the noncoordinating imidazole of His-35 extends the effective π-system from the copper center towards the above-mentioned peptide loop, and sets the upper limit of approach to N_δ of His-35 from the proposed Cr-binding site to < 4 Å. This structural element could serve as an extended relay for the resonance transfer of an electron from the surface of the protein to the Cu(II) ion. Based on these observations, the following mechanism was proposed for the affinity-labeling reaction (cf. Figure 4): $Cr(II)_{aq}$ becomes bound to Glu-91 and Lys-85 as well as to the water molecule hydrogen bonded to N_δ of His-35 (not shown in the figure). From this position the electron is transferred to the empty d orbital of the Cu(II) via the water molecule and the two parallel imidazole rings.[52]

The PMR studies of Bersohn, Hill, and associates,[46,47] which show that this noncoordi-nating imidazole titrates with a $pK \approx 7$, provide evidence for the solvent accessibility of the His-35 imidazole. The reported slow proton exchange of that residue[46,47] is then most probably related to the slow transition between the active and inactive conformer of *Pseu-domonas* Az(I) *(vide infra)*. The above-mentioned suggestion of Sykes and co-workers[45] that the two noncoordinated histidines are involved in the redox reactions of Az is also in line with this proposal. Finally, the kinetic accessibility scale of Gray and co-workers[1] can be rationalized by the presence of two different electron-transfer routes on the single blue copper proteins. One is resolved by the Cr-affinity labeling and involves the His-35 side chain, and the second is via the partly exposed imidazole of the Cu-ligating His-117.

The electron-transfer reaction between *Pseudomonas* Az and Cyt c_{551} has been intensively studied.[53-59] In the first investigation carried out by stopped-flow, it was shown that the second-order rate constant reached a limiting value at high Az concentration.[53] This finding indicated that the reaction is not a simple reversible electron-transfer process between the two partners, but rather suggested a possible complex formation. In the later temperature-

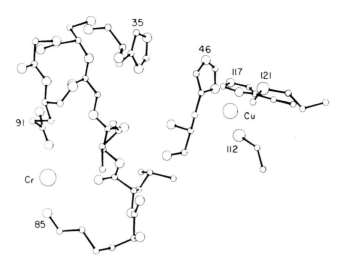

FIGURE 4. View of the area on *Pseudomonas aeruginosa* azurin proposed to be involved in electron transfer from Cr(II)$_{aq}$ to the Cu(II) site
and the residues participating in it. (Drawn from coordinates of the model
kindly provided by Dr. E. Adman.)

jump kinetic studies, two distinct relaxation times were reported, and careful analysis of
both kinetic and thermodynamic parameters led to the conclusion that complex formation
was insignificant.[54,55] The reciprocal value of the faster relaxation time showed a linear
increase up to high Az concentration with no tendency to level off, whereas the slower one
displayed a limiting dependence on Az concentration. Based on these observations it was
suggested that Az(I) exists in two different conformers, one of which is inactive and does
not participate in electron transfer.[54,55] In addition, because of an apparent discrepancy
between the enthalpy of the electron-transfer equilibrium, and the observed dependence of
this reaction on temperature, a further endothermic conformational equilibrium of the oxidized Cyt c_{551} was postulated. This yielded the following scheme:[55]

$$
\begin{array}{c}
\text{Cyt(III)}^* \\
k_{12} \;\; \updownarrow \; \text{fast} \\
\text{Cyt(II)} + \text{Az(II)} \; \overset{k_{12}}{\underset{k_{21}}{\rightleftharpoons}} \; \text{Cyt(III)} + \text{Az(I)} \\
k_{32} \updownarrow k_{23} \\
\text{Az(I)}^*
\end{array}
\tag{9}
$$

The following parameters were reported at pH 7.0 and 25°C: $k_{12} = 6.1 \cdot 10^6 \, M^{-1} \, \text{sec}^{-1}$,
$k_{21} = 7.8 \cdot 10^6 \, M^{-1} \, \text{sec}^{-1}$, $k_{23} = 12 \, \text{sec}^{-1}$, $k_{32} = 17 \, \text{sec}^{-1}$. The corresponding activation
parameters are $\Delta H^{\ddagger}$: 32.6, 57.3, 28.0, and 46.4 kJ mol^{-1} while $\Delta S^{\ddagger}$: -4.6, $+78.7$,
-126, and -65.3 J K^{-1} mol^{-1}. The standard enthalpies for electron transfer, Az(I)
isomerization, and Cyt(III) isomerization were -24.7, -18.4, and $+88$ kJ mol^{-1},
respectively.[55] The large positive $\Delta H^{\ddagger}$ and $\Delta S^{\ddagger}$ for oxidation of Az(I) by Cyt(III) is especially
noteworthy, contrasting the generally small activation enthalpies and large negative activation
entropies observed for protein-small molecule reactions. This strongly indicates that a different mechanism is involved in the two types of reaction. The observed compensational
behavior between entropy and enthalpy changes can be interpreted as reflecting displacement
of water molecules from the protein surfaces upon specific interaction between them.

The postulated equilibrium between two different conformers of reduced Az has recently
gained further support from several experimental approaches. Chemical relaxation meas-

urements of Az(I) solutions have resolved a conformational transition coupled to a proton-transfer equilibrium in the same time range as the slow relaxation observed in the Az-Cyt c_{551} electron-transfer reactions described above.[56-58] Monitoring a pH indicator it was found that protons are released upon temperature increase. This, combined with the above finding that $\Delta H < 0$ for the formation of the inactive form of Az(I),[55] leads to the conclusion that this is the protonated one.[56,59] The opposite conclusion has been presented recently. This has been reached because of the erroneous assumption that the inactive form of Az(I) is favored at higher temperature.[58] It is an interesting coincidence that both crystallographic[7] and kinetic results[24,25] show Pc(I) also exists in an inactive protonated form. However, the reasons for the formation of conformational isomers are very different in these two proteins. In the latter protein the redox site becomes inactive, whereas in the former it is the electron-transfer path that is affected.

The proposed mechanism for electron transfer from Cr(II) to Az(II) implies a close relationship between the protonation of His-35 and the conformational equilibrium between the active and inactive form of Az(I). Thus a process can be envisaged in which a fast proton exchange of His-35 in the active conformer of Az(I) is coupled to the slow conformational transition between two protonated isomers:[56,59]

$$Az^*(I)H^+ \underset{k''}{\overset{k'}{\rightleftharpoons}} Az(I)H^+ \underset{fast}{\overset{K_a}{\rightleftharpoons}} Az(I) + H^+ \qquad (10)$$

(cf. also Equation 9).

The identification of an electron-transfer route in Az by affinity labeling with Cr(II) (cf. above) has prompted the examination of the possible relevance of the Cr(III) binding site to the physiological electron-transfer reactions of Az. Temperature-jump relaxation measurements as those described above were carried out on the electron exchange between Cyt c_{551} and both native and chromium-labeled *Pseudomonas* Az.[59] In view of the instability of Cr(III) coordinated to the Az surface at the high phosphate concentrations used in the earlier studies, the medium was changed to 0.1 *M* HEPES, 0.2 *M* KCl, pH 7.0. It was found that although the reaction pattern of the native and Cr-labeled Az with Cyt c_{551} is qualitatively similar, the numerical values of the reaction parameters differ significantly.[59] Most probably the same mechanism is operative in both cases, yet Cr(III) attenuates the reaction rates. In agreement with the earlier studies,[54,55] two relaxation times were found and could be rationalized in terms of a fast electron-exchange step and the slow isomerization. A satisfactory numerical analysis of the data could only be reached when an additional conformational equilibrium in oxidized Cyt c_{551} was included in the reaction scheme.[59] This is in agreement with the earlier mechanistic proposal by Rosen and Pecht.[55] It was gratifying that numerical analysis of the data with no constraints on the parameters gave almost identical values for the equilibrium constant and ΔH for the Cyt c_{551} isomerization regardless of whether Az was in the native or Cr(III)-labeled state. A clear effect of Cr(III) was noticed on the fast electron-exchange rate between Az and Cyt c_{551}. Thus, in both directions the rates were slowed down by 30 to 50%.[59] Since the Cr(III) ion is coordinated near His-35 on Az,[52] it probably perturbs the optimal alignment between the latter protein and Cyt c_{551}. In addition, the slow Az(I) isomerization step was also found to be affected by the Cr(III) label.[59] The rate of formation of the inactive form of Cr(III)-Az(I) was slowed down to nearly one third in comparison to that of native Az(I), while the rate of the reverse reaction was unchanged. Again, the proximity between Cr(III) and His-35 suggests that the slow proton exchange of that residue (monitored by PMR[46,47]) may be influenced, either directly or indirectly, by a change in the protein structure in this region and cause the observed rate to decrease.

This conformational equilibrium was not detected in the Az(I)-Fe(CN)$_6^{3-}$ system,[43] probably because hexacyanoferrate ions employ a different electron-transfer locus on Az, namely that involving His-117. This is in agreement with the observed pH-dependent reactivity of this couple (discussed above).[45]

As stated before, the biochemical function of Az most probably involves donating electrons to *Pseudomonas* cytochrome oxidase.[42] The influence of the Cr(III) label of Az on the enzymatic oxidation of Az(I) by cytochrome oxidase has therefore been examined as well.[59] It has been found that the Cr(III) had no effect on either V or K_m of this reaction. Hence the conclusion was reached that there are at least two physiologically significant active sites for electron transfer on Az (as was also concluded for Pc, cf. Section II.A). Thus one site is apparently employed in the reaction with Cyt c_{551} (i.e., via His-35) and another for the reaction with *Pseudomonas* cytochrome oxidase (most probably on the "northern" end of Az at His-117).

The kinetics of the *Pseudomonas* Az reaction with algal Cyt 553 has also been studied[41] (cf. Table 1) and was found to exhibit a multistep equilibrium in accordance with the above. Two well-separated relaxations were observed and an analogous reaction mechanism was proposed.[41]

The above-described assignment of the slow step observed in the electron-exchange equilibrium between *Pseudomonas* Cyt c_{551} and Az to proton transfer-linked conformation change near His-35 in the latter, has been examined in an independent fashion.[60] In these experiments advantage has been taken of sequence differences among Az's from different bacterial strains. Thus, important amino-acid substitutions in the neighborhood of His-35 occur in Az of *Alcaligenes faecalis:* the preceding Ser-34 and subsequent Pro-36 in *Pseudomonas* Az are substituted by a lysine and a threonine, respectively, in the *Alcaligenes* protein.[61]

Measurements of the temperature jump-induced chemical relaxation of the *Alcaligenes* Az electron-transfer equilibrium with *Pseudomonas* Cyt c_{551} have clearly shown that the slow Az isomerization step is missing in this system.[60] These findings may suggest that His-35 is no longer involved in the electron transfer. Still, the observed rates are only slightly attenuated in the *Alcaligenes* Az-*Pseudomonas* Cyt c_{551} reaction which then points towards a common mechanism. A Cr(II) affinity-labeling study of this system would be revealing in this respect. The very recent PMR study of Mitra and Bersohn[62] monitored the differences in the pH dependence of the chemical shifts of imidazole protons in *Alcaligenes* Az and found that His-35 does not undergo the slow protonation reaction in contrast to the behavior of that residue in *Pseudomonas* Az. These findings further corroborate the early assignment of the slow step in the electron transfer of the latter protein to a conformational transition involving His-35 and its environment. Substitution of Pro-36 by threonine and Ser-34 by lysine in *Alcaligenes* Az could indeed cause major conformational changes and hence functional differences in this region. This could be due to the charge difference or more interestingly to a *cis-trans* isomerization of the proline residue present in *Pseudomonas* Az. However, *Alcaligenes* spp. which shares the Lys-34 of *A. faecalis* but still has Pro-36,[44] also lacks the slow isomerization step[41] which puts the main focus on differences in position-34.

C. Other Single Blue Copper Proteins

These are primarily found in nonphotosynthetic plant tissue. The physiological role of these proteins is not yet known and they are generally far less well-characterized than Pc and Az. Stellacyanin (St) (M_r 22,000) (Appendix 1) is isolated together with laccase from the latex of the lacquer tree *Rhus vernicifera*.[5] Umecyanin is found in horseradish roots and has a M_r of 14,600. A blue copper protein has also been isolated (M_r 20,000) from mung beans. Finally it should be mentioned that another interesting single blue copper protein has recently been isolated from a bacterial source *(Thiobacillus ferrooxidans)* with a M_r of 16,500. It is named rusticyanin because it seems to be involved in iron(II) oxidation.[63]

Out of all these proteins, electron-transfer reactions of St only have been studied.[1,19,20,64-]

[69] Significantly, the reactions of St(II) are all faster than the corresponding reactions of Az(II) and Pc(II) (cf. Table 1). The electron transfer from Fe(EDTA)$^{2-}$ to St(II) is characterized by a very small activation enthalpy ($\Delta H^{\ddagger} = 12.6$ kJ mol^{-1}), and the wide range of rates observed for the different blue proteins is due largely to differences in activation entropy (for St, $\Delta S^{\ddagger} = -88$ J K^{-1} mol^{-1}). The very high reactivity of St was interpreted as a result of the copper site being more accessible to external redox agents than in the two other proteins.[19-22] This is also supported by the observation that the apparent self-exchange rate constant for St is quite insensitive to the nature of the reductant as opposed to the two other proteins. Such a behavior again demonstrates that the reactivity of the different copper proteins with inorganic reagents is in many cases reflecting the accessibility to the copper redox center. Based on the characteristics of the redox reactions involving St, this protein was chosen as the reference for calculating electron-transfer distances described above for Pc and Az (Sections II.A and II.B).[22] Thus, it was assumed that the closest-contact electron-transfer distance for a blue copper protein was a Van der Waals contact of 1.85 Å,[22] which is the radius of an imidazole ring or a sulfur ligand. This distance is in some conflict with the 6 Å distance as determined by ENDOR measurements[70] for the closest approach of solvent water protons to Cu(II) in St.

A detailed study of the electron-transfer reactions of St has been pursued by Holwerda and co-workers.[64-67] They have examined the kinetics of oxidation of reduced St by aminocarboxylatocobaltate(III) complexes. A fast formation of a relatively stable protein-oxidant complex was found with Co(III)EDTA ($K = 149$ M^{-1}) followed by an intracomplex electron-transfer step ($k = 0.17$ sec^{-1} at 25°C).[64] The very low activation enthalpy (7.5 kJ mol^{-1}) and the high negative activation entropy (-230 J K^{-1} mol^{-1}) once more demonstrate that the requirements for inner-sphere rearrangement in the redox center of the proteins are minimal and that a poor overlap between the redox orbitals is leading to an essentially nonadiabatic electron transfer in the rate-limiting step. A variety of anions inhibit the association of Co(EDTA)$^-$ with the metalloprotein, and the oxidation becomes a simple first-order reaction in oxidant.[66] This probably reflects a competition between the oxidant and these anions for the same binding site. Substituting EDTA with propylenediaminetetraacetate (PDTA) or *trans*-1,2-diaminocyclohexanetetraacetate (CDTA) in the Co(III) complex markedly changes the kinetics of electron transfer.[65] With these oxidants no saturation behavior was observed, which sets an upper limit for precursor complex formation to <5 M^{-1}. The rate parameters for the electron-transfer reaction of Co(PDTA)$^-$ and Co(CDTA)$^-$ are identical ($k \approx 17$ M^{-1} sec^{-1}, $\Delta H^{\ddagger} \approx 36$ kJ mol^{-1}, and $\Delta S^{\ddagger} = -100$ J K^{-1} mol^{-1} at 25°C). These observations were interpreted in terms of steric effects. Thus, alkylated derivatives of EDTA apparently prevent the reactant from forming the required hydrogen bonds for complex formation.[66] For attaining a good overlap, the Co(III)-coordinated carboxylate groups must be accessible to the redox orbitals of the electron donor, whereas this is not achieved by way of the saturated methylene groups on the oxidants. A plot of $\Delta H^{\ddagger}/\Delta S^{\ddagger}$ showed a good isokinetic correlation for the reactions of St(I) with the substituted Co(III) EDTA complexes,[66] which is in marked contrast to oxidation of the same protein by substituted *tris*-(1,10-phenanthroline) cobalt(III) complexes.[19] This enthalpy-entropy compensation was rationalized as being the result of the preferred orientation of the oxidants in the precursor complex formed with St(I). The above results show that the environment of the presumed relatively exposed redox site in St must differ quite markedly from the less accessible ones in Pc and Az. The very small $\Delta H^{\ddagger}$ values and large negative $\Delta S^{\ddagger}$ values strongly suggest a poor overlap between the donor and acceptor redox orbitals, thus reflecting the nonadiabatic character of the electron-transfer process.[66]

Holwerda et al.[67] have recently proposed a physiological role for St. Measuring the kinetics of the thermodynamically favored electron transfer from St(I) to benzoquinone and benzo-semiquinone, they have found them both rather fast ($2.3 \cdot 10^4$ and $5.1 \cdot 10^6$ M^{-1} sec^{-1} at

25°C). Hence it was suggested that St acts in concert with laccase by quenching the semi-quinone produced by the latter. In other words, the proposed role is of a semiquinone reductase.[67] The question arising is how St is kept reduced under physiological conditions. This proposal is in disagreement with the proposed physiological role of laccase, where the semiquinone (of uroshiol) is formed in order to polymerize and form a shield. Here the trick is to protect the semiquinone from further reduction!

The electron transfer to St has also been studied by the pulse-radiolysis method.[68] As with Az(II) the reduction by e_{aq}^- is found to be biphasic, yet there are marked quantitative differences. The fast reduction of the 600-nm chromophore is concomitant with the decay of the e_{aq}^- absorption, but shows a slower specific rate of $2.3 \cdot 10^{10}$ M^{-1} sec^{-1} at pH 7.[68] The slower rate of St(II) reduction compared with that of Az(II) (Section II.B) may be a reflection of the former carrying a large amount of carbohydrates (40% w/w) on its surface.[71] Thus, for a diffusion-controlled process, with highly reactive radicals, the spherical angle of approach to the site may become the limiting factor. This would be in contrast with the requirements for high reactivity with inorganic reagents where orbital overlap is decisive.

The second, slow phase in the reduction by hydrated electrons was found to be a mono-molecular process and assigned to an intramolecular migration of the electron to Cu(II) from its primary site of attachment on the protein surface. The amplitude ratio of the fast to slow phases was found to depend on the protein/e_{aq}^- ratio.[68] When the latter ratio was raised to 16, the fast phase reached 90% of the reduction. This behavior is particularly interesting in view of the high carbohydrate content of this protein. Assuming these carbohydrates to be carried on the protein surface, one may rationalize it in terms of a competition between the copper site and other groups which are less reactive yet present in a large amount. This competing species might also yield the relatively weak transient absorption that appears between 400 and 450 nm.[68] This transient is formed within the decay time of e_{aq}^- and disappears monomolecularly at a rate similar to the slow phase of the 600-nm chromophore reduction.[68] Because of its relatively low intensity, it is probably not due to an RSSR$^-$ radical and it could be an electron adduct of a histidyl residue. Assuming $\epsilon_{410} = 1,500$ M^{-1}cm^{-1} (ϵ_{max} is reportedly 2,000 M^{-1} cm^{-1} at 369 nm)[72], this transient could account for the amount of Cu(II) reduced in the slow phase. The detailed analysis of the conversion yield of the hydrated electron has shown that depending on the protein:reductant ratios, one can go from 20 up to 90% conversion. No similar analysis for the e_{aq}^- conversion in the reduction of Az(II) or Pc(II) has yet been carried out. Such an analysis would be very interesting in view of the differences in reactivity between Az or Pc and St towards other low M_r reductants.

Protein-protein electron transfer between St and either horse heart Cyt c or *Pseudomonas* Cyt c_{551} has recently been investigated.[69] Both these reactions were found to follow a simple bimolecular equilibrium mechanism.

$$St(II) + Cyt(II) \underset{k_{-1}}{\overset{k_1}{\rightleftharpoons}} St(I) + Cyt(III) \tag{11}$$

with $k_1 = 1.6 \cdot 10^5$ and $k_{-1} = 1.8 \cdot 10^6$ M^{-1} sec^{-1} for Cyt c_{551} and $k_1 = 3.5 \cdot 10^2$ and $k_{-1} = 1.9 \cdot 10^3$ M^{-1} sec^{-1} with Cyt c (all at 25°C and pH 7.0). The large differences between the reaction rates of St with these two cytochromes were explained as reflecting the significantly different self-exchange rate constants of each of these two proteins, which in turn may reflect the different accessibility of the heme group in them. The activation parameters for St(I) oxidation by Cyt c_{551} were determined with the activation entropy of this process being close to zero. This was interpreted in terms of breaking of the ordered water structure,[69] as suggested by Wood,[39] or by an induced conformational change bringing the redox sites of the two proteins closer together. Once again, the relatively high reaction rates are unexpected for a glycoprotein like St carrying 40% of its M_r as surface carbohydrates.

It remains to be seen by future structural studies how another macromolecule can obtain a close access to St so as to allow for the fast electron transfer to occur.

D. Conclusion

A comparison of the reaction parameters determined for protein-small molecule electron-transfer with those for protein-protein electron-transfer makes the differences obvious. At least with hydrophilic reagents the former reactions are generally characterized by very small activation enthalpies so that the rates are essentially determined by the very large negative activation entropies. Also, complex formation between the reactants seems to play an important role. The very low $\Delta H^{\ddagger}$ values often found could then be a consequence of an activation enthalpy of the actual electron-transfer step combined with a negative enthalpy of association. This is in sharp contrast with protein-protein electron-exchange reactions which frequently exhibit quite large activation enthalpies then compensated for by favorable entropies of activation. Further, precursor-complex formation has never been demonstrated for these reactions. This illustrates that one should be very cautious in drawing parallels to the far more complex protein-protein systems from protein-small molecule reactions as the mechanisms most probably will be different for the two cases. The calculated self-exchange rate constants, k_{11}, given in Table 1, also show that there is no correlation between inter-protein reactivity and the kinetic accessibility scale which successfully describes protein-small molecule reactions.

Another important point that should be stressed again is the apparent lack of selectivity in the protein-protein reactions, as is clearly suggested by comparing the calculated self-exchange rate constants for the latter reactions (Table 1). This is contrasted by the widely accepted sequential electron transfer between some of the blue copper proteins and certain cytochromes, e.g., Pc and Cyt f.

Finally, it is significant that Pc and Az exhibit at least two different active sites for electron-transfer in their reactions with both small molecules and with their physiological partners. Evidence for that has now accumulated from several different chemical approaches. One site seems to be at the exposed imidazole of one of the copper-ligating histidines. The other one is 10 to 12 Å from the redox center, and an electron-transfer relay system involving a series of weakly coupled conjugated π-systems was inferred as connecting the copper to the protein-solvent interface. This may very well have a general significance for the mode of operation of other redox proteins.

III. BLUE OXIDASES

A. Reactivity of Copper Sites in Blue Oxidases — Electron Uptake

The reactivity of the different copper sites present in blue copper oxidases (cf. Appendix 1) is naturally determined by their different functional roles.[5] The catalytic cycle of these oxidases involves the uptake of single electrons from their respective substrates and reduction of dioxygen with four electrons to water. In this cycle the blue oxidases employ three different types of copper sites, all endowed by unique chemical or spectroscopic properties: the type-1 blue copper site described in Section I; the type-2 copper site, characterized by unusual chemical reactivity yet with spectral properties similar to low M_r copper complexes; and type-3 copper, with two copper ions in a binuclear site. In their oxidized Cu(II) state, the latter two are antiferromagnetically coupled so that they are undetected by EPR or magnetic susceptibility measurements. Three enzymes constitute the group of blue oxidases: the laccases, ascorbate oxidase, and ceruloplasmin. Each of these proteins contain all three types of copper sites. The first ones contain just one of each type, while the two other contain several.[5,73] In the following we shall consider the reactivity of each of the different types of copper sites present in these oxidases. The most intensely studied blue copper oxidase is laccase from the Japanese lacquer tree *R. vernicifera*. This extracellular, water-soluble enzyme contains four copper atoms bound to the three distinct sites, mentioned above.

In contrast to the profound specificity exhibited by all blue oxidases towards their oxidant, dioxygen, their specificity with respect to reductants is relatively low (with the exception of ascorbate oxidase). This aspect raises the question of which of the three sites are the electron-uptake site of these oxidases. While the type-2 and type-3 sites have the capacity to interact with low M_r ligands, type 1 lacks this feature. Still, mainly because of its reactivity and the role played in single blue proteins, the latter has repeatedly been considered to be the main electron-uptake site in all three blue oxidases. In more detail, two different mechanisms have been proposed for the initial electron transfer from the reducing substrates to oxidized laccase. In one of them, the type-2 site Cu(II) was assumed to be the electron acceptor from substrates via an inner-sphere pathway, followed by intramolecular electron transfer to the other sites.[74] This proposal was based on kinetic studies of *Rhus* laccase reduction by hydroquinone (H_2Q) showing that under anaerobic conditions, the type-1 and the type-3 sites are reduced at comparable rates.[74] At 25°C and pH 7.0, typical second-order rate constants are 3 to $5 \cdot 10^2$ M^{-1} sec^{-1}. The reduction also shows a $[H^+]^{-1}$ dependent pathway which suggests that HQ^- is the reactive species.[74] A detailed study of the anaerobic reduction by differently substituted hydroquinones, monitored via type-1 Cu^{2+} at 614 nm,[75,76] provided evidence for substrate binding. It was proposed that the singly ionized hydroquinones bind to the enzyme in a pre-equilibrium step before a rate limiting intracomplex electron-transfer. The transfer rate is essentially constant irrespective of the thermodynamic driving force of the particular hydroquinone ($k \approx 30$ sec^{-1} at pH 7.0, 25°C).[76] The electron-transfer reactivity therefore seems to be controlled largely by activation barriers within the protein. Unfortunately, due to absorption of the substituted hydroquinones in the near UV region, no reliable kinetic data for the type-3 site could be obtained. Excluding coordination to type-1 Cu(II), the above authors suggest that the substituted hydroquinones form a bridge between type 2 and type 3 in the rapid pre-equilibrium step.[75,76] This, however, is difficult to reconcile with the findings that anions like F^- and N_3^- bind to type-2 Cu(II) in a very slow monomolecular step[74,77] (for F^--binding the rate of complex formation is $6 \cdot 10^{-5}$ sec^{-1} at pH 5.5, independent of fluoride concentration).[77] It is also difficult to assume that the bulky HQ^- anions would be an exception to this behavior. Further, it has recently been shown that the water molecule coordinated to type-2 Cu(II) exchanges rather slowly ($t_{1/2} >$ 10 min).[78] These results suggested that the type-2 site is located in a cavity,[79] which in *oxidized* laccase is isolated from the bulk solvent, and that a redox state-dependent conformational change controls its accessibility.[78] Thus, water and extrinsic ligand-exchange processes at the type-2 site proceed efficiently only when the enzyme is *reduced*. Finally, one can assume binding of HQ^- also without the requirement of a metal ion.

An alternative scheme which involves electron transfer from substrate to the type-1 Cu(II) is widely supported by kinetic studies showing that the initial rate of reduction of the blue chromophore is proportional both to oxidized type-1 site and to substrate concentration.[80] The subsequent intramolecular electron transfer to the type-3 site is then linked to a protein conformational change.[80] The latter mechanistic suggestion may be supported by the recent finding that the rate of reduction of the type-1 Cu^{2+} is not affected by the selective removal of type-2 copper, while reduction rate of the type-3 copper(II) pair is markedly decreased.[81] Also pulse-radiolytic studies corroborate the idea of the blue copper being the primary electron acceptor (see below).

The possibility that type-2 Cu(II) can serve as a parallel entry site for electrons from substrates has recently been substantiated by kinetic studies using rapid-freeze EPR.[82] The proposal limits this role of type 2 to pH conditions lower than 7.0, and a triggering role is assigned to the prior reduction of the type-1 site.[82] The possibility that type 3 can also serve as the port for electron uptake from substrates cannot be completely dismissed either in view of the reductive titrations described below, showing a reductant-dependent electron distribution among the redox sites.[83] However, the kinetic studies reported to date lend no support to the existence of this pathway,[76,80,82] but it should also be kept in mind that neither of the reductants used in the former study[83] have been used in kinetic studies.

Early reductive titrations of native *Rhus* laccase suggested that the type-1 and type-2 Cu(II) are each one-electron acceptors, while the type-3 Cu(II) pair acts as a cooperative two-electron acceptor.[84] More recently, however, using electron donors spanning a wide range of redox potentials and differing in their chemical and structural properties, a more complex situation has been resolved. The electron distribution among the redox sites of laccase was found to be qualitatively correlated with the redox potential of the reductants' first oxidation step.[83] These findings were analysed in terms of a variable extent of cooperativity between the two type-3 copper ions, which provided a good fit to the data. The apparent variability was suggested to originate in a transition of the type-3 site from a cooperative two-electron acceptor to a pair of independent one-electron acceptors, caused by sufficiently strong reductants.[83] This nonequilibrium behavior was further rationalized by assuming two different reaction schemes: strong reductants, like $Ru(NH_3)_6^{2+}$ and durohydroquinone, may react directly with the type-3 site and bring about the single electron reduction of the type-3 site, hence uncoupling the Cu(II) pair, while weaker reductants (like benzohydroquinone) use the type-1 (or the type-2) Cu(II) as primary reduction site. The uncoupling of the type-3 Cu(II) ion pair may in principle lead to the appearance of a new EPR signal. However, no direct evidence for such a signal has been obtained in these titrations. Such signals may be difficult to resolve for native laccase, since computer simulations show that even with the strongest reductants, the uncoupled type-3 Cu(II) ions will not contribute more than $\approx 10\%$ of the total EPR intensity.[83] Furthermore, electron exchange between the two type-3 ions may also prevent detection of such a signal. Recently, however, EPR signals assigned to uncoupled type-3 Cu(II) ions have been reported.[85]

Contrary to the above situation for *Rhus* laccase there seems to be a general agreement that the type-1 Cu(II) is the first electron acceptor in fungal laccase. From stopped-flow and rapid-freeze EPR experiments, a mechanism has been proposed which involves a series of successive one-electron transfers from substrate [ascorbate, hydroquinone, hexacyanoferrate(II)] to the blue copper.[86,87] For the reduction by hydroquinone, a specific rate of $1.7 \cdot 10^7$ M^{-1} sec^{-1} was reported.[87] This mechanistic proposal has also been supported by a chemical relaxation study[88] of the electron exchange between fungal laccase and external redox couples, $Mo(CN)_8^{3-/4-}$ and $Ru(CN)_6^{3-/4-}$. Only a single relaxation time was observed at 610 nm which implied that under anaerobic conditions there is only one pathway for the electron exchange between the type-1 Cu(II) and the redox reagents. With $Mo(CN)_8^{3-/4-}$ the specific rates were $1.6 \cdot 10^5$ and $1.6 \cdot 10^4$ M^{-1} sec^{-1}, for oxidation and reduction of laccase, respectively.[88]

The pulse-radiolysis technique has also been useful in probing details of potential pathways through which the electron may be transferred from reductants to the metal sites of the blue oxidases. The exceptionally high reactivity of hydrated electrons (e_{aq}^-) allows them to react with a number of amino-acid residues exposed on the surface of the proteins at rates which reflect the electron acceptor capacity of the latter groups.[48] All blue oxidases examined to date have been shown to be reduced by hydrated electrons through an indirect pathway, where the reduction equivalent is first attached to an exposed protein residue and eventually migrates to the type-1 Cu(II).[68,89-93]

The reaction of fungal laccase with e_{aq}^- clearly showed the dramatic difference between the rate of decay of e_{aq}^- and that of the appearance of the reduction equivalent at the first redox site of the enzyme to be reduced, i.e., the type-1 Cu(II).[89] Thus, whereas the former occurs in the range of a few microseconds (at μM concentrations of reactants, i.e., at a rate of $8.5 \cdot 10^{10}$ M^{-1} sec^{-1}), the blue chromophore is reduced only in the millisecond time range. The detailed analysis of the concentration dependence and the spectra of the transients was limited by the amount of enzyme available at that time.[89] The second-order pattern of the reduction of the 610-nm chromophore would suggest that the solvated electrons are converted into electron adducts of surface residues of the protein molecules which are equilibrating by *inter*molecular transfer to the type-1 sites. The process at this stage may be analogous to electron transfer from substrates. The conclusion that the first copper site to be reduced is type 1 could be challenged since the type-2 site has not been monitored. However, earlier stopped-flow studies also lend support to this sequence of events.[87]

Recent studies by Henry et al.[94] have confirmed and extended the above observations for fungal laccase reduction by e_{aq}^-. These authors have carried out a detailed study over a wider range of concentrations and have also used CO_2^- and O_2^- as reductants, reaching essentially similar conclusions to those described above. The recent pulse-radiolysis study of the reduction of *Rhus* laccase has also been a more detailed one.[68,90-92] Again the e_{aq}^- was found to decay at a bimolecular, diffusion-controlled rate ($3.7 \pm 0.4 \cdot 10^{10}$ M^{-1} sec^{-1}),[92] and the blue Cu(II) chromophore is reduced by a monomolecular process and in a time domain which is two orders of magnitude slower. However, in *Rhus* laccase it has been clearly established that the latter reaction is an intramolecular process, independent of either protein or reductant concentrations ($1.6 \pm 0.3 \cdot 10^3$ sec^{-1}).[92] An interesting feature is the very low reduction yield of the oxidase, particularly that of *Rhus* laccase. It amounts to about 2% of the total e_{aq}^- produced per pulse. Thus, depending of the enzyme concentration and pulse length, between 0.9 and 1.8% of the total type-1 Cu(II) was reduced.

By "titration-like" experiments where the same sample of laccase is submitted to a large number of sequential pulses with a few seconds interval, it has been shown that the contributions of the pulses are additive, leading to the reduction of a large fraction ($\approx$40%) of the type-1 Cu(II).[92] Upon reoxidation of the reduced laccase by O_2 no significant radiation damage to the enzyme was found. This behavior excludes the possibility that the changes in absorption around $\approx$600 nm are due to the reduction of sites other than the type-1 Cu(II). (Such sites, e.g., type-2 Cu(II), have a much lower absorbance in this range.) Hence, the reason for the low reduction yield of the "blue" chromophore is primarily due to nonspecific reactions of e_{aq}^- with surface residues. This should indeed be predominant in *Rhus* laccase which contains 45% carbohydrates,[5] most probably on its surface.

The transient absorption bands formed concomitantly with the e_{aq}^- decay have been found to decrease in two phases. The first disappears at essentially the same rate as the reduction of the 610-nm chromophore ($1.4 \cdot 10^3$ sec^{-1}). Its maximum is at 410 nm and therefore can be assigned to the electron adduct of a disulfide bond (RSSR$^-$).[92] Using the absorption coefficient reported for this radical ion,[49] one finds close to a 1:1 stoichiometric relation between the amount of the latter species disappearing and that of the type-1 Cu(II) being reduced. These findings corroborate the suggestion made earlier that the type-1 Cu(II) is the first site to accept the electrons from external donors, and that this reduction most probably proceeds by an intramolecular process via a disulfide residue exposed on the surface. It remains to be established whether this reduction route is also employed by other electron donors. The steady-state concentration of the radical would usually be rather low and hence difficult to detect. The second transient has a maximum absorption around 360 nm and a much longer lifetime ($\gg$ 10 msec). This has, however, not been further characterized as it provides only a very minor contribution to the reduction yield.

The detailed process of electron transfer from the milder reductant, the CO_2^- radical ion, to laccase has also been examined.[92] This radical is produced by the following reactions:[48]

$$e_{aq}^- + N_2O + H^+ \rightarrow N_2 + {}^{\bullet}OH \tag{12}$$

$$ {}^{\bullet}OH + HCOO^- \rightarrow CO_2^- + H_2O \tag{13}$$

In this case a monophasic reduction of the type-1 Cu(II) is observed. It occurs synchronously $(1.4 \cdot 10^3 \text{ sec}^{-1})$ with the decay of the major transient peaking at 410 nm. Thus the same, most probably general pathway of electron transfer is also used here. Indeed, substrate independence would be expected for an intramolecular, common electron-transfer pathway.

Significantly, pulse-radiolysis studies of the reduction of human ceruloplasmin carried out earlier[93] indicated a very similar reaction pathway involving an electron adduct of both a disulfide bond and an imidazole. Particularly noteworthy is the similarity of the specific rates observed for the intramolecular electron transfer to the blue Cu(II) site in ceruloplasmin and in *Rhus* laccase.

B. Reactivity towards Dioxygen

The marked contrast between the relatively low specificity of blue oxidases towards reducing substrates and their very high reactivity and specificity to dioxygen has already been stated in Section III.A. Thus, the rate of reoxidation of fully reduced *Rhus* laccase by dioxygen,[95] $5 \cdot 10^6 \, M^{-1} \text{ sec}^{-1}$, is several orders of magnitude faster than the above reduction rates. Moreover, reoxidation rates by other oxidants are all significantly slower. Obviously, this is of great importance with respect to the catalytic cycle. The mechanism of dioxygen reduction has been investigated in greatest detail for tree and fungal laccases. The results of these studies will be reviewed below along with the analogies that can be drawn today to the other blue oxidases.

Oxidative titrations of reduced laccase showed that both type-1 and type-3 sites act as one-electron donors.[96] This was found to be independent of the nature of the oxidant used, whether it was a single $[\text{Mo(CN)}_8^{3-}]$ or multiple electron acceptor (H_2O_2 and O_2). In view of this observation and the fact that in certain cases the type-3 Cu(II) pair can behave as a strongly cooperative two-electron acceptor, it would suggest that the transition from the uncoupled reduced Cu(I)-d^{10} state to the antiferromagnetic coupling of the two Cu(II)-d^9 ions is not spontaneous. As there can only be one equilibrium pathway for a chemical process, at least one of the two states of the oxidized type-3 pair (coupled and uncoupled) must be of a nonequilibrium nature. This implies metastable conformational differences in the protein probably linked to the coupling-uncoupling of the type-3 pair.[96] Again, a new EPR signal is expected to arise from the uncoupled type-3 Cu(II). However, none was observed in these experiments, probably due to the same difficulties described above. Recently, the formation of a new EPR signal, assigned to a type-3 Cu(II) ion has been reported.[85] It has been produced by a low-temperature oxidation of native *Rhus* laccase or a partial reduction of the type-2 copper-depleted (T2D) fungal laccase. Also upon peroxy-laccase formation a decrease in the coupling between the two type-3 Cu(II) ions has been implied by susceptibility measurements.[97]

Recently, indications for the existence of metastable conformational changes associated with the type-3 site have also been observed spectroscopically (see below).[98,99]

The *inter*molecular rate of electron transfer between laccase molecules has been determined by mixing equal volumes of fully oxidized and fully reduced enzyme molecules and monitoring the time course of absorbance change at 615 nm. It was found to be a very slow process in the absence of a mediator,[100] with a second-order constant of $\approx 0.5 \, M^{-1} \text{ sec}^{-1}$ at pH 7.0 and 25°C. This finding was actually predicted by earlier Marcus calculations[3] showing that the self-exchange rate for the type-1 copper in *Rhus* laccase is several orders of magnitude slower than in the single blue copper proteins. A limited intermolecular equilibration could be the basis for the observed nonequilibrium electron distribution among the sites of different molecules described above.

From the foregoing sections it emerges that the primary role of the type-1 site and one role of the type 2 is the uptake of electrons and its intramolecular transfer to the type-3 site. The nature of the latter site clearly evolved to make it optimally suited for dioxygen binding and reduction. Originally, the type-2 site has been proposed to be the locus of binding dioxygen and its reduction intermediates.[73] This assignment was based on the relative accessibility of the type-2 Cu(II) to solvent molecules as well as to low-M_r ligands like F^- or CN^- ions. It had to be changed in view of more recent studies which have clearly established the role of the type-3 site as that of dioxygen reduction.[96-101] It was first shown [96,99] that in the oxidized state the type-3 site can form a stable peroxide complex upon reaction with H_2O_2. This complex was found to carry a marked spectroscopic resemblance to oxyhemocyanin.[99,101] This is not surprising in view of the expected isoelectronic structure of these complexes in the two proteins. It remains still to be rigorously examined, whether the type-2 copper is involved in the reduction of the enzyme-bound peroxide (see below).

Transient-state kinetic studies employing optical stopped-flow and rapid-freeze quenching EPR measurements have shown that when fully reduced laccase reacts with dioxygen, both the type-1 copper and the type-3 pair are rapidly reoxidized while the type-2 site remains undetected.[95] At the same time a new EPR signal was observed at temperatures below 25 K. Based on its characteristic features, and using ^{17}O isotope, this signal was proposed to originate from a bound O^- radical.[102] The kinetics of formation of this paramagnetic species were well correlated to the reoxidation of the type-1 and type-3 centers, while its disappearance followed the reoxidation of the type-2 site. This was interpreted as suggesting that in fully reduced laccase, three electrons are rapidly transferred to O_2.[95] Although this process is accompanied by an absorbance increase around 360 nm (which is also observed under turnover),[95,98,99] it leaves the question unanswered of what relevance the fast three-electron transfer to O_2 has to the enzymatic cycle of laccase under physiological conditions where excess O_2 prevails. From the reported rate constants for oxidation by O_2 and reduction by substrate, it is obvious that fully reduced laccase would only exist under rare conditions.

Long-lived characteristic changes in both absorption[98] and CD spectra[99] are formed upon interaction of *Rhus* laccase with O_2 and its reduction intermediates. These can be attained both in aerobic reduction experiments and from partially or fully reduced protein reacting with O_2. A transient species with extra absorbance in the 300- to 700-nm region[98] is also characterized by a negative CD band around 330 nm and a strong enhancement of the (+) 370-nm band.[99] This species was suggested to be produced through the electron transfer to O_2 from laccase containing three redox electrons.[98,99] In the absence of external electron donors it decays slowly in a monomolecular reaction ($k = 5.1 \cdot 10^{-4}$ sec^{-1}, 25°C, pH 7.0). A stable species is obtained by reoxidation of the half-reduced protein with O_2. Since most of these molecules will have the two electrons at the type-3 site,[83] it can safely be concluded that O_2 reacts with the reduced type-3 site. In the complex they form, two electrons are transferred to O_2, producing a peroxy-laccase intermediate.[98,99] This scheme is further supported by the observations that the above product has very similar chemical and spectroscopic properties to those of peroxy-laccase produced by reacting native oxidized laccase with H_2O_2 directly.[96,101]

Spectral features characteristic of peroxy-laccase were also observed in solutions under steady-state conditions, strongly suggesting that the peroxy species is involved in the catalytic cycle.[98,99] However, the above spectral features which were in all cases produced by reoxidation of laccase with either O_2 or H_2O_2 were also accompanied by other long-lived spectral changes. These observations led to the idea that the oxidized enzyme formed under these conditions assumes a metastable state, named the C state.[98,99] This form of the enzyme though fully oxidized is different from the native, freshly prepared one in analogy with other "pulsed" enzymes. The C state probably involves changes in the coordination sphere of at least one of the copper sites, but it *does not* contain any bound oxygen intermediate.[98] It may be related to the "active" isomer postulated by Andréasson and Reinhammar.[80]

Based on these observations and the fact that all four copper ions are essential for the enzymic oxygen reduction, several schemes for the catalytic cycle of laccase have been

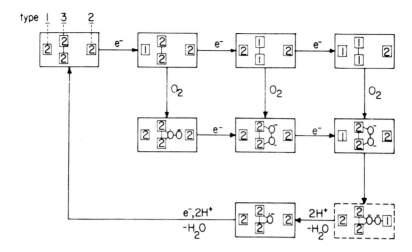

FIGURE 5.   The catalytic mechanism of *Rhus* laccase. The pre-steady-state mechanism is not included. The binding site and electronic structure of the O_2 adduct to singly reduced laccase is unknown but is formulated as type-3 bound O_2^- for reasons given in the text. The broken line is meant to indicate a short-lived intermediate or a configuration close to or at the transition state of a concerted process. (From Farver, O., Goldberg, M., and Pecht, I., *Eur. J. Biochem.*, 104, 71, 1980. With permission.)

proposed.[82,86,95,98,99] The version presented in Figure 5 illustrates the authors' present conclusions. The flexible characteristics of the depicted mechanism was suggested by the following experimental observations.

1.   Electrons can be taken up and distributed among the three sites of the enzyme in several ways and sequences, though clear thermodynamic or kinetic preferences are present.
2.   The type-3 site is the O_2 binding and reduction site; interaction with O_2 can occur as soon as one reduction equivalent is in the enzyme.
3.   Intermolecular electron equilibration is too slow to be of significance in the enzymatic reaction (cf. above).

The suggestion that reactivity with O_2 occurs already with laccase molecules containing a single reduction equivalent is based on pulse-radiolysis studies.[91] In these experiments O_2-saturated laccase solutions were pulse irradiated yielding O_2^- radicals. These radicals were found to reduce the type-1 Cu(II) at relatively fast rates followed by a slower reoxidation process. The above conclusion has been reached since this reoxidation is observed under conditions where the extent of reduction of laccase never was more than one electron per molecule.[91] This seems contradictory to the frequently stated concept that at least the initial step must be a two-electron transfer in order to avoid the highly endoergic step of O_2^- production. That argument is, however, based on the thermodynamics of free O_2^- species in solution and underestimates the effect of its binding to the enzyme.[92] Thus the relevant question is whether laccase containing one reduction equivalent can form a sufficiently stable complex with O_2 to overcome the unfavorable thermodynamics governing the formation of free superoxide.[99]

Clearly the next step, in the reaction with two reduction equivalents in the enzyme, allows for the formation of the peroxide which would occur at the type-3 site.[98,99] Agreement concerning that step is therefore quite general. The intriguing problem is now how the enzyme-bound peroxide is reduced to two water molecules. Information about this phase in the catalytic cycle stems primarily from transient-state kinetics,[82,87,95] experiments with labeled oxygen,[103,104] and more recently studies using the T2D laccase.[81,105]

The main reason for the slow reduction rate of H_2O_2 in inorganic systems is the high activation energy required for breaking the O-O bond. This could be lowered by an asymmetric coordination of O_2^{2-} to two different sites present in different oxidation states, thus leading to a polarization of the O-O bond. As shown in the scheme, we suggest cooperation between the type-2 and the type-3 sites in breaking this bond and reducing the oxygen atoms to water. This involves a bridged configuration either as a short-lived intermediate or more probably as a transition state in a concerted process involving electron transfer and bond cleavage. This suggestion has several attractive aspects. First of all it proposes a specific role for the type-2 site in the catalytic cycle. Second, a similar process probably takes place in the final step of the reoxidation of fully reduced laccase with O_2. Thus, in experiments using isotopically labeled O_2, one of the two oxygens has been shown to end up in a water molecule bound to reoxidized type 2.[103] However, as the primary O_2 interaction site is considered to be type 3, the atom transfer must take place in a controlled manner. The above-proposed bridging mechanism would naturally require appropriate proximity between type 2 and type 3. Studies on the proton and ^{17}O relaxation enhancement by laccase[78] and rapid-freeze EPR[104] have strongly suggested that the two sites indeed are situated closely in a cavity in the protein. It is assumed that the proposed binding configuration is related to the observed conformational change taking place during the catalytic cycle.[99]

Finally, the above-proposed involvement of the type-2 site in the dioxygen reduction has recently been challenged by studies of the reoxidation of the T2D enzyme. The fully reduced, T2D laccase (i.e., containing three reduction equivalents) reacts with O_2 to produce the paramagnetic intermediate assigned as an enzyme-bound O^- radical at a rate similar to that of the native enzyme.[81] However, the properties of the T2D laccase and its reactivity are still a matter of controversy,[105,106] and this issue must await more extensive examination.

Appendix 1
PROPERTIES OF SOME BLUE COPPER PROTEINS

Name	Source	M_r	Optical absorption of the blue band (absorption coefficient/ wavelength) ($M^{-1}cm^{-1}$/nm)	Oxidation-reduction potential of the blue site (mV/pH)	Ref.
Single-Cu proteins					
Azurin	*Alcaligenes faecalis*	14,000	4,000/625	266/7.0	60
	Pseudomonas aeruginosa	14,600	5,700/625	304/7.0	43
Mavicyanin	Squash	18,000	—	285/7.0	7
Plastocyanin	French bean	10,800	4,500/597	350/7.0	5
	Spinach	10,800	4,900/597	370/7.0	5
Rusticyanin	*Thiobacillus ferrooxidans*	16,500	2,200/597	680/2.0	63
Stellacyanin	*Rhus vernicifera*	22,000	4,030/608	184/7.1	5
Umecyanin	Horseradish roots	14,600	—	283/7.0	5
Multi-Cu enzymes					
Ascorbate oxidase	Cucumber, squash	140,000	3,300/610 per blue Cu	—	5
Ceruloplasmin	Human serum	150,000	4,400/610 per blue Cu	490, 580/5.5	5
Laccase	*Polyporus versicolor*	64,400	4,900/610	785/5.5	5
	Rhus vernicifera	110,000	5,700/615	395/7.5	5

REFERENCES

1. **Wherland, S. and Gray, H. B.**, Electron transfer mechanisms employed by metalloproteins in *Biological Aspects of Inorganic Chemistry*, Addison, A. W., Cullen, W. R., Dolphin, D. H., and James, B. R., Eds., John Wiley & Sons, New York, 1977, 289.
2. **Moore, G. R. and Williams, R. J. P.**, Electron transfer proteins, *Coord. Chem. Rev.*, 18, 125, 1976.
3. **Holwerda, R. A., Wherland, S., and Gray, H. B.**, Electron transfer reactions of copper proteins, *Annu. Rev. Biophys. Bioeng.*, 5, 363, 1976.
4. **Farver, O. and Pecht, I.**, Electron transfer processes of blue copper proteins, in *Copper Proteins, Metal Ions in Biology*, Vol. 3, Spiro, Th. G., Ed., John Wiley & Sons, New York, 1981, chap. 4.
5. **Fee, J.**, Cooper proteins, *Struct. Bonding*, 23, 1, 1975.
6. **Colman, P. M., Freeman, H. C., Guss, J. M., Murata, M., Norris, V. A., Ramshaw, J. A. M., and Venkatappa, M. P.**, X-ray crystal structure analysis of plastocyanin at 2.7 Å resolution, *Nature (London)*, 272, 319, 1978.
7. **Freeman, H. C.**, Electron transfer in blue copper proteins, in *Coordination Chemistry*, Vol. 21, Laurent, J. P., Ed., Pergamon Press, Oxford, 1981, 29.
8. **Adman, E. T., Stenkamp, R. E., Sieker, L. C., and Jensen, L. H.**, A crystallographic model for azurin at 3 Å resolution, *J. Mol. Biol.*, 123, 35, 1978.
9. **Adman, E. T. and Jensen, L. H.**, Structural features of azurin at 2.7 Å resolution, *Isr. J. Chem.*, 21, 8, 1981.
10. **Adman, E. T.**, A comparison of the structures of electron transfer proteins, *Biochim. Biophys. Acta*, 549, 107, 1979.
11. **Wiesenfeld, J. M., Ippen, E. P., Corin, A., and Bersohn, R.**, Electronic relaxation in azurin: picosecond reverse charge transfer, *J. Am. Chem. Soc.*, 102, 7526, 1980.
12. **Basolo, F. and Pearson, R. G.**, *Mechanisms of Inorganic Reactions*, 2nd ed., John Wiley & Sons, New York, 1967.
13. **Marcus, R. A.**, Chemical and electrochemical electron tranfer theory, *Annu. Rev. Phys. Chem.*, 15, 155, 1964.
14. **Hopfield, J. J.**, Electron transfer between biological molecules by thermally activated tunneling, *Proc. Natl. Acad. Sci. U.S.A.*, 71, 3640, 1974.
15. **Brill, A. S.**, Activation of electron transfer reactions of the blue proteins, *Biophys. J.*, 22, 139, 1978.
16. **Chance, B., DeVault, D. C., Frauenfelder, H., Marcus, R. A., Schrieffer, J. R., and Sutin, N.**, Eds., *Tunneling in Biological Systems*, Academic Press, New York, 1979.

17. **Boulter, D., Haslett, B. G., Peacock, D., Ramshaw, J. A. M., and Scawen, M. D.,** Chemistry, function, and evolution of plastocyanin, *Int. Rev. Biochem. Plant Biochem. II,* 13, 1, 1977.

18. **Freeman, H. C., Norris, V. A., Ramshaw, J. A. M., and Wright, P. E.,** High resolution proton magnetic resonance studies of plastocyanin, *FEBS Lett.,* 86, 131, 1978.

19. **McArdle, J. V., Coyle, C. L., Gray, H. B., Yoneda, G. S., and Holwerda, R. A.,** Kinetics studies of the oxidation of blue copper proteins by tris(1,10-phenanthroline)cobalt(III) ions, *J. Am. Chem. Soc.,* 99, 2483, 1977.

20. **Cummins, D. and Gray, H. B.,** Electron-transfer protein reactivities. Kinetic studies of the oxidation of horse heart cytochrome *c, Chromatium vinosum* high potential iron-sulfur protein, *Pseudomonas aeruginosa* azurin, bean plastocyanin, and *Rhus vernicifera* stellacyanin by pentaamminepyridineruthenium(III), *J. Am. Chem. Soc.,* 99, 5158, 1977.

21. **Holwerda, R. A., Knaff, D. B., Gray, H. B., Clemmer, J. D., Crowley, R., Smith, J. M., and Mauk, A. G.,** Comparison of the electron-transfer reactivities of tris(oxalato)cobaltate(III) $(Co(ox)_3^{3-})$ and tris(1,10-phenanthroline)cobalt(III) $(Co(phen)_3^{3+})$ with metalloproteins, *J. Am. Chem. Soc.,* 102, 1142, 1980.

22. **Mauk, A. G., Scott, R. A., and Gray, H. B.,** Distances of electron transfer to and from metalloprotein redox sites in reactions with inorganic complexes, *J. Am. Chem. Soc.,* 102, 4360, 1980.

23. **Segal, M. G. and Sykes, A. G.,** Stopped-flow kinetic studies on electron-transfer reactions of blue copper proteins. Evidence for an initial association in reactions of plastocyanin with inorganic complexes, *J. Chem. Soc. Chem. Comm.,* 764, 1977.

24. **Segal, M. G., and Sykes, A. G.,** Kinetic studies on 1:1 electron-transfer reactions involving blue copper proteins. I. Evidence for an unreactive form of the reduced protein (pH $<$ 5) and for protein-complex association in reactions of parsley (and spinach) plastocyanin, *J. Am. Chem. Soc.,* 100, 4585, 1978.

25. **Lappin, A. G., Segal, M. G., Weatherburn, D. C., and Sykes, A. G.,** Kinetic studies on 1:1 electron-transfer reactions involving blue copper proteins. II. Protonation effects and different binding sites in the oxidation of Parsley plastocyanin with $Co(4,7\text{-}DPSphen)_3^{3-}$, $Fe(CN)_6^{3-}$, and $Co(phen)_3^{3+}$, *J. Am. Chem. Soc.,* 101, 2297, 1979.

26. **Cookson, D. J., Hayes, M. T., and Wright, P. E.,** Electron transfer reagent binding sites on plastocyanin, *Nature (London),* 283, 682, 1980.

27. **Cookson, D. J., Hayes, M. T., and Wright, P. E.,** NMR study of the interaction of plastocyanin with chromium(III) analogues of inorganic electron transfer reagents, *Biochim. Biophys. Acta,* 591, 162, 1980.

28. **Handford, P. M., Hill, H. A. O., Wing-Kai Lee, R., Henderson, R. A., and Sykes,A. G.,** Investigation of the binding of inorganic complexes to blue copper proteins by ¹H nmr spectroscopy. I. The interaction between the $[Cr(phen)_3]^{3+}$ and $[Cr(CN)_6]^{3-}$ ions and the copper(I) form of parsley plastocyanin, *J. Inorg. Biochem.,* 13, 83, 1980.

29. **Farver, O. and Pecht, I.,** Identification of an electron-transfer locus in plastocyanin by chromium(II) affinity labeling, *Proc. Natl. Acad. Sci. U.S.A.,* 78, 4190, 1981.

30. **Katoh, S., Shiratori, I., and Takamiya, A.,** Purification and some properties of spinach plastocyanin, *J. Biochem. (Tokyo),* 51, 32, 1962.

31. **Taube, H.,** Experimental approaches to electronic coupling in metal ion redox systems, in *Tunneling in Biological Systems,* Chance, B., DeVault, D. C., Frauenfelder, H., Marcus, R. A., Schrieffer, J. R., and Sutin, N., Eds., Academic Press, New York, 1979, 173.

32. **Farver, O., Shahak, Y., and Pecht, I.,** Electron uptake and delivery sites on plastocyanin in its reactions with the photosynthetic electron transport system, *Biochemistry,* 21, 1885, 1982.

33. **Bouges-Bocquet, B. and Delosme, R.,** Evidence for a new electron donor to P-700 in *Chlorella pyrenoidosa, FEBS Lett.,* 94, 100, 1978.

34. **Haehnel, W., Hesse, V., and Pröpper, A.,** Electron transfer from plastocyanin to P700. Function of a subunit of photosystem I reaction center, *FEBS Lett.,* 111, 79, 1980.

35. **Olsen, L. F., Cox, R. P., and Barber, J.,** Flash-induced redox changes of P700 and plastocyanin in chloroplasts suspended in fluid media at sub-zero temperatures, *FEBS Lett.,* 122, 13, 1980.

36. **Kunert, K.-J., Böhme, H., and Böger, P.,** Reactions of plastocyanin and cytochrome 553 with photosystem I of *Scenedesmus, Biochim. Biophys. Acta,* 449, 541, 1976.

37. **Kunert, K.-J. and Böger, P.,** Absence of plastocyanin in the alga *Bumilleriopsis* and its replacement by cytochrome 553, *Z. Naturforsch.,* 30C, 190, 1975.

38. **Wood, P. M. and Bendall, D. S.,** The kinetics and specificity of electron transfer from cytochromes and copper proteins to P700, *Biochim. Biophys. Acta,* 387, 115, 1975.

39. **Wood, P. M.,** Rate of electron transfer between plastocyanin, cytochrome *f,* related proteins and artificial redox reagents in solution, *Biochim. Biophys. Acta,* 357, 370, 1974.

40. **Davis, D. J., Krogmann, D. W., and San Pietro, A.,** Electron donation to photosystem-I, *Plant Physiol.,* 65, 697, 1980.

41. **Wherland, S. and Pecht, I.,** Protein-protein electron transfer. A Marcus theory analysis of reactions between *c* type cytochromes and blue copper proteins, *Biochemistry,* 17, 2585, 1978.

42. **Horio, T.,** Terminal oxidation system in bacteria. I. Purification of cytochromes from *Pseudomonas aeruginosa, J. Biochem. (Tokyo),* 45, 195, 1958.

43. **Goldberg, M. and Pecht, I.,** Kinetics and equilibria of the electron transfer between azurin and the hexacyanoiron (II/III) couple, *Biochemistry,* 15, 4197, 1976.

44. **Rydén, L. and Lundgren, J. -O.,** Homology relationships among the small blue proteins, *Nature (London),* 261, 344, 1976.

45. **Lappin, A. G., Segal, M. G., Weatherburn, D. C., Henderson, R. A., and Sykes, A. G.,** Kinetic studies on 1:1 electron-transfer reactions involving blue copper proteins. III. Protonation effects, protein-complex association, and binding sites in reactions of *Pseudomonas aeruginosa* azurin with $Co(phen)_3^{3-}$, $Co(4.7-DPSphen)_3^{3-}$, and $Fe(CN)_6^{3-}$ (oxidants) and $Fe(CN)_6^{4-}$ (reductant), *J. Am. Chem. Soc.,* 101, 2302, 1979.

46. **Hill, H. A. O., Leer, J. C., Smith, B. E., Storm, C. B., and Ambler, R. P.,** A possible approach to the investigation of the structures of copper proteins: ^{1}H-NMR spectra of azurin, *Biochem. Biophys. Res. Commun.,* 70, 331, 1976.

47. **Ugurbil, K. and Bersohn, R.,** Nuclear magnetic resonance study of exchangeable and nonexchangeable protons in azurin from *Pseudomonas aeruginosa, Biochemistry,* 16, 3016, 1977.

48. **Klapper, M. H. and Faraggi, M.,** Applications of pulse radiolysis to protein chemistry, *Q. Rev. Biophys.,* 12, 465, 1979.

49. **Adams, G. E., Redpath, J. L., Bisby, R. H., and Cundall, R. B.,** The use of free radical probes in the study of mechanisms of inactivation, *Isr. J. Chem.,* 10, 1079, 1972.

50. **Faraggi, M. and Pecht, I.,** The reactions of *Pseudomonas* azurin with hydrated electrons, *Biochem. Biophys. Res. Commun.,* 45, 842, 1971.

51. **Pecht, I. and Faraggi, M.,** The reduction of cytochrome *c* by hydrated electrons, *FEBS Lett.,* 13, 221, 1971.

52. **Farver, O. and Pecht, I.,** Affinity labeling of an electron transfer pathway in azurin by Cr(II) ions, *Isr. J. Chem.,* 21, 13, 1981.

53. **Antonini, E., Finazzi-Agrò, A., Avigliano, L., Guerrieri, P., Rotilio, G., and Mondovì, B.,** Kinetics of electron transfer between azurin and cytochrome 551 from *Pseudomonas, J. Biol. Chem.,* 245, 4847, 1970.

54. **Wilson, M. T., Greenwood, C., Brunori, M., and Antonini, E.,** Electron transfer between azurin and cytochrome *c*-551 from *Pseudomonas aeruginosa, Biochem. J.,* 145, 449, 1975.

55. **Rosen, P. and Pecht, I.,** Conformational equilibria accompanying the electron transfer between cytochrome *c* (P551) and azurin from *Pseudomonas aeruginosa, Biochemistry,* 15, 775, 1976.

56. **Bersohn, R. and Corin, A.,** private communication.

57. **Wherland, S. and Pecht, I.,** unpublished results.

58. **Silvestrini, M. C., Brunori, M., Wilson, M. T., and Darley-Usmar, V. M.,** The electron transfer system of *Pseudomonas aeruginosa:* a study of the pH-dependent transitions between redox forms of azurin and cytochrome $c_{551}, J. Inorg. Biochem.,$ 14, 327, 1981.

59. **Farver, O., Blatt, Y., and Pecht, I.,** Resolution of two distinct electron transfer sites on azurin, *Biochemistry,* 21, 3556, 1982.

60. **Rosen, P., Segal, M., and Pecht, I.,** Electron transfer between azurin from *Alcaligenes faecalis* and cytochrome c_{551} from *Pseudomonas aeruginosa, Eur. J. Biochem.,* 120, 339, 1981.

61. **Ambler, R. P. and Brown, L. H.,** The amino acid sequence of *Pseudomonas fluorescens* azurin, *Biochem. J.,* 104, 784, 1967.

62. **Mitra, S. and Bersohn, R.,** Proton NMR of the histidines of azurin from *Alcaligenes faecalis:* linkage of histidine-35 with redox kinetics, *Proc. Natl. Acad. Sci. U.S.A.,* 79, 6807, 1982.

63. **Cox, J. C. and Boxer, D. H.,** The purification and some properties of rusticyanin, a blue copper protein involved in iron(II) oxidation from *Thiobacillus ferrooxidans, Biochem. J.,* 174, 497, 1978.

64. **Yoneda, G. S. and Holwerda, R. A.,** Kinetics of the oxidation of *Rhus vernicifera* stellacyanin by the $Co(EDTA)^-$ ion, *Bioinorg. Chem.,* 8, 139, 1978.

65. **Yoneda, G. S., Mitchel, G. L., Blackmer, G. L., and Holwerda, R. A.,** Oxidation of cuprous stellacyanin by aminopolycarboxylatocobaltate(III) complexes, *Bioinorg. Chem.,* 8, 369, 1978.

66. **Holwerda, R. A. and Clemmer, J. D.,** Isokinetic relationship in the oxidation of cuprous stellacyanin by cobalt(III) complexes, *J. Inorg. Biochem.,* 11, 7, 1979.

67. **Holwerda, R. A., Knaff, D. B., Gray, G. O., and Harsh, C. E.,** Reactivity of cuprous stellacyanin as a quinone and semiquinone reductase, *Biochemistry,* 20, 4336, 1981.

68. **Pecht, I. and Goldberg, M.,** Electron transfer pathways to and within redox proteins: pulse radiolysis studies, in *Fast Processes in Radiation Chemistry and Biology,* Adams, G. E., Fielden, E. M., and Michael, B. D., Eds., John Wiley & Sons, London, 1975, 277.

69. **Wilson, M. T., Silvestrini, M. C., Morpurgo, L., and Brunori, M.,** Electron transfer kinetics between *Rhus vernicifera* stellacyanin and cytochrome *c* (horse heart cytochrome *c* and *Pseudomonas* cytochrome c_{551}) ,*J. Inorg. Biochem.,* 11, 95, 1979.

70. **Rist, G. H., Hyde, J. S., and Vänngård, T.,** Electron-nuclear double resonance of a protein that contains copper: evidence for nitrogen coordination to Cu(II) in stellacyanin, *Proc. Natl. Acad. Sci. U.S.A.,* 67, 79, 1970.

71. **Peisach, J., Levine, W. G., and Blumberg, W. E.,** Structural properties of stellacyanin, a copper mucoprotein from *Rhus vernicifera,* the Japanese lac tree, *J. Biol. Chem.,* 242, 2847, 1967.

72. **Faraggi, M. and Bettelheim, A.,** The reaction of the hydrated electron with amino acids, peptides, and proteins in aqueous solutions. III. Histidyl peptides, *Radiat. Res.,* 71, 311, 1977.

73. **Malkin, R. and Malmström, B. G.,** The state and function of copper in biological systems, *Adv. Enzymol.,* 33, 177, 1970.

74. **Holwerda, R. A. and Gray, H. B.,** Mechanistic studies of the reduction of *Rhus vernicifera* laccase by hydroquinone, *J. Am. Chem. Soc.,* 96, 6008, 1974.

75. **Holwerda, R. A., Clemmer, J. D., Yoneda, G. S., and McKerley, B. J.,** Reduction of *Rhus vernicifera* laccase type 1 copper by substituted hydroquinones, *Bioinorg. Chem.,* 8, 225, 1978.

76. **Clemmer, J. D., Gilliland, B. L. Bartsch, R. A., and Holwerda, R. A.,** Substituent effects on the electron transfer reactivity of hydroquinones with laccase blue copper, *Biochim. Biophys. Acta,* 568, 307, 1979.

77. **Branden, R., Malmstrom, B. G., and Vanngard, T.,** The effect of fluoride on the spectral and catalytic properties of three copper-containing oxidases, *Eur. J. Biochem.,* 36, 195, 1973.

78. **Goldberg, M., Vuk-Pavlovic, S., and Pecht, I.,** Proton and oxygen-17 magnetic resonance relaxation in *Rhus* laccase solutions: proton exchange with type 2 copper(II) ligands, *Biochemistry,* 19, 5181, 1980.

79. **Branden, R. and Deinum, J.,** Type 2 copper(II) as a component of the dioxygen reducing site in laccase: evidence from EPR experiments with ^{17}O, *FEBS Lett.,* 73, 144, 1977.

80. **Andreasson, L.-E. and Reinhammar, B.,** Kinetic studies of *Rhus vernicifera* laccase. Role of the metal centers in electron transfer, *Biochim. Biophys. Acta,* 445, 579, 1976.

81. **Reinhammar, B. and Oda, Y.,** Spectroscopic and catalytic properties of *Rhus vernicifera* laccase depleted in type 2 copper, *J. Inorg. Biochem.,* 11, 115, 1979.

82. **Andreasson, L.-E. and Reinhammar, B.,** The mechanism of electron transfer in laccase-catalysed reactions, *Biochim. Biophys. Acta,* 568, 145, 1979.

83. **Farver, O., Goldberg, M., Wherland, S., and Pecht, I.,** Reductant-dependent electron distribution among redox sites of laccase, *Proc. Natl. Acad. Sci. U.S.A.,* 75, 5245, 1978.

84. **Reinhammar, B. R. M.,** Oxidation-reduction potentials of the electron acceptors in laccases and stellacyanin, *Biochim. Biophys. Acta,* 275, 245, 1972.

85. **Reinhammar, B., Malkin, R., Jensen, P., Karlsson, B., Andréasson, L.-E., Aasa, R., Vänngård, T., and Malmström, B. G.,** A new copper(II) electron paramagnetic resonance signal in two laccases and in cytochrome *c* oxidase, *J. Biol. Chem.,* 255, 5000, 1980.

86. **Malmström, B. G., Finazzi-Agrò, A., and Antonini, E.,** The mechanism of laccase-catalyzed oxidations: kinetic evidence for the involvement of several electron-accepting sites in the enzyme, *Eur. J. Biochem.,* 9, 383, 1969.

87. **Andréasson, L.-E., Malmström, B. G., Stromberg, C., and Vänngärd, T.,** The kinetics of anaerobic reduction of fungal laccase B, *Eur. J. Biochem.,* 34, 434, 1973.

88. **Pecht, I.,** Chemical relaxation study of the electron transfer between laccase and external redox couples, *Isr. J. Chem.,* 12, 351, 1974.

89. **Pecht, I. and Faraggi, M.,** Reduction of copper(II) in fungal laccase by hydrated electrons, *Nature (London), New Biol.,* 233, 116, 1971.

90. **Pecht, I., Farver, O., and Goldberg, M.,** Electron transfer pathways in blue copper proteins, *Adv. Chem.,* Ser. No. 162, American Chemical Society, Washington, D.C., 1977, 179.

91. **Goldberg, M. and Pecht, I.,** The reaction of "blue" copper oxidases with O_2. A pulse radiolysis study, *Biophys. J.,* 24, 371, 1978.

92. **Goldberg, M.,** Tree Laccase: Its Mechanism of Action, Ph.D. thesis, The Weizmann Institute of Science, Rehovot, 1979.

93. **Faraggi, M. and Pecht, I.,** The electron pathway to Cu(II) in ceruloplasmin, *J. Biol. Chem.,* 248, 3146, 1973.

94. **Henry, Y., Guissani, A., and Gilles, L.,** Radical scavenging and electron transfer reactions in laccase, *Biochimie,* 63, 841, 1981.

95. **Andréasson, L.-E., Bränden, R., and Reinhammar, B.,** Kinetic studies of *Rhus vernicifera* laccase. Evidence for multi-electron transfer and an oxygen intermediate in the reoxidation reaction, *Biochim. Biophys. Acta,* 438, 370, 1976.

96. **Farver, O., Goldberg, M., Lancet, D., and Pecht, I.,** Oxidative titrations of *Rhus vernicifera* laccase and its specific interaction with hydrogen peroxide, *Biochem. Biophys. Res. Commun.,* 73, 494, 1976.

97. **Farver, O. and Pecht, I.,** Magnetic susceptibility study of the laccase-peroxide derivative, *FEBS Lett.,* 108, 436, 1979.

98. **Goldberg, M., Farver, O., and Pecht, I.,** Interaction of *Rhus* laccase with dioxygen and its reduction intermediates, *J. Biol. Chem.,* 255, 7353, 1980.

99. **Farver, O., Goldberg, M., and Pecht, I.,** A circular dichroism study of the reactions of *Rhus* laccase with dioxygen, *Eur. J. Biochem.,* 104, 71, 1980.

100. **Farver, O., Goldberg, M., and Pecht, I.,** unpublished results.

101. **Farver, O., Goldberg, M., and Pecht, I.,** Circular dichroic spectrum of the laccase-peroxide derivative, *FEBS Lett.,* 94, 383, 1978.

102. **Aasa, R., Brändén, R., Deinum, J., Malmström, B. G., Reinhammar, B., and Vänngård, T.,** A ^{17}O-effect on the EPR spectrum of the intermediate in the dioxygen-laccase reaction, *Biochem. Biophys. Res. Commun.,* 70, 1204, 1976.

103. **Brändén, R., Deinum, J., and Coleman, M.,** A mass spectrometric investigation of the reaction between ^{18}O$_2$ and reduced tree laccase. A differentiation between the two water molecules formed, *FEBS Lett.,* 89, 180, 1978.

104. **Brändén, R. and Deinum, J.,** The effect of pH on the oxygen intermediate and the dioxygen reducing site in blue oxidases, *Biochim. Biophys. Acta,* 524, 297, 1978.

105. **Graziani, M. T., Morpurgo, L., Rotilio, G., and Mondovì, B.,** Selective removal of type 2 copper from *Rhus vernicifera* laccase, *FEBS Lett.,* 70, 87, 1976.

106. **Morpurgo, L., Graziani, M. T., Finazzi-Agrò, A., Rotilio, G., and Mondovì, B.,** Optical properties of Japanese-lacquer-tree *(Rhus vernicifera)* laccase depleted of type-2 copper(II). Involvement of type-2 copper(II) in the 330 nm chromophore, *Biochem. J.,* 187, 361, 1980.

ABBREVIATIONS AND SYMBOLS

Az	azurin
CD	circular dichroism (dichroic)
CDTA	*trans*-1,2-diaminocyclohexanetetraacetate
CT	charge transfer
Cyt	cytochrome
EC	Enzyme Commission
ENDOR	electron nuclear double resonance
EPR	electron paramagnetic resonance
EXAFS	extended X-ray absorption fine structure
Hc	hemocyanin
H_2Q	hydroquinone
K_m	Michaelis constant
L	ligand (intraligand)
LCAO	linear combination of atomic orbitals
LEFE	linear electric field effect
LMCT	ligand to metal charge transfer
M-L	metal-ligand
MO	molecular orbital
M_r	relative molecular mass
NCA	normal coordinate analysis
NMR	nuclear magnetic resonance
Pc	plastocyanin
PDTA	propylenediaminetetraacetate
phen	1,10-phenanthroline
PMR	proton magnetic resonance
py	pyridine
RR	resonance Raman
RREP	resonance Raman enhancement profile
SD	standard deviation unit(s) of the random distribution
SDS	sodium dodecyl sulfate
SOD	superoxide dismutase
St	stellacyanin
T2D	type-2 copper depleted
TWT	traveling wave tube
V	rate of enzyme-catalysed reaction at infinite concentration of substrate (maximal velocity, limiting rate)
XAS	X-ray absorption spectroscopy

One-Letter Symbols for Amino Acids

A	Ala	I	Ile	S	Ser
B	Asx	K	Lys	T	Thr
C	Cys	L	Leu	V	Val
D	Asp	M	Met	W	Trp
E	Glu	N	Asn	X	unknown
F	Phe	P	Pro	Y	Tyr
G	Gly	Q	Gln	Z	Glx
H	His	R	Arg		

INDEX